A FRAMEWORK FOR POST-PHYLOGENETIC SYSTEMATICS

Richard H. Zander

Zetetic Publications, St. Louis

3p

Richard H. Zander
Missouri Botanical Garden
P.O. Box 299
St. Louis, MO 63166
richard.zander@mobot.org

Zetetic Publications in St. Louis produces but does not sell this book. Any book dealer can obtain a copy for you through the usual channels. Resellers please contact CreateSpace Independent Publishing Platform of Amazon.

ISBN-13: 978-1492220404
ISBN-10: 149222040X

The image on the cover and title page is a stylized dendrogram of paraphyly (see Plate 1.1). This is, in macroevolutionary terms, an ancestral taxon of two (or more) species or of molecular strains of one taxon giving rise to a descendant taxon (unconnected comma) from one ancestral branch.

The image on the back cover is a stylized dendrogram of two, genus-level speciational bursts or dissilience. Here, the dissilient genus is the basic evolutionary unit (see Plate 13.1). This evolutionary model is evident in analysis of the moss *Didymodon* (Chapter 8) through superoptimization. A super-generative core species with a set of radiative, specialized descendant species in the stylized tree compromises one genus. In this exemplary image; another genus of similar complexity is generated by the core supergenerative species of the first.

TABLE OF CONTENTS

PREFACE

This book is an attempt to find common principles and intellectual continuity in addressing today's problems in systematics. Certain difficulties endemic in human thought, and often faced in the past in other fields, are now evident in systematics. This has been perceived by many workers and science is self-correcting, however tardily. This book suggests a needed correction that deals with several problems at once, and its particular solution will be accepted or fail as weighed in the marketplace of reason.

A new paradigm should present an acceptable solution to a problem by addressing it in a new way. Phylogenetics has redefined the problem of devising an evolution-based classification by presenting evolutionary relationships not as descent with modification of taxa but as descent with modification of traits. According to the Web home page of the phylogenetically oriented Society for Systematic Botany (December 2012): "Systematics is the study of biological diversity and its origins. It focuses on understanding evolutionary relationships among organisms, species, higher taxa, or other biological entities, such as genes, *and the evolution of the properties of taxa including intrinsic traits, ecological interactions, and geographic distributions.* An important part of systematics is the development of methods for various aspects of phylogenetic inference and biological nomenclature/classification." [Italics mine.]

Phylogenetics eliminates any hint of progenitor-descendant relationships in evolutionary analysis, and relies on algorithmic clustering data from descriptions or specimens to provide a "hard science," mathematically non-trivial, statistically based, integrable (fully calculable) solution that has the appearance of an evolutionary tree but lacks identification of the nodes of the tree as being any extant taxon beyond the name of that taxon including all specimens or descriptions used as data distal to that node. That all nodes are treated as pseudoextinction events never budding evolution totally vitiates any responsible macroevolutionary inferences in sister-group analysis.

Phylogenetics imposes a classification on the results of cladistic analysis without a process-based explanation of those results. The sister-group structure is taken to be a classification itself. Evolution is not clustering, classification is. Evolution is not nesting, classification is. Phylogenetics leaps from the clustering and nesting of cladistic analysis straight to classification without explanation of the analysis in terms of serial transformations of one taxon into another, which is the nut of macroevolutionary theory.

This book rejects classification informing evolution rather than the other way around. "The map is not the territory." Given that historical reconstruction cannot be directly verified, and will remain forever notional not actual, mere precision will never make up for natural limits on accuracy, particularly if precision is obtained at the sacrifice of a total evidence approach involving discursive logic and macroevolutionary theory. Phylogenetic attempts at "reconstruction" try to reconstruct evolutionary nesting, not a process in nature. Yet, with application of a pluralist approach involving classical techniques, morphological cladistics, and phylogenetic analyses, satisfying advances can be made within such natural limits.

The proposed Framework will probably not change the methods of career phylogeneticists who may feel loyal or responsible to sunk-cost professional investments. The story goes that the Buddha, after enlightenment, went into the world to teach. The first person he came upon was a holy man, a fellow seeker of enlightenment. The Buddha cried, "Wait! Listen! I have found enlightenment!" The holy man paused and looked at the Buddha a moment. He said "Maybe so...," and walked on. If I can obtain a "maybe so" from the phylogenetic establishment, I will be well satisfied.

This book is largely intended for students and uncommitted professionals in systematics and evolution, and for those in other fields, such as philosophy, physics and psychology, that deal with scientific and decision theory. The basic ideas and methods presented here are a pluralistic means to correct the difficulties in which modern systematics has found itself. The reader will find the same basic concepts presented often in this book, but this is defensible because the concepts are sometimes difficult, relate to other fields, and require a familiarity with both classical and modern methods in systematics. In addition, judging from reviewers' comments of previous manuscripts, I have decided it is necessary to present certain novel concepts each in several ways and in different contexts to (1) clarify what is meant, (2) hammer past intransigent preconceptions, and (3) dispel through reasoned discourse and perhaps a little humor the fog now shrouding classical systematics. Repetition of logical argument is often the only way to break through or reprogram hard-set mental viewpoints.

Practitioners of evolutionary systematics are methodologically diverse, and this book does not try to represent the field. Instead, presented here are my ideas on how systematics as a whole might develop. I do not claim to have generated all the central ideas presented in this book as many or even most are already in the literature, albeit dispersed, or they have coalesced from the suggestions and discussions of others. As I was completing the work, however, certain ideas arose as obvious given the method, e.g., self-nesting ladders, and supergenerative ancestral taxa, so I expect that future workers may find this pluralistic view of systematics a complex new field for their own novel concepts and solutions. The method presented here is complex because it taps the research programs of both classical and phylogenetic taxonomies, is process-based, not structuralist, and requires reason, judgment, and insight to be successful.

In this book, many examples concern the plants, notably the bryophytes (mosses), particularly the family Pottiaceae, which I have studied for 40 years. I make no excuses for such stress, in part because expertise is one of the elements of the pluralistic methodology I describe. Reviewers have sometimes remonstrated over my occasional reference to matters philosophical or how things are done in the fields of physics and mathematics. These references, however, serve to remind readers that systematics, though a historical science creating results not directly confirmed or supported, is nevertheless part of the scientific endeavor and cannot be excused from shirking rigorous scientific method or logical reasoning.

In (very) short, phylogenetics chooses an analytic aspect of evolution, sister-group nesting, that can be precisely modeled. It bases classification on that, which requires strict phylogenetic monophyly (holophyly) to work well. Evolutionary systematics attempts classifications from all data, including those that are not informative of such nesting, or precisely measured, or which are intuitively inferred, and rejects strict monophyly because it masks information on macroevolutionary transformations. This book is not a continuation of the grand remonstrance of many recent authors against phylogenetic rejection of paraphyly, but is an attempt to consiliate, conciliate and consolidate the two schools of systematics, classical and phylogenetic.

August 20, 2013

Richard H. Zander
Missouri Botanical Garden
St. Louis, Missouri, U.S.A.
richard.zander@mobot.org

ACKNOWLEDGMENTS

This book is dedicated to Richard K. Brummitt, who first formally resisted (Brummitt 1996, 1997) the classification principle of strict phylogenetic monophyly, and has continued to defend this position (see Bibliography) in the face of much opposition.

I thank Patricia M. Eckel for renderings of Latin descriptions, discussions of the subject, and continued encouragement. Bernard Goffinet is acknowledged for suggesting at least obliquely the taxon mapping concept in a review. Tod Stuessy and Elvira Hörandl have provided significant support and feedback. Users of the Bryonet and Taxacom listservers are again thanked for their helpful criticism and in some cases (and understandably usually off-line) support, including M. Blanco, S. Brady, C. Clark, D. Colless, P. Deleporte, E. De Luna, T. Dickinson, J. Enroth, K. Fitzhugh, A. D. Forrest, B. Goffinet, J. Grehan, R. Hastings, M. Heads, L. Hedenäs, P. Hovenkamp, R. Jensen, K. Kellman, K. Kinman, W. Kipling, J. Kirkbride, D. Lahr, T. G. Lammers, J. F. Mate, R. Mesibov, B. Mishler, P. J. Morris, A. Nicholas, R. Pyle, D. Quandt, C. Rothfels, F. W. Schueller, R. Schumacker, R. Seppelt, J. Shaw, J. Shuey, P. Stevens, S. Thorpe, K. van der Linde, D. Wagner, P. Wilson, and D. Yanega. Many of these persons argued contra my suggestions, yet provided valuable feedback.

Of particular importance to me is the work of members of the University of Murcia Pottiaceae group, who have contributed many papers of high quality research on this moss family, including molecular analysis. I have based many of my examples on their publications.

I salute my advisor Harold Robinson, who taught me much about systematics during a pre-doctoral internship at the Smithsonian Institution in 1966–1967. He also wrote the delightful and infamous "A key to the common errors of cladistics" (Robinson 1986).

I appreciate the generosity of the Missouri Botanical Garden in providing well-appointed shelter for this research. I also value the decision of certain journal editors who shared with me some objurgatory reviews of my papers that were more revelatory of the reviewer's closely held beliefs than solvent of my own mistakes or misconstruals. These reviews confirmed my intention to publish this book as samizdat.

CHAPTER 1
Introduction

Précis — The fundamental premise in phylogenetics is that two of every three taxa are more closely related evolutionarily because of pseudoextinction, or demise of a shared ancestral species upon speciation. Any instance of paraphyly implies no pseudoextinction and two cladogram nodes that are the same taxon. An instance of paraphyly then cannot be analyzed cladistically because of three clades there is only one descendant and two clades are most closely related because they are the same taxon. If paraphyly is common, particularly extended paraphyly with many descendants from one ancestral taxon, either presently or in the past or both, then the cladistic method fails or is only of superficial value. Phylogeneticists mask this problem with (1) the principle of holophyly, that is, strict phylogenetic monophyly, and (2) treating changes in DNA tracking sequences as primary evolutionary events. A pluralistic method can address this affair by using macroevolution, distinguishing pseudoextinction from budding evolution, as an explanatory process. Macroevolution and associated linear taxic transforms can replace general relationships due to pseudoextinction as a fundamental premise in evolution as reflected in classification.

Cladistics and phylogenetics are often distinguished in that the former is simply dichotomous clustering by synapomorphies, while the latter introduces evolutionary elements such as the time dimension and "shared ancestors." Both are considered much the same in this book because both are limited by rejecting the naming of ancestral nodes and therefore crippling inferences of macroevolutionary transformations. Phylogenetics implies macroevolution through transformation from an unnamed "shared ancestor" via pseudoextinction (see Glossary). Both the cladistic assumption of maximum parsimony and the phylogenetic assumption of *universal* pseudoextinction are essentially the same thing, one with an evolutionary explanation. Both render a sister-group analysis non-operational in that neither distinguishes pseudoextinction and budding evolution. This is in the face of much evidence that commonly one of a sister-group will be easily inferred as evolving in expressed traits from the other. Such evidence is not informative, according to phylogeneticists, because it is not evidence about sister groups; see discussion of super-optimization in Chapter 8.

To start, consider the following extreme simplification of modern competing views for solving a complex problem, namely that of making an evolution-based classification. Phylogenetics chooses an aspect of evolution, nesting of sister groups, that can be precisely modeled, and bases classification on that.

Evolutionary systematics attempts classifications from all data, including those that are not precisely measured or are informedly intuitively inferred, data that are reflect evolutionary uniqueness rather than relationship, and rejects strict monophyly as masking information on macroevolution (generation of one taxon from another). "Total evidence" in phylogenetics (Allard & Carpenter 1996; Eernisse & Kluge 1993; Nixon & Carpenter 1996) only means total evidence about sister-group relationships. This book attempts to conciliate and consolidate the two schools of systematics (phylogenetics and evolutionary systematics) through a Bayes' Solution (Kendall & Buckland 1971), which reconciles all sources of uncertainty involved in the various methods used, and incorporates additional certainty from neglected information, this in light of risk if wrong. A simple example is choosing a low-stakes poker game if you have little money and hope to play all night. At stake, for systematists and conservationists, is a correct and workable classification of the world's fauna and flora.

This is a reframing of the evolutionary element in systematics from nested exemplars or taxa to serial transformations of taxa, that is, from tree-thinking of the cladogram to stem-thinking of the caulogram (or commagram or Besseyan cactus). Above all, this book rejects the idea that a cladogram is automatically a monophylogram, and that a clade is necessarily monophyletic. Such ideas are purely definitional and have no place in science.

The difference in tree-thinking and stem-thinking may be exemplified in the modern analysis of the evolutionary position of *Amborella* as the evolutionary root of the angiosperms. Although *Amborella* may be the lowest diverging lineage (or maybe not) (Goremykin et al. 2013) this says nothing of the taxon from which *Amborella* and its sister lineage diverged. Was it *Amborella* or perhaps Nymphaeaceae? Or a lineage diverging even higher in the present angiosperm tree of life? Tree-thinking cannot even approach dealing with this question yet it can be address by discovery of heterophyly after dense sampling of molecular strains of both taxa.

The six-element Framework presented below eliminates inconsistencies that contribute to lowered posterior probabilities by using a theory, macroevolution, that places apparent inconsistencies in a context in which they are consistent and contributory to a higher posterior probability. Macroevolution cannot be analyzed without distinguishing pseudoextinction from budding evolution. All evolutionary data then are relevant to inferences of a shared macroevolutionary structure addressed, no matter how indirectly at times, by all three major methods, classical systematics, morphological cladistics, and molecular analysis. This structure, which shows descent with modification of taxa as best possible from the data, is considered here the proper basis for evolutionary classification.

The method in a nutshell — A new, syncretic method is needed to address the central problem of cladistics, that if cladogram nodes are not named, cladistics alone cannot demonstrate monophyly. In addition, molecular phylogenetics deals only with extant molecular strains while other molecular strains of the same taxon may be extinct or unsampled but potentially scattered on the cladogram in a manner similar to extant paraphyly.

The main *problem* in a nutshell is that cladogram analysis treats *all* tree splits as sister groups—in cladistics as dichotomous synapomorphic pairs, and in phylogenetics as pseudoextinction events (with a disappearing shared ancestor). The solution is to recognize both pseudoextinction and budding evolution using both phylogenetically and non-phylogenetically informative information. That is, using both sister groups diagrammatically as in (AB) and taxon transformations as in **C > D**. Although there are other evolutionary scenarios, these are the two basic choices for systematists who use evolutionary trees to help classify.

Presently, taxonomic analysis represents evolutionary relationships as a set of nested taxa. There is commonly contradiction, that is, lack of congruence, in nesting patterns derived from classical taxonomy (Linnaean classification is essentially hierarchical), morphological cladistics, and molecular phylogenetics. This is because serial transformation patterns are interpreted as nestings in different ways by different methods. If an ancestral taxa generates two or more descendant taxa at the same rank and does not go

itself extinct, which is apparently not uncommon, then the fundamental assumption of phylogenetics, that of three taxa at the same rank, two are always more closely related to each other, fails. Using the Framework methods, estimates of, not nesting, but serial macroevolutionary transformations of taxa, are derived both separately and through reciprocal illumination from the three methodological sources of analysis. Where such serial transformations overlap when superimposed is considered well supported. Where they do not is evaluated in the Bayesian context using coarse priors (Chapter 8). We end up with an estimate, based on all available information and standard theory, of a natural process, macroevolution, for a particular group. This may be used in classification, but is not expected to result in the myriad classification changes now associated with molecular phylogenetics and its classification principle of holophyly (strict phylogenetic monophyly).

Discussion of macroevolution in this book reflects a contest between two concepts: (1) in standard cladistic methodology all speciation events are considered pseudoextinction (see Glossary), and (2) in evolutionary systematics pseudoextinction and budding evolution (see Glossary) are carefully distinguished for each node in an evolutionary tree (= "superoptimization," see Chapter 8). Inasmuch as only superoptimization actually reviews and defines speciation events based on both phylogenetic and non-phylogenetically informative information, the term "macroevolution" is used here for a transformation of one taxon from another, either by pseudoextinction or by budding evolution, neither of which is successfully modeled by the assumption of universal pseudoextinction in cladograms (see Plate 5.3).

The caulistic macroevolutionary methodological concept is easy to conceive and support theoretically, but difficult to execute given the need to integrate several dimensions of data and analysis with varying degrees of precision and accuracy, as well as to deal with historical burdens of preconceptions and presumptions on the part of practitioners in differing schools. The reader is asked here to suspend, for a time, his or her present-day assumptions of the proper way to do systematics.

Paradigm change — The cladistic revolution in systematics (Stuessy 2009) imbued a new way of perceiving or modeling evolution (i.e., tree thinking, Baum & Smith 2012) among many systematists. This was a logical extension of phenetics (Heywood & McNeill 1964; Sneath 1976; Sokal & Sneath 1963; Yablokov 1986). Consider the proverb, "If at first you don't succeed, change the rules" (i.e., from serial to nesting models). This may have both good and unfortunate results; sometimes the rules are too restrictive, sometimes not restrictive enough, and one can be wrong either way. Burke (1985)—of television "Connections" fame—offered the idea that the universe, reality itself, actually changes whenever new scientific paradigms (Kuhn 1970) of perception, interpretation, and thought are accepted. People "see" the heavens, for instance, as something quite different post Newton. Yet, after many changes in scientific paradigms, nature remains, and only human viewpoints and mental methods have changed with each new paradigm. The switch from one scientific paradigm to another may be occasionally difficult for humans, but the universe does not change.

Revolution — Was the phylogenetic revolution actually a revolution? It indeed had many of the traits of revolutions as recognized by Brinton (1952), an authority on revolutions. Revolutions, according to Brinton, are characterized by a struggle between what become essentially two governments, between the ins and the outs. There is an eventual overthrow of the revolutionary moderates by the much more fully committed and disciplined extremists, including taking control of centers of power, press, banks, and ministries. A ruling elite is required to illuminate through education, rules, and censorship the masses who are slow to learn new ways of thinking. Pure democracy is considered "mob rule" (for instance, the leaders of the American Revolution gave us a representative-based republic). There is also a "Terror," which I was going to avoid mentioning to duck accusations of sensationalism, but supportive comments from other scientists on my relevant contributions to listservers Taxacom and Bryonet were always sent to me offline, by private email. This could be explained by widespread mystification of the "black box" phylogenetic classification methods—who wants to admit ignorance? On the other hand, United States federal funding for systematics is much influenced by a review panel of scientists largely selected from the phylogenetic establishment.

Brinton goes on to say that revolutionaries strive to "achieve a reign of virtue on earth." He suggests (p. 193) that revolutionaries have many of the traits of the religious, not the theistic dimension, but certainly "the important thing about a religious belief is that under its influence men work very hard and excitedly in common to achieve here or somewhere an ideal, a pattern of life not at the moment universally—or even largely—achieved." The fervor of cladists seems part and parcel of the revolutionary phenomenon Brinton describes. Religious attitudes also extend to resistance to criticism. According to Grant (1977: 198), historians of Christianity may unacceptably maintain that "no one but a believer in Jesus' divinity is entitled to write a single word about him," or that the burden of proof has passed from the true believer to the historian. A similar stance is detectable in a phylogenetist reviewer's rejective comment on one of my papers that "we discussed and dealt with that back in the 1980's."

Dyson (1999) has made the point that scientific revolutions are commonly associated with the development of new research tools. That phylogenetics is powered by computer analysis is incontrovertible. Falling into the mechanical knowledge fallacy of Gigerenzer et al. (1989: 288), the phylogenetic revolution seems to have bypassed the need for insight and judgment. But insight and judgment are never more necessary than today, when faced with mountains of data that support very conflicting explanations.

The history of the struggle of cladists against pheneticists related by Felsenstein (2001), extended and criticized by Farris (2012), presented in the general historical context (Vernon 1993), and more deeply analyzed by Scott-Ram (1990), is that of conflict between groups of partisans of two new methods of numerical, computer-driven methodology. The cladists won. Felsenstein suggested that the even newer statistical approach is something more scientific, and as such it has overwhelmed early axiomatic cladism. A glance, however, at the list of phylogenetic axioms he gave indicates that statistical phylogenetics is little more than an extension of elementary cladistics, with inference limited to statistically amenable data sets rather than philosophical justification of simplicity through parsimony. Felsenstein said little about the "overthrow of the moderates" that Brinton emphasized as an essential part of revolution. The moderates, in my opinion, were the young systematists who wanted to use all the new methods, including computer analysis and biosystematy (common gardens, reciprocal transplants, cytology, autecology, and the like). They (we) are still around but the dedication of adherents of cladistics plus the magic of DNA has fueled their present hegemony in scientific culture. That cladists now control positions in universities, funding through granting agencies, and publication in major journals cannot be gainsaid. Vernon (1993) pointed out that "The history of taxonomy in the twentieth century, then, could be viewed as a response to its perceived low status."

Major changes in infrastructure (e.g., revamping herbaria to reflect the APG III classification system) are not left to a democratic vote by staff or users. There are cogent arguments against the biases of cladistics, pointing out associated often negative effects on biodiversity research. Such arguments against phylogenetic classifications cannot prevail, however, without offering an alternative that will catch the imagination of a new generation of systematists. This book is an attempt at such an alternative.

Many of the problems of systematics are not associated with the cladistic or molecular revolution, but with a kind of Cultural Revolution in business methods. The systems of Drucker (management by objective) and Deming (total quality, teamwork, customer focus) rededicated institutions towards financial prosperity, which led eventually to the moral hazard of monetizing anything collateralizable for debt and speculative leverage of publicly attractive products. The dislocations in systematics regarding a paucity of positions in classical systematics may be largely caused by academic institutions and natural history museums refocusing on a new business model monetizing popular aspects of science. (I remember the anthropologist at the museum I once worked at being forced to make popcorn for a horde of screaming kids. He soon left.) Although, at least anecdotally, there are more taxonomists than ever before, the "taxonomic impediment," given the biodiversity emergency, is real and urgent.

Classical taxonomy versus evolutionary systematics — Another, deeper view of the historical background of the phylogenetic revolution is provided by Vernon (1993). In the late 1950s, there were two contending factions: (1) classical taxonomists, who felt that taxonomy could exist on its own and produce, using standard methods, classifications that evolutionists could use in their own work, and (2) evolutionary systematists, who, by "putting evolutionary issues as the primary focus of taxonomy, ... sought to connect it to one of the most important biological questions of the time." The problem was that although some groups of birds and mammals had good fossil records and known breeding behavior, and some beetles, mollusks and butterflies were amenable to the evolutionary analysis of the day, most invertebrate zoologists, most botanists, and all microbiologists were not counted among practitioners of the cutting edge of evolutionary systematics. Classical taxonomy uncommonly involves direct inference of macroevolution but is intended to present a hierarchical classification.

So who won the mid-Century contest described by Vernon (1993)? Given the present-day hegemony of phylogenetics, it would seem that the evolutionary systematists won. But consider this—a cladogram is much like a dichotomous key in classical taxonomy, with similar nested state changes (including reversals if classical descriptions are polythetic). A cladogram can be viewed as a classification as long as the classification principle of holophyly is used to reject putting names of higher rank under names of lower rank, or the same rank nested in another rank; that is, the principle of holophyly may be used to reject any results that are not like a classification. A classification as imbued in a dichotomous key is presupposed and imposed on evolutionary evaluations in systematics. There are many ways a tree of life (Gontier 2011) depicts relationships. A phylogenetic Tree of Life (Pennisi 2003) is not an evolutionary tree, it is a classification based on a dichotomous key, although use of outgroups and clustering of state changes introduce an evolutionary dimension. Phylogenetics short-circuits a deep evolutionary analysis because sister-group trees are a ready-made hierarchical classification. Classification and classification principles (e.g. holophyly) are effectively treated as a natural process to be modeled in analysis. This leads to the fact that all evolutionary analyses done in phylogenetic systematics must fit a classificatory dichotomous key as a basic structure. In fact, modern cladistic systematics is the triumph of classical taxonomy over evolutionary systematics. Its methods are clearly attractive to classical taxonomists needing a philosophically and statistically impressive justification for their classifications. R. Feynman (1985: 313) wryly observed, "The easiest person to fool is yourself."

The basic analytic format imposed on evolutionary information in both hierarchic classical classification and cladistics is ((A, B) C), while that of evolutionary systematics is $A \rightarrow B$, or occasionally $? \rightarrow (BC)$ using information from both classical and cladistic analysis. The methods of phylogenetics are powerful, however, and much information on evolution can be derived from them as long as one can keep the cart behind the horse, and derive classifications based on evolutionary relationships, not evolutionary relationships from hasty classifications. The Framework attempts to remedy this.

A test for paradigm change — In most cases, I believe, paradigm change does not involve the stress of deprogramming, abreaction, and indoctrination; in fact, it may be scarcely noticed. The change may take many years, or be as simple as recognizing something as now "obvious" where before it was unthought or unthinkable. As an instructive exercise, the cladistically inclined reader might, before reading further, examine a complex cladogram in the literature and mark his or her way of interpreting it. After reading this book or some substantial part of it, examine again the same cladogram, and see if perception and interpretation of that cladogram are changed to an important extent. If so, this is a paradigm change in the small. Reviewers of this book might note that this subjective but quite real measure is the basis on which the author feels the success of this book depends. But will such a change spread across the field? Only the future marketplace of ideas will determine this.

Six elements — The proposed new context for biological systematics obviates inconsistencies in modern phylogenetic analysis with generation of an overarching theory of macroevolution through time that is particular for each group studied. It is pluralistic in using both classical and phylogenetic analytic techniques. There are six elements (as previously proposed, Zander 2010b): (1) Alpha taxonomy is a hard-won set of genetic-algorithm-based heuristics that in large is accurate in clustering relationships. (2) Cladistic analysis of morphology aids in developing a natural key to taxa by assessing transformations of weighted conservative characters. (3) Molecular systematics establishes genetic continuity of tracking sequences and order of isolation events of exemplars (but not necessarily speciation events) and may determine deep ancestors by taxa split apart on a molecular cladogram. (4) Taxa low in the morphological tree but high in the molecular tree are theoretically ancestral taxa of all lineages in between. (5) Superoptimization by maximizing theoretical ancestor-descendant hypotheses eliminates hidden causes as unobservable superfluous postulated shared ancestors, while biosystematic and biogeographic studies provide biological evidence, often experimental or quasi-experimental, that supports or modifies alpha taxonomy. Dollo evaluation at level of the whole organism allows inferences distinguishing progenitors and descendants. Implied reliable credible interval calculation and the use of coarse priors for a Bayes' Solution (allowing for statistical preselection) leads to consolidation of all evolutionary information into an evolutionary tree of serial (as opposed to nested) macroevolutionary transformations. (6) Classification by diagnosable macroevolutionary constraints requires a generalist Linnaean classification capable of representing to the most simplified degree all taxonomic

concepts. A methodological pluralistic analysis of evolution is here considered essential for a modern systematics based on all evidence relevant to theory of descent with modification of taxa.

Evolutionary systematics, or evolutionary taxonomy, is the science of apprehending nature through a naming system of nested groups, with the species as basic unit of classification (but see Chapter 8 on superoptimization), using the Linnaean system and, to the extent possible, what is known about evolutionary relationships. Evolutionary systematics uses both sister-group relationships and ancestor-descendant relationships as recommended by Darwin (1859: 420) in his "natural system" to present a classification and evolutionary tree reflecting descent (splitting or budding of lineages) with modification (macroevolutionary change or speciation and generation of higher taxa).

This is in contradistinction to the now popular Hennigian phylogenetic system, which focuses exclusively on sister-group relationships (splitting of lineages). Rieppel (2012) pointed out that Hennig's method "renders phylogenetic systematics a search for sister group relationships, not for ancestor-descendant relationships." According to Korn and Reif (2003: 688) "As the phylogenetic analysis in cladistics is based on the search for sister-groups only, real ancestral species cannot meaningfully be dealt with and also behave as 'noise'." Phylogenetics now incorporates powerful analytic tools, including statistical analysis of molecular data, but the elimination of representations of macroevolution in phylogenetic trees has led to various problems that have been pointed out in recent literature.

Evolutionary systematists generally agree that recognition of paraphyletic groups (ancestral groups denied recognition at separate and taxonomically equal rank to that of their descendants by phylogeneticists) contain important evolutionary information that should be represented in classifications. Well-known examples are the sinking or attempted sinking of Aves into Reptilia (particularly as discussed by Hörandl & Stuessy 2010), or the polar bear into the brown bears, or Cactaceae into Portulacaceae. Detailing macroevolution in classification through taxonomic recognition of paraphyletic groups is fundamental to evolutionary systematics. Macroevolution is a real scientific concept supported by plenty of data showing macroevolutionary transformations of derived lineages arising from the midst of paraphyletic lineages.

Phylogenetics is methodologically inconsistent. For example, one taxon may be represented by two exemplars that are distant from each other on a molecular tree (phylogenetic paraphyly or apparent polyphyly). Given the structuralist justification for evolutionary classification, this is explained away by "convergence" or not yet attaining "reciprocal monophyly" (the phylogenetic desideratum). But a cladogram does not diagnose exactly what different ancestors the two distant OTUs came from. The explanation is inconsistent with the method because it uses a different method. In addition, after years of cladistic emphasis on the importance of determining morphological homologies, this rule is thrown out the window. An alternative evolutionary systematics explanation discussed at length here is the scientific theory of a deep shared ancestral taxon with the same diagnosis inclusive of both OTUs, which is consistent. The *point* of classical systematic is to present information such that complete morphological convergence is quite improbable, certainly not as common as molecular analyses seem to make it.

Given the commonness of apparent "convergence" at the taxon level, if it were in fact true at that level, then the fabric of classical taxonomy fails. Because molecular analysis depends on classical analysis to name its exemplars, molecular analysis must also fail (there are no facilities in nestings for distinguishing taxa, only clustering them). Clearly what seems to be convergence at the taxon level (two different species or genera are molecularly the same or closely clustered) in molecular systematics is a phenomenon different from what is usually accepted as evolutionary convergence. Taking "convergence" as an explanation of multiple salting of exemplars among other taxa will take one to the nihilist position that there are no taxa, only lineages of OTUs. Examples of researchers willing to countenance such are Mishler (1999) and Fisher et al. (2007), among others.

A morphological cladogram may be different from a molecular cladogram, and both may be intuitively convincing, the one clarifying morphological

.

changes, and the other DNA sequence changes. Phylogenetics cannot explain this inconsistency, but tries to conflate them in "total evidence" analysis, lumping all data together. This combines inconsistent results into a jumble where multiple data points of one process (mostly non-coding changes in DNA) overwhelm, usually, apparently fewer data points of another, different process (fixation of expressed traits). Through Simpson's Paradox (support for alternative clades adds, in a combined data set tree, to a better supported branch order than for either alone), branch orders are often generated that are not in any cladogram from any one data set of a partitioned data set. Properly, if Bayesian analysis is done, inconsistencies must be addressed with Bayes' Formula, where low support *for* the molecular clade *from* the morphological tree may radically reduce support for the molecular clade. An alternative (evolutionary systematics) over-arching theory is that taxa basal in a morphological cladogram but terminal in a molecular cladogram signal the status of that taxon as surviving ancestor of possibly many lineages (a kind of cladogrammatic coelacanth). This last theory renders the differences between morphological and molecular analyses consistent. It is only a scientific theory, not a fact, but this is far better as a basis for classification than the apparent axiomatic perfection of a phylogenetic cladogram. A pattern is not an explanation, it needs an explanation.

At times cladists may accuse evolutionary systematists of "confusing pattern with process," deriding evolutionary theory as metaphysics while lauding pattern in science, encouraging "systematists to study patterns of relationship rather than to tinker with algorithmic models that specify evolutionary processes...." (Brower 2009).

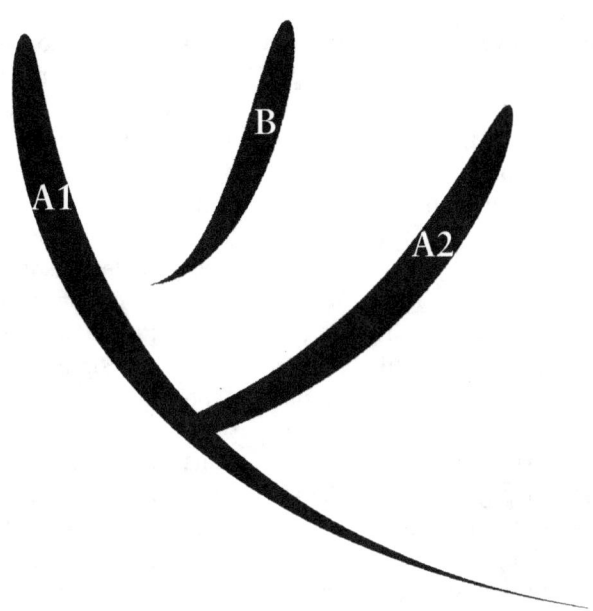

Plate 1.1 – A stylized dendrogram demonstrating paraphyly. When translated into macroevolutionary theory, paraphyly signals, in the simplest case (above), an ancestral morphological taxon "A" of two (or more) *species*, or of molecular *strains* "A1" and "A2" of one taxon "A" giving rise to a descendant taxon "B" from one of them ("A1"). Paraphyly on a cladogram can *therefore* be evolutionarily informative. Morphological paraphyly simply implies one taxon (species, genus or family) emerging from another of the same or lower rank. Molecular paraphyly directly implies a caulistic ancestral taxon inclusive of all paraphyletic terminal taxa. A caulogram of macroevolutionary transformations superimposed on a molecular cladogram may be considered a kind of Feynman graph (Watzlawick 1976: 234), with distance in space on one axis, and back and forth through time on the other axis.

Paraphyly — Brummitt (2006) emphasized that "paraphyly is the most important issue debate in taxonomy today." Paraphyly is, however, much condemned in both standard works on cladistics, and in specific by a number of cladist apologists (e.g. Schmidt-Lebuhn 2011). Paraphyly (Plate 1.1) is a somewhat disparaging phylogenetic word for what is generally known as ancestors involved in macroevolution, that is, a label for a group from which one or more other groups at the same (or higher) taxonomic rank have apparently evolved. "Para" implies faulty, wrong, amiss, or merely similar to the true form. Evolutionary systematists, on the other hand, celebrate that which is presently known as phylogenetic paraphyly.

Papers from a symposium on paraphyly are soon to be published in Annals of the Missouri Botanical Garden (last issue of 2013). I have not yet seen this issue, but it will doubtless be of relevance.

Molecular systematists restrict the term paraphyly to molecular paraphyly. They state (Rosenberg 2005: 1474) "paraphyletic genealogies are most frequent for only a short period of time" before reciprocal monophyly takes place. The present book asserts that although molecular changes occur and are fixed in all products of speciation, paraphyly as two lineages in morphological (and essential evolutionary) stasis is common, even though they gradually accrue different molecular tracking traits and become reciprocally monophyletic by such molecular traits. According to Vanderpoorten and Shaw (2010),

"We argue that there are events of major biological import that occur when a new divergent taxon is 'budded off' from within an ancestral widespread species; however, the point at which both species become reciprocally monophyletic can simply reflect the stochastic process of gene coalescence and is of no real biological significance in and of itself. Reproductive isolation through one mechanism or another is necessary, though not necessarily sufficient, for the development of [true] reciprocal monophyly. Thus, the evolution of reproductive isolation is of critical importance evolutionarily, whereas the development of [molecular] reciprocal monophyly is biologically trivial."

Thus, Aves is an apophyletic (= autophyletic, derived) product of the paraphyletic Reptilia; while Cactaceae is apophyletic to the paraphyletic Portulacaceae; and in my own field of specialization Cinclidotaceae, Ephemeraceae (Holyoak 2010), and Splachnobryaceae are all apophyletic to the large, extended paraphyletic moss group Pottiaceae (see discussion of Zander 2007, 2008).

According to the blurb on the home page of the Willi Hennig Society: "Hennig's idea that groups of organisms, or taxa, should be recognized and formally named only in cases where they are evolutionarily real entities, that is 'monophyletic,' at first was controversial. It is now the prevailing approach to modern systematics." I reject this simplistic, mechanical definition of an evolutionary real entity (and the egoist declaration that such is now largely accepted).

Tobias et al. (2010) argued in favor of a relaxation of phylogenetic monophyly:

"We take the...view, that a distinctive, reproductively isolated lineage can be classified as a species even though it is nested within a phenotypically homogeneous ancestor. To clarify, if subspecies A and B are phenotypically similar, but genetically and geographically interposed by a third divergent and reproductively isolated taxon C, it does not follow that the classification of C as a separate species must necessarily trigger the splitting of A and B....It is clear that lumping non-sisters in this way results in a mismatch between species and clades. However, we concur with Lee (2003), who argued that 'this mismatch is precisely what makes the species category worthy of special recognition: species are not merely another type of clade, but a different type of biological entity altogether.' From this perspective, useful information is lost when taxonomy is forced to reflect gene trees by either over-lumping daughter and parent species, or over-splitting inherently paraphyletic taxa, and thereby ignoring the evolutionary reality of the nested lineage...."

A natural taxon is any group that is probabilistically the best representation of the expressed traits as an evolutionary trajectory at that taxonomic rank. This is like the "natural taxon" of Gilmour (1940: 468), which he characterized by being the most highly predictive, in that what is predicted is the evolutionary trajectory. By evolutionary trajectory I mean whatever species concept an author uses that has an evolutionary dimension. When a taxon is split to fit a molecular tree, for a split to be a natural taxon it must be significantly more robustly supported by expressed traits than any other split in any other way of splitting. This is because some of many expressed traits (a combination by chance alone) may support any split that may have appeared in a molecular cladogram. The traits should include autapomorphic traits and any traits that particularly fit the organism to a particular environment. Finding some expressed traits by chance alone that support a taxon in any particular molecular clade is a problem in statistics called "multiple tests" or "multiple comparisons." An entire chapter of this book is devoted to this important subject.

Other methods have been proposed to reinsert the phyletic or divergence element back into classification (reviewed by Stuessy 2009). Unfortunately, these largely deal with atomized ("taxon agnostic") traits lacking the integration of expertise (as in phenomics and ontologies: Balhoff et al. 2013; Burleigh et al. 2013) in sometimes quite pruned data sets, rather than whole taxon. Atomized traits are now used as tracking traits (whether morphological or molecular) not as contributions to understanding adaptive strategies.

Egan (2006) pointed out that the total evidence paradigm of cladistics stems from Popper's writings, and that "exclusion of evidence could render the hypothesis untestable (by protecting the hypothesis from conflicting evidence, since testability (falsifiability) is central to corroboration...." Szalay et al. (2008) recommended a strong *a priori* weighting of morphological traits through biologically well-founded character analysis. The massive plant study (73,060 taxa) of Goloboff et al. (2009) combined molecular data with morphological data, the latter arbitrarily upweighted to be each equivalent to three base pair substitutions.

The recent patrocladistics (Stuessy & König 2008) method, for instance, agreeably recommends character weighting of states of evolutionary importance, and largely identifies apophyletic lineages that differ by large numbers of apomorphies and autapomorphies (assuming these are included in the data set) plus a measure of cladistic (number of nodes) and patristic (number of apomorphies) distance between taxa. Using autapomorphies is valuable but again reduces to a few major traits a taxon that is better viewed as a gapped cluster of traits with central tendencies and a biological dimension associated with selection to a particular habitat, in short a biological entity.

Using a taxon or exemplar as an outgroup in morphological cladistics has much to recommend it over a contrived hypothetical outgroup, yet in some cases some of the traits of the outgroup are clearly advanced. These traits bias the analysis, which after all is founded on classical systematics. Cladistic analysis merely presents in morphological cladistics the relationships of classical systematics as a easy-to-understand tree. Traits of an outgroup should be changed to hypothetical plesiomorphic trait states if clearly advanced in the outgroup's own taxonomic grouping.

It has been long recognized that phylogenetic analysis does not model evolution of branching series of named ancestral and descendant taxa, i.e., genealogies, but demonstrates the evolution of characteristics as branching lines of trait changes (e.g., Bowler 1989: 345–346; Farjon 2007; Hörandl 2006, 2007) for exemplars of named terminal taxa. Of three particular taxa, two are more likely to share an ancestor (Williams 2002), which seems a good deduction from theory. That ancestor, however, is generally not identified as a taxon different from its descendants; it is simply represented in phylogenetics by an unnamed node, or "common ancestor" of descendant lineages. This is allowable only when there is evidence that pseudoextinction has occurred.

When fossils are at hand, however, they are potentially more informative of evolution as descent with modification (Hall 2003) of taxa because bioroles may be inferred from expressed trait combinations. In phylogenetic analysis, on the other hand, ancestral mapped morphological or molecular traits, though presented as sequential, remain atomized. Attempts to infer soft tissues in geologic fossils also deal with individual traits. For instance, in extant phylogenetic bracketing (Witmer 1995, 1998), a fossil lineage bracketed by two lineages each sharing one particular trait in their extant taxa would be expected to also have that trait, but features not present in both bracketing lineages would be expected to be

absent in the fossil. This method does not infer taxic ancestors, which the present method of heterophyly (i.e., non-monophyly, Zander 2008) in molecular trees theoretically can do.

It should be noted that phylogenetic paraphyly exists only because shared ancestors (nodes) are not named. It could very well be, and probably often is, that an ancestral taxon of a number of clades of one name extends below the set of phylogenetically monophyletic clades to include one or more clades of a different name at the same rank or higher. Thus, monophyly is hidden in the case of trees that only show nesting of exemplars and do not inferentially name shared ancestors. *Phylogenetic analyses cannot estimate evolutionary monophyly.*

Although paraphyly usually implies that the apophyletic (derived) taxon is a descendant of the paraphyletic taxon, this may be reversed in the case when a taxon found to be near the base of a morphological cladogram (or from other data) and is therefore probably primitive (see Glossary) is perceived (by the analytic software) as an apparently deeply nested taxon. This is a case of self-nesting laddering (see chapter on superoptimization), and the more primitive taxon, though deeply nested in a molecular cladogram, is the progenitor of all lines between its place on the morphological cladogram and its place in the molecular cladogram. This explanation may seem mechanical, yet the explanation is solidly based on macroevolutionary theory, that is, descent with modification of taxa. An example of macroevolutionary transformation that is a much extended paraphyly is that of the moss genus *Erythrophyllopsis,* see extensive discussion elsewhere in this book. The detection and elucidation of macroevolution at the taxon level in the transformation implied by paraphyly is an example the need for "discursive reasoning" in pluralist systematic analysis.

Textbook examples of identification of paraphyletic taxa as ancestral are given by Futuyma (1998: 456, 470), citing Moritz et al. (1992) where coastal and Sierran Californian subspecies of the salamander *Ensatina eschscholtzii* "appear to have been derived from" subsp. *oregonensis,* and citing Hey and Kliman (1993) and Kliman and Hey (1993) for the *Drosophila melanogaster* species group where the paraphyletic *D. simulans* "gene copies are traced back to a 'deeper' common ancestor than in any other species." Rieseberg and Brouillet (1994) discuss mechanisms for evolution of monophyletic daughter taxa from paraphyletic parental taxa through geographically local models of speciation. All this assumes that the molecular analysis has accounted for any homoplasy introduced into the analysis by inappropriate technique, e.g. wrong model (Alfaro & Huelsenbeck 2006) or inappropriate data, e.g., incomplete concerted evolution (Doyle 1996).

Evolutionary paraphyly versus phylogenetic paraphyly — In cladistics, by definition, every clade is monophyletic. This is a misuse of the term monophyly (all terminal groups derive from one ancestor) and phylogenetic monophyly should be retermed "cladophyly." Consider the following cladogram:

$$((((((A, B) C) D) E) F) G) H, I$$
where I is outgroup

By definition, phylogenetic monophyly occurs at every close parenthesis in this example of a pectinate cladogram, that is, encompassing every clade. Every set of exemplars distal to a close parenthesis is a phylogenetically monophyletic group. This introduces a initial amount of certainty in estimation of monophyly. On the other hand, suppose A, B, and C were directly derived from the taxon A; also, D, E, and F were directly derived from the taxon E. If so then *evolutionary monophyly* would include only those groups distal (to the left of) to the close parenthesis between C and D, and the close parenthesis between F and G. We thus have only three monophyletic groups given this additional information. The additional information must necessarily add uncertainty to any statistical estimation of those three evolutionarily monophyletic groups because such analysis is based on scientific method, not definition.

Eliminating science-by-definition, we have the strange case of cladists recognizing *evolutionarily* paraphyletic groups (e.g., those distal to every close parenthesis but the three mentioned), while evolutionary systematists themselves have found themselves defending phylogenetic paraphyly (as an expected effect on a cladogram of macroevolution in molecular cladograms). There are in the example above six close parentheses and only three of these imply evolutionary monophyly, thus in this example cladistics has only a 0.5 chance of establishing true monophyly in an analysis of the data set.

Other than for this section, to be clear, paraphyly when discussed means phylogenetic paraphyly unless otherwise noted. Cladists have never been identified, to my knowledge, as recognizing paraphyletic groups, but they do, and it is wrong.

Explicit and implicit paraphyly — In molecular systematics, there is simple paraphyly, e.g. ((A1, B)

A2) C, D. That is, one lineage of another taxon (B) between two lineages of the same taxon A). There is also extended paraphyly, e.g., ((A1, B) C) A2) D, E. In this case, two or more lineages of different taxa B and C between two lineages of the same taxon A).

Explicit paraphyly has extant heterophyletic lineages that signal paraphyly, e.g. A1 and A2. Implicit paraphyly is missing one of these two lineages. Any indication in a cladogram that one of a sister group is clearly the ancestral taxon of the other of the pair signals implicit paraphyly. How common is implicit paraphyly? Well, if explicit paraphyly is common, one might expect implicit paraphyly to be even more common.

Extended paraphyly has the distinction of being able to scramble branch ordering in molecular cladograms. When explicit, one can make allowances. When implicit, the switching of branch ordering in the cladogram is hidden. It is possible that analysis with traits other than phylogenetic that the correct branch order might be detected, but this would never be easy or sure.

Inferring monophyly solely by maximizing parsimony of trait transformation leads to evolutionary paraphyly. Suppose we have a number of taxa terminal on a clade. Two are most terminal with shared synapomorphies. The next taxon down, however, is clearly derived from the same ancestral taxon as the two most terminal taxa. Asserting that the two most terminal taxa are monophyletic splits the ancestral taxon.

Try it with a group of your specialization. Many subgenera commonly have some one wide-ranging species of generalized morphology with some closely related species specialized into more recent habitats. A theory could be developed (by yourself) that these are all daughter species of the more generalized species.

It is a fallacy that all daughter species must occur as polychotomies, since, in morphological analysis, given few important traits, some will reverse and some will be duplicated given false (aleatory) resolution. In molecular analysis, extinct and unsampled molecular strains confound resolution totally when one ancestral species gives rise to two or more other taxa (lineages may survive from any point in the cladogram the ancestral taxon has occupied).

The fact that cladistics has promoted recognition of evolutionary paraphyly (while damning phylogenetic paraphyly) has been one of those things right in front of us for thirty years. We've ignored it.

Definitions for variations on a theme — *Pseudoex-*

tinction is the preferred manner of speciation in phylogenetics. In pseudoextinction, the shared ancestral taxon disappears fairly rapidly by anagenesis after generation of descendant species, and becomes the second of a sister-group pair. Grant (1971: 48) dissects stages in divergence of this phenomenon. Pseudoextinction is doubtfully as common as expected by phylogeneticists (Raup 1986; but see Hegde et al 2006) but forms the analytic basis of phylogenetics. If universal, it would imply that any two taxa are then more closely related to each other than to a third, also basic to cladistic analysis. If much less than universal, cladistic analysis and simplicity as an analytic method becomes doubtful. In phylogenetics, pseudoextinction is automatically treated as universal.

The idea of pseudoextinction is now so prevailing that some new and interesting methods are inadvertently biased by it. For instance, Shaban-Nejad and Haarslev (2008) introduced category theory as a way to analyze evolutionary relationships. Unfortunately, pseudoextinction is made central: "...category theory is capable of solving problems related to reverse analysis (mentioned in cladistics method) through recursive domain equations [33]. In order to analyze the bifurcating pattern of cladogenesis, which states that 'new organisms may come to exist when currently existing species divide into exactly two groups' [6], we have used two categorical constructors: pushouts and pullbacks." They cite a phylogenetics Web site at Berkeley, and a paper on category theory by Smyth and Plotkin (1982). This indicates that multifield attention to scientific problems may be initially stymied by misunderstandings or superficial knowledge of specialized theory.

Many or even most nodes in cladograms can be assigned by evolutionary systematists to a speciation process different than pseudoextinction, using judgment and insight with non-phylogenetically informative data. That process is speciation without dissolution of the ancestor. An example of superoptimization of *Didymodon*, see Plate 8.1 below, demonstrates only a single node among more than 20 as evidence of pseudoextinction in that genus.

Readers may feel they detect a bit of circular reasoning, that is, ancestral taxa are such because they have descendants, and descendants are such because, well, they have ancestors. Not so! There are clear criteria for identifying ancestral taxa (that is, probable ancestral taxa) as discussed in the treatment of *Didymodon* in this book. In the most simple case, a taxon closely related to another may be its descendant if isolated, specialized, and recent, while the other, the potential ancestral taxon, may be widely spread in ancient geography and habitats, polymorphic, and unspecialized (following Mayr 1954). Postulating a new ancestral taxon at each cladogram node is unparsimonious when an ancestral taxon is often extant and identifiable. Consider polar bears and brown bears. Did polar bears evolutionarily generate brown bears as glaciers retreated? Probably not, since prey seals do not live in glaciers, and there are many bears in a variety of ancient habitats more like the brown bear than the highly specialized polar bear (but see Hailer et al. 2012).

The ancestor may be extant or at least last long enough to generate additional descendants. *Paraphyly* is generation of a descendant from one branch of two extant lineages (usually molecular lineages) of the *same ancestral species* or taxon. *Extended paraphyly* is generation of two or more lineages from one of two extant molecular lineages of the same ancestral species. *Pseudopolyphyly* is generation of descendant lineages of the same taxon (say two species of the same genus) from both branches of an extant ancestral taxon of the same name. It is much like parallelism but is restricted to molecular analyses. *Heterophyly* is simply a general term for all these processes.

True parallelism and polyphyly is when two or more same-name descendants are each generated from a taxon of a *different name* but at the same taxonomic level. One might term this phenomenon anastasis (resurrection), and has been discussed as evolutionary Lazarus taxa (Zander 2006), as opposed to the geologic Lazarus taxa of Jablonski (1986) that simply have large gaps in the fossil record. One must be able to demonstrate, however, that the ancestral taxon is indeed different from the anastatic descendants; otherwise the apparent parallelism or polyphyly (often ascribed to "massive" convergence) is more likely due to heterophyly.

The generation of two different descendants from one ancestral species or taxon certainly may be called extended paraphyly if two branches of the ancestor are extant and bracket the descendant(s). But they have the same effect on evolutionary analysis if only one branch of the ancestral species or taxon is extant, or even if no ancestral populations are extant and leave a legacy of two or more nodes of a cladogram having the same ancestral taxon name and character.

Limits to tree resolution by potential for unsampled or extinct extended paraphyly — "Phylogenetically informative" may prove somewhat of an oxymoron. This is because empty precision leads to aleatory classification. This is how—in parsimony analysis of morphology, traits are not necessarily tacked onto a taxon as speciation gradually continues, but an initial linked set of traits may be necessary for selection into a new environment. Thus, if A and B share three traits that are selectively linked, and A and C share two traits that are not (maybe neutral or sequentially added as the environment changes over time), then A and C probabilistically share the latest ancestor, not A and B. Although when dealing with masses of shared traits, main clusters of a parsimony cladogram may be okay or acceptably approximate, parsimonious decisions about relationships of small groups of OTU's may need additional information, but are for now cladogrammed by chance.

In a theoretic discussion of the effect of unobserved extinction on modeling macroevolution, Stadler (2013) used a model based entirely on phylogenetic splitting, and did not discuss heterophyly. In molecular analyses, any sister group pair may have had an extinct lineage identical in phenotype to one of the sister groups occurring below the split. If so, then this is not a sister group relationship but ancestor-descendant relationship instead. If the extinct lineage identical in phenotype to one of the sister groups is even farther down in the tree (phylogenetic polyphyly, or if within reasonable patristic distance then extended paraphyly), then the molecular tracking of splits in the gene history is further compromised. This is further gone into elsewhere in this book, but here I report an email exchange I had recently with a phylogeneticist. We were discussing offline an exchange of views held publically on the listserver Taxacom, and I wrote to him, "If two or three or four nodes in a row are the same taxon (which can be estimated with non-phylogenetically informative data), the import of the 'shared ancestor' is lessened. If pseudoextinction is very rare, the import of pseudoextinction is close to zero." He replied, "Two or three nodes in a row can't be the same taxon. Cladogenesis implies speciation. If you are working within the limits of cladistic reconstruction, this scenario is non-existent." I'm not sure how many phylogeneticists agree with this, but the sentiment is doubtless typical.

The results of macroevolution can often give statistically near-certain sets of nested lineages of present-day specimens (exemplars). But should we use these nested patterns for classification? Because macroevolution involving progenitors in stasis shuffle lineages of taxa, even trying to "fix" the pattern by renaming taxa that are out of order does not give a classification that reflects evolution well. Only by going beyond pattern, and using phylogenetic pattern to help infer evolutionary process, can we create a classification that is not often plain wrong.

The evolutionary story has been lost to reductionism in ignoring all information on evolution not in a database of phylogenetically informative traits, and to irredeemably faulty methods of analyzing evolution and assessing classification (e.g., sister groups in, sister groups out). Phylogenetic analyses can be important if interpreted in the pluralist context of information from chromosome counts, ecology, biogeography, phyletic weighting of traits, and genomics, among other information. The phylogenetic practice of renaming taxa that occur in two or more different lineages or of lumping paraphyletic groups with their autophyletic macroevolutionary products is just ignoring significant evolutionary information to preserve assumptions that are contrary to reality (e.g., the false notion that "a taxon cannot be in two molecular lineages at once") and save the hyperexact Method.

Extinct or otherwise unsampled paraphyly is a problem with resolution of sequence of molecular lineage splitting. The resolution of a molecular tree depends on distinguishing extended paraphyly, i.e., a reasonable inference of a deep shared ancestral taxon (evolutionary monophyly) from evolutionary polyphyly (no reasonable inference of a deep shared ancestral taxon). The question remains whether any particular sister group is or is not the remnant of a paraphyletic ancestor, which would affect accuracy of mapping of expressed traits or taxa on the molecular tree. Without additional information like relative age of the groups involved, the best guide is the extent of paraphyly or extended paraphyly, by some measure of patristic distance, of related extant natural taxa.

Without other data, a cladogram with 10 percent of the nodes exhibiting paraphyly in extant taxa may indicate that 10 percent of the ancestral nodes at any past time were also paraphyletic. Individual lineages that are well-supported by bootstrapping or credible intervals are in no way immune to this problem. Other data possibly of value in evolutionary analysis preliminary to classification include various autapomorphic (phylogenetically uninformative) traits, paleontology, chemistry, ecology, biogeography, chromosome numbers, and any other information that might throw light on ancestor-descendant relationships of accepted or natural taxa. Also relevant here is recent work on irreversible traits (Bridgham et al. 2009).

Molecular systematics alone cannot determine branch order of taxa, not even with dense sampling, because of the possibility of extended paraphyly of extinct or unsampled molecular strains of the same taxon. This last is doubtless common given the prevalence of paraphyly among extant taxa.

A perspective — The way phylogenetic analysis has itself evolved apparently parallels recent changes in the way history proper is studied. Fischer (1989) pointed out that three generations ago, there was a standard paradigm for doing (non-science) historical work consisting of narrative reports of a fairly narrow class of variables (authority and power in politics through time), based on thorough, Gestalt knowledge of the literature, with major findings offered as interpretations discovered by intuition underlain by testimony. Early in the 20th Century, the topics of history expanded greatly and historical relativism became central (1930–1960), though unsatisfactory because such was static. In the 1960's the French school of the "Annales" invented a radical new method that examined change in all of social history, requiring rigorous methods of logic and empiricism. As a synthesis, however, in the 1980's this newest paradigm failed by devolving into competing special fields with narrow focuses and philosophies of study. Fischer's solution was to combine as well as possible the best elements of all previous syntheses. Both interpretation and empiric evaluations contribute to a more broadly based interdisciplinary view of history, combining fact-based, interpretive story-telling and rigorous empiric problem-solving as a "braided narrative."

A history of phylogenetics (e.g., as related by Felsenstein 2004) follows approximately this nutshell historiography. Originally, evolutionary work was based on thorough, Gestalt evaluation of the facts, resulting in reasonable scenarios. Then, emphasis on data from crossing experiments, common gardens, reciprocal transplants, cytology, and other fields supported a "New Systematics" with more robust, empirically based narratives. Phenetic analysis of the 1970's introduced a rigorous mathematical method emphasizing similarity, with prediction focused on predicting phenetic similarity. Cladistics, with a competing new rigorous method based on maximum parsimony, then gained popularity, and proponents ridiculed the older descriptive methods (e.g., Crowe 1994) as overly subjective and similar to the "just so stories" of Kipling (1966), e.g., "How the Leopard Got His Spots." Dennett (Dennett 1984; Brockman 1995: 180), however, lauds such narratives as "intuition pumps."

The fundamental phylogenetic presupposition—that of any three species two are more closely related—fails totally in two cases: (a) paraphyly, including nesting of genera among species of other genera (Plates 6.1, 7.3, 13.2); and (b) when any one generalist, wide-ranging extant species can be easily hypothesized as ancestral to two or more derived, highly specialized, and possibly evolutionarily dead-end descendant species (Plate 8.1, 8.2). Both cases are common. Ergo phylogenetic resolution of certain stretches of branch order is commonly random in both morphological and molecular analyses. Extensive explanation is given below.

Today, well-known methods such as maximum parsimony, maximum likelihood, and Bayesian Markov chain Monte Carlo methods all have their own partisan schools and somewhat different results with the same data (but see Rindal & Brower 2011), while new techniques both complex, e.g., codon substitution (Ren et al. 2005) and simplistic, e.g., DNA bar-coding (Hebert & Gregory 2005; Will et al. 2005), vie for researchers' attention, and all commonly relegate morphology into the background or at best include it submerged in total evidence studies. There is no return to systematics of the past because all new methodological practices and viewpoints contribute positive analytic aspects. A pluralistic Newer Systematics must include the most powerful features of phylogenetic systematics and downplay or exclude the contradictory or biased. That researchers are now willing to countenance pluralistic analyses rather than the mechanistic phylogenetic model is exemplified by the work of Hörandl & Emadzade (2012) who used several methods in an attempt to balance evolutionary viewpoints in a taxonomic study of the plant genus *Ranunculus* (Ranunculaceae). There criteria for using monophyly and paraphyly were not as critical and restrictive as those presented in this book. A scientific pluralism, capable of combining all methods, is here predicated on pursuit of evidence of macroevolution. Using the best methods and theory from multiple fields and experiential vantages to address a problem in a new way is typical of advancement in science, and is clearly a "positive sum game" (Wright 2001).

CHAPTER 2
Pluralism versus Structuralism in Phylogenetic Systematics

Précis — Mechanical knowledge based on a selected subset of information and a simple model intended to reveal relational structures in nature is pitted against process-based theory meant to explain all information.

Phylogenetics is a form of structuralism, offering a precise, statistically well-supported pattern of present-day relationships as a fundamental axiomatic ground plan to which all other data are relegated (Zander 2010a,b). Pattern cladism seems to match Chakravartty's (2004) definition of the ontic form of scientific realism (there are no real objects, only structure), while less dogmatic forms of cladism match his "epistemic" form (relationships are based on real but unknowable objects. According to de Queiroz (1992), evolution is a "central tenet from which the principles and methods of taxonomy are to be deduced." Even the most simplistic central tenet ("evolution happens") for a topic as complex as evolution, however, is not axiomatic or empirically simple, particularly if evolution is defined as nested relationships rather than serial macroevolutionary transformations, or these relationships based on characters evolving rather than taxa. Deductions of taxonomic methods and principles from the argumentative evolutionary literature must be limited to suggestions for approaching general study and specific ideas for analyzing the distinctive evolution of particular groups. Structuralism as used here is a more general term than developmental structuralism, i.e., evidence of design limitations in body form or "ontogenetic trajectories."

According to Korn and Reif (2003), cladists regard the nested hierarchy of taxa and the nested hierarchies of characters as caused by a "natural law," in particular a stance of pattern cladists who believe morphological data is more important than explanations, in a "theory-free" analysis (discussed by Kearney & Rieppel 2006).

A cladogram, even a molecular tree, is, however, not a fundamental pattern, being rife with inconsistencies, e.g., paraphyly within molecular results, and between morphological and molecular results. Paraphyly is actually what macroevolution looks like on a molecular cladogram. Phylogenetic inference is wedded to an artificial classification principle, *holophyly* (strict phylogenetic monophyly), that determines whether taxa merit names. Holophyly refers to a named group containing the common ancestor, all organisms descended from the common ancestor, but no other organisms; this means that a named group cannot have another named group of the same or higher rank nested in it because the first group then would not include all organisms. Although paraphyly is the main clue to macroevolution in cladograms, phylogenetic methodology completely rejects representing macroevolution as distinguished by pseudo-extinction or budding evolution in cladistic evaluation of evolution. This is inconsistent with evolutionary theory. Dismissal of the relevance (other than as ordered by the cladistic pattern) of information other than that which is phylogenetically informative of sister groups means that structuralist systematics not only rejects empiricism but also scientific inference itself in the context of the scientific method (Cleland 2001).

A new Framework is offered here that explains inconsistencies and conciliates classical and phylogenetic systematics with an over-arching diachronic (through-time) theory of macroevolution as a basis for a robust classification. It is pluralistic in using both classical and phylogenetic analytic techniques. Stevens (2008), in an explanation of his phylogenetic classification system wrote that "Evolutionary classifications in general try and combine phylogeny and morphological gaps, although that is no easy thing to do - it is akin to combining chalk and cheese…." The pluralistic context of the present book, however, it is like two barrels of a pair of binoculars (morphology and DNA), the two images being combined in the brain to reveal a third dimension (well-explained macroevolutionary transformation). Given that methodologically pluralistic inference of serial macroevolutionary transformations of taxa (stem-thinking leading to a caulogram) is analytically orthogonal to nesting sets of taxa by most parsimonious, maximum likelihood, or Bayesian Markov chain Monte Carlo trait transformations (tree-thinking leading to a cladogram), it may be difficult for phylogeneticists to suspend belief long enough to consider the former seriously. The idea of a certain fundamental structure against which all else may be measured and guided has an explosive psychic power, whatever its objective validity.

Taxonomic and evolutionary studies are closely intertwined. It is advanced here that the methodological fundamentals on which a theoretically pluralistic systematics (Padial et al. 2010) should be based are the following basic precepts: (1) alpha taxonomy as a hard-won set of informal genetic algorithms as rather successful heuristics, often termed "expertise"; (2) rigorous statistical re-evaluation of published molecular cladograms (Cohen 1994; Gigerenzer et al. 1989; Zander 2007a); (3) recognition that possible surviving ancestral taxa (Lewis 1962, 1966; Lewis & Roberts 1956; Mayer & Beseda 2010; Vasek 1968) may introduce uncertainties in cladistic analyses, as may extinct or otherwise unsampled paraphyly (Zander 2007b, 2008b); (4) ancestral taxa may be mapped on a molecular tree through discursive reasoning and inference from paraphyly or phylogenetic polyphyly on molecular cladograms (Zander 2008a, 2010a); and, (5) additional mapping of taxa to tree nodes is possible through cross-tree heterophyly (superimposition of morphological and molecular cladograms) refereed by Dollo's Rule at the taxon level. At the present time, evolutionary study is strongly pressured by phylogenetic ways of thinking. According to Butler (2011), "Without a robust phylogeny in place upon which to base biological classifications, i.e., the raw data sets which palaeobiologists analyze and draw their conclusions from, model-based analyses of diversity and evolution will have questionable empirical justification, especially if the underlying taxonomic framework is riddled with paraphyletic groups and taxa of dubious validity." Although evolution as a theoretic study is not as vulnerable to phylogenetic methods as is systematics, its data are.

Pluralism in systematics is not dialtheistic, where two conflicting phenomena do coexist (e.g., one atomic particle traveling through both slits of a graticule at the same time), or paraconsistent, where two theories are necessary for complete description and explanation of a subject (e.g., particle and wave theory of light), but it means that although facts and patterns may be diverse they can be held together by postulating through abduction an overarching, unifying scientific explanatory process-based corrigible theory (macroevolution) involving both induction and deduction. Sober (1991: 20) pointed out that restricting analysis to deduction is conservative in avoiding reaching false conclusions from true premises, while inductive analysis has more theoretical power but chances false conclusions from true premises. Although phylogenetics avoids induction, its deductions are not just conservative but commonly also extremely biased. "It is always dangerous to reason from insufficient data," cautioned Sherlock Holmes. Skepticism in science follows the dictum that knowing is the enemy of learning. According to Gigeren-

zer et al. (1989: 288), discussing mechanized knowledge:

"Of course, this escape from judgment is an illusion. All inference techniques depend on a modicum of good judgment to guide their application. Once applicability has been decided, judgment must intervene again to set the decision criterion, in the case of Neyman-Pearson theory, or the level of significance in Fisherian null hypotheses testing, or the prior probabilities in Bayesian inference. No amount of mathematical legerdemain can transform uncertainty into certainty, although much of the appeal of statistical inference techniques stems from just such great expectations. These expectations are fed by ignorance of the existence of alternative theories of statistical inference, by the conflation of calculated solutions with unique ones, by the reduction of objectivity to intersubjective consensus, and above all by the hope of avoiding the oppressive responsibilities that every exercise of personal judgment entails. It would be unjust to blame the mathematical statisticians for these false hopes, although some of their number have shared them. Rather, the fascination with mechanized inference stems from more widespread yearnings for unanimity in times of strife, and for certainty in uncertain circumstances."

Carnap (1967: 29) wrote that "...each scientific statement can in principle be transformed so that it is nothing but a structure statement." This is limited to logic (including math), although Carnap does address the "intersubjective world," but probabilistic science requires that structure be less strictly regulated than by theorems and lemmas. In the context of statistical science, when scientific statements are mutually incomplete, a joint structure should be theorized.

One may note with little surprise that the U.S. National Science Foundation recently (2009) awarded a grant (number 0928772) of US$498,813.00 for research on "Developing an axiomatic theory of evolution." From the abstract, the goal is to "develop a single body of mathematical evolutionary theory that is based only on assumptions that we know to be true...." Hmmm. It will "illuminate the underlying mathematical unity of evolutionary theory" and solidify "evolutionary biology as a science grounded in universal mathematical rules." Given that mathematics is grounded in axioms, this is clearly a return—a well-funded return—to axiomatic, rule-driven science. Balhoff et al. (2013) and Burleigh et al. (2013)

present rule-driven taxonomic analysis for morphological data, which have positive aspects yet much may be lost if a GenBank-like MorphoBank (O'Leary & Kaufman 2007) is presented as a substitute for examination and re-examination of collections.

There are three kinds of knowledge. *Positive* (empirical) deals with what is. *Normative* (of norms, such as culture and ethics) deals with what should be. *Artistic* (skilled presentation of what could be) draws a target around where the arrows have landed. How the arrows got there is the process-based question.

Phylogenetics deals with *three patterns*, traditional classification, morphological cladistics, and molecular trees, now eliminating the first two by mapping or otherwise relegating them atomistically to the molecular tree. The suppression of two disparate but informative patterns in favor of the one that is most precise and involves the most data is not good logic, since it merely suppresses or relegates (in "integrative systematics") contradictions and does not explain them.

This is structuralism (Zander 2010b), a rejection of empiricism, not merely bounded rationality (Gigerenzer & Selten 2001; Martignon 2001) or reasoning under the constraint of not quite sufficient knowledge (Tversky & Kahneman 1974). One may evaluate a scientific construct by (1) proving it to be a fundamental pattern in nature, as was done in the work of Euclid, Archimedes, Newton, and Einstein, but not for systematics where multiple patterns must be accounted for; (2) falsify every other relevant theory or at least avoid the "vicious ambiguity" (van Deemter 2004) of almost-as-good alternatives; or (3) corroborate (match as "not incompatible"), (4) or support (with data adequate for stand-alone evidence) the theory. Note that many cladists apparently (Laurin 2010: 701) dispute that ancestor-descendant relationships "can ever be proven," while the reader would doubtless agree that proof has never been more than a welcome but rare surprise in science. Phylogenetics as a field has been fascinated with the potential of the computer to deal with large data sets, and has been trapped into a few computer-friendly methods of analysis. A succeeding systematics that uses computer methods, heuristics, and discursive reasoning, avoiding complete focus on mechanical methods, should be a major advance. (It is the second mouse that gets the cheese.)

Tobias et al. (2010) offered a statistical method of extending species limits determined for well-known avian faunas to birds of relatively unknown faunas, particularly of tropical areas, using mean divergence among known sympatric species to gauge taxonomic status of allopatric forms. The potential is vast, considering it is not presented as an absolute criterion, although it is somewhat limited by assumption of the biological species concept. A paper on a similar topic by Paradis (2005) also uses powerful tools but is more limited in restricting results to clades.

Reason, judgment, insight. — The antidote to structuralism in systematics is a return to scientific reason, judgment, and insight. *Reason* is simply inference, including abduction, deduction, induction and analogy. A nested set of taxa in a software-generated cladogram does not necessarily equate to a nested set of taxa in a classification based on known or well-supported theoretical processes in evolution that may affect each taxon and using (and explaining) all relevant information. *Judgment* includes decision theory, particularly the heuristics and biases program (Tversky and associates) and fast and frugal methods (Gigerenzer and associates) as reviewed in this book. It means deciding what features of morphological cladistics and molecular systematics are relevant and which are aleatory or otherwise irrelevant. Classical taxonomists must make well-considered use of phylogenetic methods and vice versa. *Insight* means informed scientific imagination, such as having the courage to examine the ideas and theories underlying a research program when systemic problems occur, and of inventing new approaches.

The importance of descriptive taxonomy — Fitzhugh (2012) suggested, in a philosophical treatment of the logical basis of biological systematics, that support for phylogenetic analysis is little more than the premises used to found the hypotheses, and thus there is actually little testing involved because test evidence cannot be supplanted by character evidence used to create the hypothesis. He wrote:

"At its best, systematics enhances descriptive understanding, and within limits the pursuit of proximate causal understanding. Where it has been especially remiss is in elevating the importance of specific and phylogenetic hypotheses beyond what they usually are—initial, very vague explanation sketches—as well as claiming increases in evolutionary understanding where none exists."

I suggest, in the same vein, that molecular exemplars are not selected randomly but are preselected from clusters based on classical study. If molecular data sets are entirely determined by preselection of morphologically defined taxa embedded in a natural key, then agreement of molecular and morphological cladograms means little, and disagreement *may* be due to wrong morphology *or* to some bias inherent in phylogenetic analysis. If reexamination of classical taxonomy still supports the classical analysis fully, then the molecular analysis is biased or a new analytic viewpoint is needed. Several biases are described in this book, and additional are doubtless to be discovered. In other words, molecular systematics does not create a separate taxonomy from dense sampling that stands on its own, and which is comparable to that of classical systematics. It therefore does not provide real support when the two agree, nor real refutation when they do not agree. This is true even after accounting for biases in molecular systematics, including here detailed self-nesting ladders, these explained later. This is not to say that classical taxonomy does not have intrinsic biases, like morphological convergence, yet these are well-documented and there are standard work-arounds. Biases and other problems associated with classical heuristic methodology are addressed in Chapter 12.

In the present book, classical taxonomy is used to inject a separate, well-supported set of data and inferred relationships into molecular systematics, which does, given allowance for known biases, contribute information on macroevolutionary transformations. Molecular analysis is not a true test of classical and morphological cladistic analyses of relationships. The only way to use molecular evidence alone is to redefine evolution as any phylogenetically informative changes in molecular sequences. Ignoring all other evidence is not scientific.

Precision and accuracy — Precision is a hallmark of mathematics. One plus one always equals precisely two. Precision is also accurate in mathematics. Yet when measurements of things in nature are involved, precision does not imply accuracy. For instance, sawing a stick into 100 1 cm lengths does not provide a 100 cm length when you glue them together.

Philosophers Millstein et al. (2009), discussing mathematical models of evolutionary drift, wrote:

"As the Hardy-Weinberg Principle, the equation $[(p + q)^2 = p^2 + 2pq + q^2 = 1]$ represents a physical process: the maintenance of genotype frequencies in a population of randomly mating, sexually reproducing organisms from one generation to the next. The physical process can be represented mathematically, but it is not purely mathematical.

As the sum of areas, however, the equation does not represent a physical process. It is purely mathematical; it is simply a geometrical relationship. Thus, it is a mistake to derive definitions from mathematics alone, as DOA [Drift as Outcome Alone] seems to do, since many, very different definitions can be derived from the same equation. Moreover, it is problematic to think that ontological questions about the causality (or lack thereof) of terms appearing in equations can be gleaned from the equations alone. On some interpretations, a physical process is represented, but on others, it is not. There is no way to tell from the mathematics alone. So, one reason that it is a mistake to think that we can glean definitions of drift from mathematics alone, as DOA seems to do, is that the same mathematics can give rise to radically different interpretations."

Empty precision in mathematics is paralleled by empty precision in statistics. The over-precise is matched by the over-focused. For example, the average measure of an anatomical element might be 10 units. Averaging can be highly precise, but the use of estimated ranges in classical taxonomy, e.g. the paradigm (a–)b–c(–d), is far better because the nature we describe is seldom definitive, far more often probabilistic and dimensionally fuzzy. With morphometrics, an average plus information on standard deviation is also more informative. With Bayesian credible intervals, trimming data to that which is easily and precisely analyzed can produce a credible interval far more narrow than can any more realistic analysis taking into account all data. Also, the idea that cladograms must be more accurate if better resolved is dependent again on rejection of data that reflects a commonly less accurate resolution level because all relevant data must be accommodated. This is discussed at greater length in Chapters 8 and 15. According to Cobley (2012: 28) the classic researcher's mistake is "impressing one's own expectations upon dataless voids," in this case on the void once filled by rejected contrary or less precise or not phylogenetically informative data.

In another view, mathematically one plus one equals two, which is true, but does mathematics genuinely lend assurance to the postulate that one hobbit plus one hobbit equals two hobbits? If pseudoextinction is not common, then patristic distance as measured by numbers of nodes between extant taxa is compromised in that most or all of the nodes may be of one extant taxon. This is an important caveat for conservation analyses.

Three different views — The three patterns mentioned above are reconciled by pluralistic evolutionary systematics. These patterns give three different views of the evolutionary process, and are not equivalent, thus one is not necessarily better at charting evolution than the others. (1) Classical systematics produces classifications of taxa distinguished by overall similarity of locally conservative and apparently homologous traits, apprehended heuristically by a "naïve form of analysis of variance" (Littlejohn 1978: 234). *Conservative traits* are those stable within a taxon (and also often its close relatives) and also at different collecting sites and across different habitats. (2) Morphological cladograms present nested sets of *trait* transformations away from an apparent primitive (basal on a caulistic tree like a Besseyan cactus diagram and similar to other basal taxa)

state usually represented in a related (outgroup) taxon, and produce charts of synchronic (present-time) relationships. Macroevolution should be represented in classification, however, only by transformations among *taxa* named through classical systematics unless pseudoextinction is well determined by non-phylogenetic data. The traits in phylogenetics that are considered transforming from one morphological state to another are, usually, characters from classical descriptions. (3) Molecular trees are variously assembled through parsimony analysis (as in morphology), or maximum likelihood or Markov chain Monte Carlo Bayesian methods. It will be shown that the order of nesting of either morphological or molecular cladograms may not be the same as the order of macroevolutionary transformation. See the discussion of self-nesting ladders.

CHAPTER 3
THE FRAMEWORK

Although phylogeneticists commonly aver that molecular trees imply relationships between taxa, actually such trees only represent genetic continuity of molecular strains and isolation events associated with the specimens analyzed but not the taxa. This not necessarily reveals speciation events because taxa in morphological stasis may generate daughter species but remain static in expressed traits (Frey 1993; Tobias et al. 2010), while continuing to change in non-coding molecular traits. The traits are mostly non-coding sequences usually assumed to be unbiased by differential selection. In terms of information theory (Shannon & Weaver 1949; Weaver 1949), systematics based on morphology and other expressed traits, and systematics based on mostly non-coding DNA sites are not each based on information redundant in the other, and one cannot be eliminated in favor of the other without sacrifice of information. All science seeks to maximize information in a simplifying explanation that addresses all relevant information. The process of analysis using a single unifying concept, macroevolution, leaves less uncertainty than analysis through classical systematics, morphological cladistics, or molecular systematics each alone. Less uncertainty lowers informational entropy and any disagreement increases entropy. Unfortunately the decrease in entropy (i.e., increase in "negentropy") due to lessening of uncertainty by eliminating relevant data that is not phylogenetically informative is artificial.

One can interpret molecular heterophyly (paraphyly and phylogenetic polyphyly) as an indicator of progenitor-descendant diachronic (caulistic, through time) evolution (Zander 2008b), and make an evolutionary tree (plates 6.1, 6.2, 8.1) with named nodes or series of nodes, leading to, e.g., a commagram or "Besseyan cactus" (Plate 8.2). For a modern example of a Besseyan cactus of macromolecular transformation see Denk and Grimm (2010).

Macroevolution is used here in Jablonski's (2007) sense as evolution at and above the species level, as opposed to *microevolution* being minor genetic and phenetic changes within a species, although microevolution may lead to speciation through infraspecies in gradualist scenarios. Two kinds of macroevolution are emphasized here, (1) pseudoextinction where a descendant species is produced from a progenitor, which itself changes into another species by anagenesis (gradual change), and (2) budding evolution (ex stasis speciation) where a descendant species is produced from a progenitor species but without change in the progenitor species. In the former, the caulis is broken, and maximum parsimony is appropriate, while in the latter the caulis remains intact, and a less parsimonious solution is necessary.

Systematic pattern and evolutionary process are nowadays commonly divorced in the development of phylogenetic classifications (Rieppel and Grande 1994), beyond acknowledgement that shared advanced traits imply shared immediate ancestry. This leads to well-defined sister-group cladograms that are amenable to mathematical and statistical manipulation, the "mechanized knowledge" of Gigerenzer et al. (1989: 211). But, because ancestor-descendant relationships are ignored (Dayrat 2005; Grant 2003; Mayr & Bock 2002; O'Keefe & Sander 1999), it also leads to empty precision (Rieppel 2010), including the assumption that dating geologically one sister group necessarily gives the same date to the other. Introduced in the present work is the structuralist concept of isomorphism (Giere 2009), i.e., a structure-preserving map of hidden relationships, which, in systematics, can be the identification of progenitor-descendant-based matches or mismatches between cladograms based on morphological data and those based on molecular data.

This is equivalent to the search for "hidden variables" in physics, such as a non-obvious classical explanation for the nonsensical non-deterministic rules of quantum mechanics. These structural isomorphisms empirically (van Fraasen 2007) support the consilient (see Glossary) and consistent historical structure hidden in both data sets through a theory of joint cause (common ancestry), and this is true whether the data sets achieve congruent results or not. Reconciliation of morphological and molecular cladograms is not needed for the structure they share (but see wrong corroboration from two identical self-nesting ladders, below), namely that due to genetic continuity; otherwise a theory that provides for general caulistic macroevolutionary agreement is of interest.

Curiously, the search for reconciliation of classical and phylogenetic classifications, and morphological and molecular cladograms, is quite like that of Hegelian dialectics, where two opposites (thesis and antithesis) are both explained by finding some shared process (synthesis) that obviates a perceived necessarily excluded logical middle (e.g. A or B is right, but not both). The Framework synthesis is, of course, not a theory of class struggle in a historical context (Engels (1989: 86), but simply macroevolution in a scientific context.

A methodologically and theoretically pluralist Framework is presented here in an effort to explain all available evolutionary information by an over-arching, unifying theory of macroevolution, representing the results in a classification of value to workers in many fields. The present contribution is a methodological technique for standard evolutionary systematics based on pluralist scientific philosophy. As suggested by Kitcher (1988: 172), the virtue of Darwin's theory of evolution is that it promised to unify a host of biological phenomena, even though much was not understood at the time. The present book advances the idea that a reevaluation of evolutionary principles in systematics will do the same.

FRAMEWORK FLOW CHART

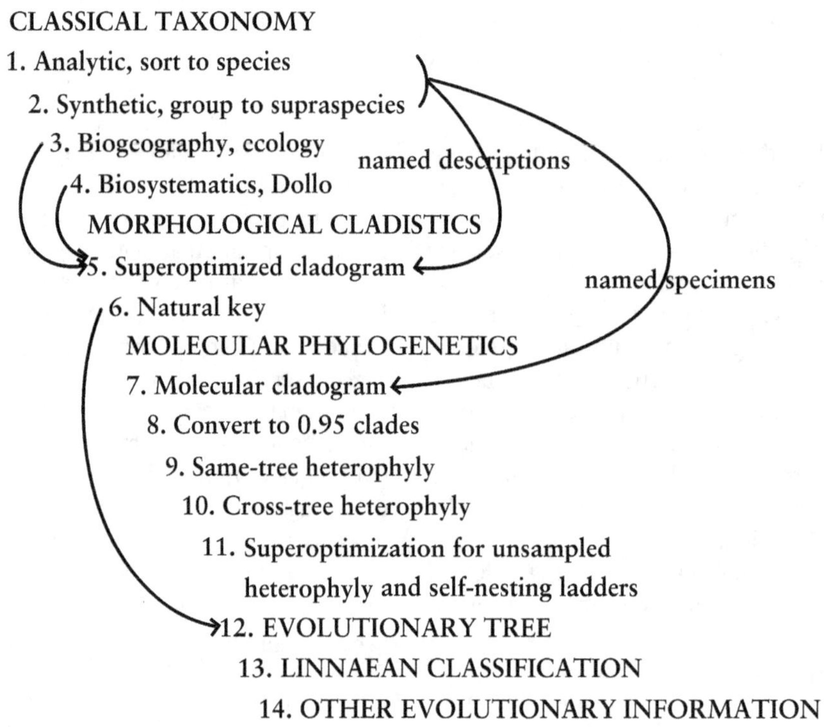

CLASSICAL TAXONOMY
1. Analytic, sort to species
2. Synthetic, group to supraspecies
3. Biogeography, ecology
4. Biosystematics, Dollo
MORPHOLOGICAL CLADISTICS
5. Superoptimized cladogram
6. Natural key
MOLECULAR PHYLOGENETICS
7. Molecular cladogram
8. Convert to 0.95 clades
9. Same-tree heterophyly
10. Cross-tree heterophyly
11. Superoptimization for unsampled heterophyly and self-nesting ladders
12. EVOLUTIONARY TREE
13. LINNAEAN CLASSIFICATION
14. OTHER EVOLUTIONARY INFORMATION

named descriptions

named specimens

Plate 3.1. — Idealized Framework flow chart for combining classical taxonomy, morphological cladistics and molecular phylogenetics to generate an evolutionary tree of serial (as opposed to nested) macroevolutionary transformations. This results in a Linnaean classification plus additional information that allows interpretation of the classification. Classical taxonomy provides descriptions as data sets for morphological cladistics, and provides named specimens as exemplars for molecular phylogenetics. It also contributes information on geographic distribution, ecology, (ideally) biosystematics, and possible directions for unidirectional Dollo transformation at the taxon level. When molecular phylogenetics is not available or unresolved, an evolutionary tree may be constructed from the natural key developed from the superoptimized morphological cladogram, or from a superoptimized classification alone.

CHAPTER 4
Element 1 - Contributions of Classical Systematics

Précis — Classical taxonomy distinguishes species and taxa using 250 years of highly developed heuristics in identifying conservative traits for use in establishing taxa by differences and for grouping the distinguished taxa by similarities. Classical taxonomy demonstrates its utility by matching in major features the results of morphological and molecular cladistics. It is superior in creating classifications because morphological cladistics simply parrots back the original descriptions and their implied relationships, in part due to preselection. Traits valid very locally (clearly strongly affected by adaptation or epigenetics as judged by covariance with environmental variables) are not treated the same in classical taxonomy as conservative traits stable across a larger group or groups and associated with a larger group of stable traits. It is superior to molecular systematics because it samples geographically both densely and widely the complete taxon as opposed to one or a very few specimens, and is not confounded by tracking one of many strains that may be paraphyletic.

Alpha taxonomy deals with the distinguishing of species, their traits and distribution, commonly by evaluating gaps between ranges of expressed traits, particularly morphology, that are perceived as coherent. *Classical systematics* groups species into assemblages of apparent evolutionary close relationship using whatever information is available, consistent with evolutionary theory. Both are here referred to as *classical taxonomy.* Even the "morphological" species concept of Cronquist (1978): "Species are the smallest groups that are consistently and persistently distinct, and distinguishable by ordinary means" has a significant element of evolutionary theory embedded. Also, compare Sonneborn's (1957, in Heywood 1963), "minimal irreversible evolutionary and morphological divergence that yields constant and readily recognizable difference."

Systematics today requires an initial evolutionary analysis, however perfunctory, contributing to and building on a predictive classification that is, as such, useful to many fields. Scientific knowledge is based on facts and the testing of predictive rules, in taxonomy rules developed over 250 of increasing expertise. Phylogenetic analyses, even though very precise, poorly predict classical results. Alpha taxonomy, in the modern context, involves the analytic discernment of groups of organisms in nature developed by evolutionary processes promoting distinction rather then similarity, while classical systematics focuses on the synthetic shared-ancestry-based similarity between groups. Because classical systematics at least in the past has included a morphological cladistic analysis, this, among other inferences of relationship including phenetic clustering, provides at least a first theoretical glimpse of the dimension of macroevolutionary transformation.

Given that phylogenetic classifications are biased by the artificial classification principle of holophyly (strict phylogenetic monophyly), the Framework requires that any taxonomic group being addressed be initially restricted to classical taxa plus any past true advances in understanding, i.e., restoration of taxa phylogenetically lumped merely because they make another taxon at same rank paraphyletic, and restoration of taxa split because they are paraphyletic.

Basic textbooks in taxonomy (e.g., Stuessy 2009; Stuessy & Lack 2011) review classical methods. These include such basic techniques as clustering by similarity, gaps, intergradation, and homology, identification of conservative traits, and characteristic geographic and ecological distributions. "Natural" in this book follows Darwin's (1859: 404) definition of natural taxa as based on both genealogical relationships *and* degree of divergence. A modern revision may also involve information from ecology, paleontology, evo-devo, population genetics, chemistry, cytology, ethology, and other biosystematic indicators of descent with modification of taxa, particularly inferences through application of Dollo's Rule to help group taxa by homologous, conservative traits that organize (i.e., predict) other traits that are less conservative and more easily lost or reversed. Standard taxonomic methods build on both differences and similarities to produce a first pass at evolution-based classificatory guides to biodiversity (Zander 2007b). This parallels Linnaean classification where differences are noted by distinguishing species, and similarities by organization into higher ranks. Alpha taxonomy deals with conservative traits "locally" because some traits may be conservative only for particular subgroups, while weighting traits in morphological cladistics is biased by being globally applied.

Sample size — In statistics, required sample size for a particular level of confidence can be estimated by recourse to the Central Limit Theorem (Ross 2009: 393) or through the Bayesian context (Winkler & Hays 1975: 598). A fundamental but largely unrecognized feature of taxonomic heuristics is the general statistical rule-of-thumb that about 30 samples are sufficient to ensure a normal distribution of samples from distributions that may be skewed but are not very complex, e.g. are not multimodal (Games & Clare 1967: 247–248; Yamane 1967: 146). Testing of this rule by Smith and Wells (2006) demonstrated a complete spectrum of reliability, from 15 samples being sufficient in most normal data sets, and 30 for bimodal well-behaved data sets, to not even 300 samples being able to deal with heavily skewed distributions. Consistent following of the normal sampling distribution in a real data set did not begin until 175 samples were made. Alpha taxonomy commonly expects sampling of specimens for each species at this ballpark level.

Sample size can be dismissed easily. Morrison (2013) wrote: "...there is an exponential relationship between sample size and precision, so that doubling the precision of an estimated quantity requires the sample size to be squared, which leads to rapidly decreasing return for effort. Any sample size beyond $n = 30$ is, for practical purposes, little different from n = infinity." This assumes a rather simple distribution of the sampled data.

Sample size in classical systematics is usually excellent, with many specimens of each taxon examined, but sample size in molecular systematics is usually one specimen per species or even genus. In molecular systematics, large groups like families can be adequately well-sampled taxon-wise by summing molecular samples of species such that the large groups can be represented by known variation in the *molecular* samples rather than just the *morphological descriptions* assigned to the exemplars. Zhang et al. (2011) averred that they were able to infer correct species delimitations with a single sample and 50 DNA loci, or 5 to 10 samples and 50 loci, based on assumptions of pseudoextinction, a relaxed biological species concept, and that concordance of gene trees across multiple loci indicate a distinct, stable species. Walsh (2009) was apparently able to confirm a lepidopteran as a distinct species because its CO1 sequence DNA barcode had a greater than three percent difference from that of its closest relative. Although the moth was morphologically distinctive, both the above methods may identify molecular species that have no morphological identifying features. Phyloge-

netic analysis uses heuristics (e.g., Hastings-Metropolis sampling) to sample multimodal data spaces of DNA sequences. Assessing adequate molecular sample size to determine distinct phylogenetic units has not been much studied but the literature indicates that 20 to 59 individuals (Crandall et al. 2000; Walsh 2000) is a minimum number. Adequate sample size does not mean there are no unsampled specimens (e.g., extinct lines) that may render a taxon paraphyletic, but does help search for extant isolated populations with informative divergent molecular lineages.

The normalized sampling distribution allows a good estimate of the mean of the sampled, potentially non-normal distribution. For this reason, though perhaps only implicitly recognized by taxonomists, dozens or hundreds of specimens are examined in classical taxonomy to establish diagnostic measurements. Many more specimens are examined to establish and describe bimodal distributions of traits, such as when two species have been wrongly conflated, or infraspecies are diagnosed. The point of doing taxonomy is to understand variation globally for a group of species such that when a possibly new species in the group is detected based on only one or a very few examples, reasoning by *analogy* (Kline 1985: 48; Lim et al. 2012) can provide a guide or prediction as to whether it should be described or not. Ranges and modes of variation of similar, related taxa are assumed to be similar (as per Vavillov's "Law of Homologous Series," Vavillov 1951). Molecular analysis uses analogy, too, when representing morphological taxa with single molecular samples, but arbitrarily rejects the analogy in instances of paraphyly.

Because molecular systematics must stand alone if it is to be supportive of morphological studies, or withstand being refuted by them, the small sample sizes in molecular analysis makes reasoning by analogy difficult or impossible in groups in which all taxa are each represented by one or few samples. Families and higher ranks of organisms are commonly well-sampled molecularly, and these may provide an analogy, yet the prevalence of paraphyly among families indicates, by analogy, that genera and species should also be predicted as commonly paraphyletic whenever they are ultimately well-sampled.

Whittaker (2009) emphasized the importance of intraspecific sample size in conservation and biodiversity analysis. Goldstein et al. (2002) pointed out that inferences are facilitated by an amplification of relationship detection when working with small sample sizes due to overestimation of Pearson correlations, while Aron et al. (2008: 225) indicated that small samples of high Bayesian credibility have high power of discrimination. Such analogy is common, and has been found (at least in modern taxonomy) generally predictive of estimated features when additional specimens become available. The benefits are may not be worthwhile, however, if chances of detection failure are high.

Reasoning — Scientists use deduction, induction, abduction and reasoning by analogy, if they are organized in thinking. Abduction, the devising of a hypothesis, is a central feature of the scientific method. A reason is posited as an explanation for a given observation (Fitzhugh 2012; Niiniluoto 1998; Pierce 1903). There may be many abduced explanations, yet, for hypothesis testing, one is singled out as the more worthwhile to test. Selection can be as simple as educated guesswork or there may be rules for selecting explanations for testing. Popper (1959: 128) did not seriously investigate abduction beyond pointing out that the simplest hypotheses are better testable because the epistemic content is greater. A most impressive example of abduction is that of physicist Edward Tryon (Parker 1988: 190) who realized that the universe's gravitational potential energy was exactly that of its mass energy, but negative, so the net energy of the universe was zero. Thus, probabilistic vacuum quantum pair fluctuation could well have been the origin of the universe, and "where did the energy come from? is no longer a question. A neat hypothesis like this is based on expertise and hundreds of years of standing on the shoulders of giants.

The following illustrates the need for discursive reasoning to put in context and transform into coherent theory the "discovered" evolutionary relationships in molecular phylogenetics. Real things are different in some way, that is why we can tell they are different even if at minimum the only detectable difference is their position in space. Mathematics may add one and one and get two, but the two items are always different. Thus mathematics is a first approximation for the spectrum of fuzziness in the real world. It takes discursive reasoning to see if the mathematical analysis actually applies to the situation. For example, one apple and one apple make two apples, even if one is a mackintosh and the other a granny smith. Why? Because we decide this category "apple" is mathematically transigent because important for our purposes in future calculation. One apple and one orange make two fruit, and "fruit" is an acceptable transformation because the category has value in future calculation; you are sampling similarities or likenesses. One apple and one automobile are at best two "things" because this category has no apparent. That is, there is no apparent value to the category "things" for reasoning purposes.

In phylogenetics, taxa may be combined even if evolutionarily unalike by substituting axiomatic phylogenetic monophyly for reasoning about evolution. The category "phylogenetic monophyly" has value to phylogeneticists interested in dealing with sister groups but not to those who need information about serial macroevolutionary transformations. Taxa may also be dismembered in phylogenetics merely because of position, being not contiguous (being heterophyletic) on a cladogram. This is the complement or obverse of mechanical addition and may yield categories that are, for phylogenetic purposes, practically identical and so mathematically transigent if combined.

Scientific intuition — The "creative act" associated with intuition is well discussed by Springer and Deutsch (1993: 312), who give examples of major scientific discoveries associated with hypnogogic semi-dream states. Such eureka events, however, are cited by them as not "accidental or purely intuitive discovery." Most are in a scientific context associated with a long-term problem, with a background "set by years of rigorous work," often with a latent period in which, apparently, the unconscious attends to the puzzle.

Classical taxonomy has been accused of being antique, subjective, "merely" intuitive, or even instinctual (Hey 2009; Scotland et al. 2003; Yoon 2009), a product of "authority figures" (Mooi & Gill 2010) invoking a personal *nous*. Classical taxonomists are likened to red-daubed feathered shamans dancing in fitful firelight in smelly, smoky, dank caves, their only analytic tools being a bull-roarer, some popping bladders, and the occasional scry from a fresh liver. When a range extension is published, such a paper is criticized as having no theoretical framework, no experiment, and no results. In fact, taxonomy is a 250-year research effort whose communal beginning is attributed to Linnaeus, and which seeks to document (and explain if possible) the distinctions, groups, and distributions of the world's plants and animals. Any little distribution record is a part of this research context. The project goes back farther than Linnaeus, of course, through the Greek and Roman naturalists and physicians, straight back to the dancing shaman. But it is an integral project with a noble end, a clearly stated basic corrigible scientific method, a receptiveness to advances in theory and methods, and a proven practical dimension.

The words intuitive and subjective must not be conflated. Intuition is a bright idea grounded in thorough familiarity with data and theory, while subjective means existing only in the mind or illusory. Intuition is fundamental to hypothesis generation, which is part of an objective scientific endeavor. Subjective is, by definition, not objective.

An evolutionary systematist might question phylogeneticists as to their own intuitive act of choosing cladistic analysis as a method in the first place. Is it the similarity with a dichotomous key that is decisive? A dichotomous key is the central feature of classical taxonomy, and has much manna. Combining transformation series with a dichotomous key seems attractive at first consideration, yet there is no reason for evolution to occur in such a pattern. According to principles of human magical thought, the law of similarity means like causes like, or, alternatively, appearance equals reality (Rozin & Nemeroff 2002). A cladogram is clearly a tree, isn't it? The shaman dances for everyone.

Additional intuitive elements of phylogenetics include choice of exemplars, which outgroup(s), which traits. These are intuitive scientific choices and can be easily defended logically, yet phylogenetics is like a mathematical proof for which one part is invalid, yet the remainder is constantly perfected with great industry and zeal.

Although Felsenstein (2001) has asserted that phylogenetics is now on a firm statistical basis, his list of early intellectual phylogenetic axioms still holds for the most part. The result is that phylogenetics, especially in its use of new statistical ways of working directly with the genome, and has not only become the darling of university deans and the staff of granting agencies, in fact phylogeneticists have become the deans and have staffed the granting agencies. There would be no problem with this if phylogenetics did not have fatal flaws, discussed in detail in this book. Phylogenetics has reserved the sobriquet of "systematics" to itself, redefining the word to mean "phylogenetics and its applications to classification," while the remainder of plant and animal natural historians are generally relegated to "taxonomists"—being service people who collect, name, and curate specimens. To paraphrase Adams (1980: 38), "The [molecular cladogram] is definitive. Reality is frequently inaccurate."

This book suggests that it is taxa that evolve, not characters divorced from the taxa. According to Arendt and Resnick (2007), because genomic analysis has demonstrated that the same genes may be involved in the same phenotypic adaptation in quite distant groups of animals, while different genes are apparently the source of the same phenotypic adaptation in related groups, the usual distinction between parallelism and convergence (parallelism expected to be based on the same genomic pathways, and convergence on different) breaks down. The authors recommend that "convergence" should be the general term. In the present work, evolution of the taxon is paramount. There should be no confusion between using the same or similar genomic pathways to help estimate evolutionary relationships, and discovering that the generation of the same adaptive expressed traits may be through different genes by chance alone responding to a particular selective regime at the genomic level. Evolution of the phenotype may be quite disconnected from "evolution" of the genotype though remaining based on it. The phenotype evolves to establish a new taxon, not the genotype, which only changes and may generate the same adaptive or neutral trait in many ways as a willy-nilly service to evolution. A good review of non-genetic inheritance is given by Danchin et al. (2011). It must be pointed out here that epigenisis may be cited as the reason for *any* unusual phenomenon that is not easily ascribed to genetics. Evidence in favor of epigenetic reversal of individual traits is known for many groups (summarized by Zander 2006, and see discussion of Dollo evaluation in Chapter 8), yet clear support for reversal of entire taxa is not at hand and may never be.

Phylogeneticists have indicated (Grant & Kluge 2004: 23) that if data are not phylogenetically informative of a transformation event, they are irrelevant, while optimality theory (van Deemter 2004) has it that a concept is "ineffable" if it cannot be expressed in a language (e.g., phylogenetic trees) through expressible in another language. Although it has been said that "nothing needs to be known about evolution to classify phenetically" (Ridley 1996: 372), as a kind of "theory-free" philosophy, modern classical taxonomists are fully cognizant of the importance of homology in expressed traits when evaluating similarity (Mooi & Gill 2010; Sneath 1995), even at the alpha level.

Additionally, alpha taxonomists distinguish between artificial and natural keys, a distinction similar to that between phenetics as optimization of number of character state identities, and cladistics, as optimization of state transformations (following Brower 2009), although of course such transformations are assumed to occur between similar taxa. A natural key may have to be trichotomous or polychotomous to reflect multiple daughter taxa from a single progenitor taxon. Laurin's (2010) evaluation of evolutionary trend detection used only simulations of character change "using known evolutionary models" of character change. Alpha taxonomists use geographical and ecological correlations as helpful in distinguishing organisms, and group (as least in modern times) species that have an apparent evolutionary relationship into higher taxa, either through overall similarity in important (homologous, conservative) traits or through some theory of taxic coherence in evolutionary divergence. Molecular systematists use exemplars that purportedly represent an entire taxon, but alpha taxonomists think in terms of taxon-areas, the range of known morphological variation in the world plus other, biosystematic traits when known. Connecting molecular and morphological thinking is what this book is about.

Heuristics — Classical taxonomy, morphological cladistics, and molecular phylogenetics all deal with what is basically an np-hard problem (Semple 2007: 299, 308). Np-hard means "not to be completed in polynomial time," i.e., full optimization involves generation and analysis of greater and greater sets (Martignon 2001). A maze is an np-hard problem, requiring examination of all paths until the exit is found although there are weak heuristic search methods (Pullen 2011). Both morphological and molecular analyses use heuristics to simplify searches to get results that are not guaranteed optimal but are at least close. Heuristic search in parsimony analysis is one kind, and Markov chain Monte Carlo analysis involving Hastings-Metropolis sampling in Bayesian analysis is another.

In the absence (or in alpha taxonomy the methodological difficulty) of clear cut and robust sampling, such as is involved in standard hypothesis testing, heuristics must be used. Evolutionary relationships are, in classical alpha taxonomy, addressed and inferred through informal genetic algorithms for rule production (Gigerenzer 2007; Hutchinson & Gigerenzer 2005) as a heuristically based expert system in systematics (Zander 1982), and the quasi-optimal results allow generation of descriptions of taxa. Goldstein et al. (2002) describe the "Take the best" heuristic, in which inferences and predictions are based on only a part (say, one-third) of the information until a stopping rule ends the search and decisions are made on basis of the cue that ends the search. Such predictions are better than those made by multiple regression, and are based on not allowing less important data to overwhelm (compensate for) highly weighted data. The parallel in systematics is the difference between classical taxonomy in which a

small set of conservative traits inform evolutionary relationships, while in morphological cladistics modeling of microevolutionary transformations from the whole data set unfortunately allows multiple weakly conservative or labile traits to overwhelm the conservative ones, resulting in a somewhat artificial evolutionary model. In Bayesian terms, weakly conservative traits are noise that overfit the training set (Martignon 2001).

Another heuristic is "Take the first," in which an expert examines a series of alternatives and stops when an adequate solution is recognized. This depends on recall and fairly similar sets of problems, but saves considerable time and effort. All "fast and frugal" heuristics exploit regularities in the environment, including those in data, but may not be entirely generalizable (Gigerenzer and Selten 2002). Aerts et al. (2010) demonstrated a go-no go theorem involving quantum-style analysis for dealing with manifest data based in part on hidden variables, and Aerts (2009) discussed the well-structured mechanics of the double layer of human thought that figures in the balance between logic and Gestalt apprehensions of reality, basing heuristics on entirely rational processes. In both papers, however, an over-arching theory (e.g., Aerts 2009: 22) can reconcile the apparently nonclassical disjunctions and conjunctions associated with apparent quantum phenomena in mesocosmic cognition. Rules of thumb are also important in statistical psychology (Wilkinson et al. 1999). Inasmuch as there is considerable matching of the groupings of classical revisions with molecular analytic results, informal genetic algorithms are apparently successful, and formalization of taxonomic heuristics, as done for decision heuristics by Goldstein et al. (2002), would be important. One can note the Darwinian evolution itself is heuristically based.

It is quite possible that simple heuristics such as the above will eventually be formalized mathematically. A good place to start is adaptation of the Fitz-Hugh-Nagumo equations (Stewart 2011: 164) that describe threshold excitability in neurons.

A 250-year scientific enterprise — As noted above, classical taxonomy is not merely descriptive, but is a 250-year joint scientific enterprise distinguishing and grouping the kinds of organisms. This, at least in modern times, is based on guidelines from and deductions about theoretic evolutionary processes that affect the history and groupings of organisms in the natural world. Non-trivial falsifiable null hypotheses are basic and abundant: e.g., groups cannot be distinguished; if false then groups have no ecological and historical traits; if false then there are no discernable sequential, tree-like, or reticulate patterns of evolution; if false then no evolutionary theories can be devised and tested; if false then such groups have no value to other sciences; if false then … and so on. Like parsimony algorithms, classical taxonomists conceive of and discard many suboptimal evolutionary models and classifications before settling on a solution that either best explains and represents the data or, even what is more desirable, one that is much better than any other in explaining and representing the data, avoiding the vicious ambiguity (van Deemter 2004) of nearly-as-good alternative explanations. What does "much better" mean? This depends on the number of alternatives that must be considered. If there are only two alternatives, and one wants a 0.95 credible interval, then the alternative should not be more than 0.05 probability. Much better in this case is 0.95 / 0.05, or 19 times better. If there are more than two alternatives, the *sum* of the probabilities of each of the alternatives should not exceed 1/19 of the probability of the main hypothesis. So if we have a case where the main hypothesis is of 0.50 probability, and the highest of many alternatives is 0.026 probability, then although 0.50 is 19 times 0.026 this is not "much better" than the sum of the alternative hypotheses (which should be 0.50) even though much better than the best alternative.

A formalization (see detailed discussion in Chapter 12) of dimensional heuristics in taxonomic descriptions of mosses (Bryophyta) revealed that, for the format (a–)b–c(–d) metric, the low range "(a–)b" is usually about 0.25 of the high range "c(–d)", while "c(–d)" is usually 0.85 of the mid range "b–c". The geometric mean of the mid range "b–c" is quite near the geometric mean of the full range from "a–d", while the arithmetic means (averages) of "a–d" and "b–c" match less well. This is because "b–c" is often a large proportion of the range zero to "c". The low range "a–b" is crowded between zero and "b". This involves the geometric mean, such as is used for problems involving proportions, or across large parts of magnitudes, e.g. "a" of 0.1 units to "d" of 10 units, or in solutions to certain "Fermi Questions" (Morrison 1963; Weinstein & Adam 2008). Gould's (2002: 893) speciational reformulation of macroevolution involving minimum structural constraints on size is also relevant and explanatory, i.e., although there are mutations promoting variation towards both small and large size, there is a developmental wall to small size for particular organisms, e.g., horses, which only seem to evolve only towards large size. A number of heuristics may be enhanced or be made more under-standable by formalization. Alpha taxonomists should not doubt their methods as they are well founded on rather basic relationships in physics and good statistical sampling.

Bayesian reasoning and multiple tests — With a change during the 1700's and 1800's towards probabilistic thinking in science (Pap 1962), philosophical or logical support for certainty or relative certainty about "truth" in science began to be replaced by a pragmatic attitude that science does not establish truth, but identifies theories that are so well supported by facts (well-corroborated observations) that they may be acted upon (used as a basis for additional research) and alternative theories may be ignored, even though some facts support them. The new probabilistic science is centered around a phenomenon of distribution of random events in physics described by the Central Limit Theorem, and is the basis for probabilistic theory and statistics. Statistics as a field is well established today and, although there are conflicts between the different schools (Gigerenzer et al. 1989) of frequentist, Bayesian, and hypothesis testers, the basic requirements for assessing reliability are clear. There are, however, commonly today many phylogenetic studies that fail to provide adequate statistical evaluation, often focusing on only one or two of several requirements for establishing reliability. Such short-cut solutions to prediction should be avoided in science, as it is in more mundane matters, such as, say, horse-race handicapping. Beyer (1975: 10), for instance, after much early failure, opined: "The systems and the gadgets are all based on the same assumption: that the complexities of horse racing, which have baffled men for centuries, which involve hundreds of factors, can be resolved by the application of a few simple rules or calculations. The assumption is a seductive one." And that "...it will be much easier [for one] to operate with an inflexible method that dictates what he should do rather than use his intelligence and judgment at every stage of the handicapping process." (Note: Do *not* bet the ponies. This is only a colorful example.)

Although the result of Bayes' Formula is the posterior probability and that is commonly regarded as the chance of a solution being correct, the Bayesian philosophy also requires no bet (i.e., no confidence in the pragmatic value of the results) except after evaluation of risk. Bayesian betting in the phylogenetic context means taking the study out of the realm of speculation and deciding whether to view the solution as sufficiently reliable to base additional work (biogeography, etc.) on it. The risk of one's science being wrong also depends on the chance of any assumptions being wrong, which affects the final probability involved in the decision to make the bet or not. The posterior probability of a branch arrangement of interest must be modified by the chance that the arrangement is wrong because an assumption may be wrong. This is not the same as the "anti-quant" arguments that exact analysis, such as standard deviation, is merely a hyperexact measure of risk, which is better estimated intuitively (Brown 2011). Risk can be estimated in many ways, but it must be estimated, exactly or as best possible.

Therefore, many studies must be made to establish a chi-square distribution, and low scores retained (not discarded if below 0.50). Consider a 4-taxon branch arrangement with the branches labeled A, B, C, and D, and possible branch lengths AB, AC, and BC when D is outgroup. With total evidence, in the 4-taxon case with random data or in the case of real data with a hard polytomy, it is possible to find or contrive an arrangement supported by chance alone with, for instance, a 0.98 CI (credible interval) reliability of a branch length of 7 steps where AB + AC + BC = 10; doubling the data will reduce the chance down to 0.96 in a branch length of 11 with AB + AC + BC = 20. This assumes the additional data is random from additional studies of the hard polytomy (or from another contrived data set) and shared about equally by AB, AC and BC. Thus, for branch lengths of about 10 steps, rather highly reliable scores that may have occurred by chance alone are not corrected by total evidence involving a hard or soft polytomy until the data set relevant to that branch arrangement is more than doubled in size.

A totally artificial data set of 50 taxa and 50 random 2-state characters was contrived with RANDSET (Zander 1999). Analysis with PAUP* under maximum parsimony (hs with 20 random sequence additions) produced 159 equally parsimonious trees from this totally random data set, and a largely unresolved strict consensus tree with, however, two distinct lineages (A, B) and ((C, D) E). An analysis of a subset of only these 5 taxa, under maximum parsimony (bandb) with PAUP*, produced one lineage of ((C, D) E with 0.58 BP support for (C, D) and 0.81 BP for (C, D, E). Bayesian MCMC analysis (MrBayes 3.1, datatype = standard, ngen = 500000) of the 5 taxon data set provided 0.74 BPP and 0.94 BPP for the same groups with their random data. Clearly, preselection of a subset on the basis of a reliability measure for further analysis introduces multiple test problems.

Multiple test problems (Felsenstein 2004, and see perfectly justified lengthy discussion of Chapter 15) occur when selection is made on the chance of being correct. For example, flipping many coins many times to determine if any are loaded will result in several coins coming up heads several times in a row even if all coins are fair. In the context of a large number of coins analyzed, this is to be expected. But selecting only that group of coins and their associated data that generates a high reliability measure and re-analyzing from that data will falsely show high reliability out of context, and the high possibility of this being random data is hidden. In phylogenetic analysis, preselecting taxa for study based on morphological analysis and a natural key gives molecular results that should match the molecular key. This is not a random analysis that discovers support from separate data since the molecular data are dependent on the phylogenetic structure of the preselected exemplars.

This dependence is also found in molecular data sets alone. If the first molecular analysis resulted in one clade with a BPP of 0.95, then the second analysis using a different molecular sequence based on just the exemplars of this clade, to reach 0.95 BPP, requires Bonferroni correction (i.e. a BPP of 0.975 is needed), such that both analyses are correct (show acceptably non-random variation) at the same time. Any preselection of taxa is a candidate for examination for introduction of multiple test problems and empty agreement.

Given the emphasis in the present book on decision theory, one should note that aircraft manufacturers often insist on "Six Sigma" precision. The Sigma level indicates the minimum number of standard deviations tolerated for acceptance, and Six Sigma means 3.4 defects per million opportunities for error, or 0.999997 error-free. Five Sigma limits defects to 230 per million, or 0.998 error-free; Four Sigma to 6,200 per million, or 0.994 error-free; Three Sigma to about 66,810 per million, or 0.93 error-free; while Two Sigma limits defects to about 308,500 per million, or 0.69 error-free. In phylogenetic analysis, tolerable error limits are generally set at Three Sigma or above. Requiring a Four Sigma level of precision would limit statistical power (of discrimination) drastically, but many clades often reach a high Sigma level of precision because there *is* such a thing as "statistical certainty," meaning precision as close to certain) given the data and analytic method) as makes no difference for decision making. Of course, precision is not the same as accurate or correct.

Preliminary macroevolutionary hypotheses — Classical taxonomy is both analytic and synthetic, a practice represented in the binomial by species and genera, respectively. The first synthetic effort is to make initial groupings of species using whatever genus concept is favored by the author for the particular group, and the second is to make a macroevolutionary hypothesis of one taxon generating another in a series or demise of the ancestral taxon. The method of intermediates is valuable for a first pass at a taxic sequence for testing, where primitive taxa are placed first in the sequence, leading gradually to taxa that are advanced in that series. The method is similar to outgroup analysis in cladistics, where traits "evolve" from plesiomorphic to apomorphic, and morphological cladograms of the taxa, including outgroup selection, can offer some guidance.

Information in classical taxonomy may be divided into three fairly well distinguished kinds: (1) that which clearly allows grouping of similarities, a simplistic form of phylogenetics or phenetics; (2) that which clearly allows splitting by differences, particularly evolutionary import of autapomorphies at various taxonomic levels of analysis; and (3) that which seems to be the magic of expertise, namely the "look and feel" of taxa, consisting of traits difficult or nearly impossible to characterize or measure but which are valuable in assessing similarity or differences. According to R. Heinlein, "One man's magic is another man's engineering." All cladistic morphological data sets lack information of kinds 2 and 3.

In the moss family Pottiaceae, a fairly obvious sequence from a similar but non-Pottiaceae outgroup of *Ptychomitrium* or *Timmia* would be *Timmiella,* *Erythrophyllopsis, Trichostomum, Barbula, Tortula,* with a side branch or split from *Barbula,* of *Pseudocrossidium,* and *Syntrichia.* This corresponds with major trait changes from plane leaf margins, to unistratose leaves, to recurved leaf margins, then one split by a side branch from *Barbula* to ovate leaves with single round-stereid-banded costa, to red leaf coloration, and another split to single, flattened stereidbanded costa and then to ovate leaves. Much of this parallels in a simplified manner the cladogram and suprageneric classification given by Zander (1993) for the family, but is couched in terms of taxic macroevolutionary sequences. not a nested classification or nested phylogeny.

Wastebasket taxa — Classical taxonomy ideally distinguishes groups that are well defined with clearly observed conservative traits, then commonly sweeps poorly understood or highly modified and reduced species into a common pool or "wastebasket" taxon. In my own field, the moss genus *Gyroweisia* of much reduced species is well known as such, and in a revision of *Leptodontium* (Zander 1972) I assigned a number of puzzling, probably unrelated, and small-statured species to the section *Verecunda.* This was before morphological cladograms were well understood by students like myself. Careful analysis may now redistribute the species of sect. *Verecunda* by determining through superoptimization the possible transformational relationships of those taxa. The restudy is in progress.

CHAPTER 5
Element 2 - **Contributions of Morphological Cladistics**

Précis —Because morphological traits in the data set are usually taken from classical taxonomy studies, one should retrieve a better organized natural key from cladistic study than from intuition alone. One commonly does not find such a result in cladistics because descriptions of species and higher groups in classical study are grouped by non-global, independent conservative traits, while cladistic studies involve global weighting of traits or all traits are equally weighted, nor is there a distinction between traits fixed independently or as a complex in evolution. In this case, labile traits bias the clustering. Conservative morphological traits are as valuable for tracking genetic continuity as are DNA traits. Careful weighting of traits in small taxonomic groups together with biosystematic evaluation of ancestor-descendent relationships help make a natural key of a cladogram, although such is limited to dichotomous splits. Phylogenetic "shared ancestors" are contrivances to ensure full resolution of a cladogram. Evolutionarily primitive taxa can be tentatively identified on a morphological cladogram if other taxa of similar morphology are also scattered about the base of the cladogram.

"Morphological" is used in this book to mean any scorable trait of the phenome (Burleigh et al. 2013) but usually limited to morphology in cladistic studies. Any good taxonomist can, with application, develop a natural key (emphasizing conservative traits) to a group. Why is a natural key often different from a morphological cladogram of the same taxa? The cladogram (e.g., Plate 5.1) can often correctly evaluate the primitive and advanced traits and group taxa accordingly. This is a benefit when groups may be complex and sorting by omnispection difficult. On the other hand, cladograms may be based on traits weighted arbitrarily (e.g., the same) and non-phylogenetically informative traits that signal distinctive individual taxa are excluded from the data set. Morphological cladograms with traits equally weighted—because they imply an unbiased mathematical analysis that supposedly leads to accuracy—are a distraction.

Morphological maximum parsimony analyses have fallen into disfavor in this day of the molecular data set, yet Schneider et al. (2010) and Mooi and Gill (2010) detail good arguments in favor of evolutionary analysis using morphological data (Radinsky 1985), particularly in search of congruence as corroboration for molecular studies, and for reciprocal illumination.

The method of parsimony with morphological data is well-known (e.g., review by Zander 1995). Simplicity itself is a good criterion for selecting hypotheses for further analysis, but not because "scientists always accept the simplest theory." In the case of phylogenetics, the simplest model is that constrained to reflect data sets to show both pseudoextinction and budding evolution where appropriate. Empiric support for simplicity alone might come from some future demonstration that there is a bias towards the least number of linked traits fixed in selection during speciation, and/or the fixation of the least number of conservative quasi-neutral traits to minimize physiological burden. Surely it seems likely that of three terminal taxa, an arrangement that minimizes the number of traits fixed per speciation event would reflect an assumption that the more traits fixed places a greater insult on a newly evolved species. If demonstrable, this would be a scientific reason for weighting some traits greater than others. Note that Kluge and Farris (1969) recommended weighting by degree of variability of a morphological character within taxa, this being an estimate of the rate of evolution of that character. Fitzhugh (2012: 61) has pointed out that a cladogram is itself not a test but a graph of the evidence. Abduction of explanatory hypotheses remains to be developed.

Thus, if selection trims the number of traits that survive speciation, then parsimony is a good way to group taxa using selection. The cladogram nests taxa, however, it does not arrange taxa in a macroevolutionary transformation series. Although there is little data on this, parsimony should reflect evolutionary constraint (signaled by conservative traits) as an influence on classification. If pruning of new trait complexity is of selective advantage, then more complex, less parsimonious combinations of new and conservative traits may initiate macroevolutionary changes at higher ranks in the tails of a statistical distribution of such pruning.

Given that morphological cladograms are based on both unitary, independent conservative traits and on linked traits from descriptions of groups previously inferred as uniquely, evolutionarily coherent, then such parsimony analysis is not an exact discovery process but an aid to generating a natural key or cladogram equivalent. It is important, if possible, to distinguish between independent conservative traits that are not involved directly in evolution and can be used as tracking traits (like molecular base changes) and traits involved in adaptation that may be fixed as a single multitrait group (and thus should be lower weighted). The stricture of a minimum of 30 (to 175 in difficult cases) samples for statistical evaluation of distributional variation has also been met by using the descriptions of taxa in classical systematics as exemplars rather than single specimens.

Effect size — Plate 5.2 presents an explanation of effect size (difference between distances between population means divided by standard deviation of populations), and how it influences interpretation of hypothetical cladograms with all splits with high Bayesian credibility, i.e., the order of splits is sure. (This is different than the "effect hypothesis" of Gould and Vrba (1982), the adaptive functions of a trait presently different from those for which the trait was originally selected.) Note that in this book all exemplary, hypothetical cladograms are rooted or are (more usually) a terminal clade of a larger, rooted cladogram not given.

Morphological cladogram: Six taxa in Plate 5.2 are well documented with many samples (vertical arrows) establishing nonoverlapping distributions with large effect sizes. The tree is a natural key or cladogram equivalent, the ancestral taxa D and E are inferred from Dollo evaluations or from heterophyly in the molecular study.

DNA cladogram: The distribution of morphological traits are the same as in the morphological taxa but molecular sampling is sparse and effect size can-

not be determined on the basis of molecular sample distributions. Taxa A, B and C are each assumed to have narrowly distributed molecular variation, in part corroborated by the match of morphology and DNA analyses. Taxon C has three exemplars within its variation and since all splits are at 95% credibility, two of these might be described as new molecular taxa, cryptic because of the narrow range of morphological variation. Lacking multiple samples, we cannot be sure, however, that the distribution of molecular data are as narrow as the distributions of morphological traits. If superoptimization using non-phylogenetic information fails, then (and only then) can taxa A, B and C be theoretically generated through pseudoextinction (note no special symbol at branch split). The D1, E and D2 lineages have high credibility (branch order is sure for these exemplars), but because D1 and D2 are of the same taxon, an ancestor is implied by the heterophyly and the molecular data cannot be used in distinguishing the populations (e.g., as cryptic molecular taxa) because E retains the molecular signature of a portion of the ancestral lineage. D and E may thus be postulated as progenitor-descendant pair (ellipses). For taxa F, G and H, continued molecular sampling may demonstrate that DNA variation overlaps (through hybridization, introgression, lineage sorting, paralogy and other factors decoupling expressed and non-coding DNA traits), or internal structure may be found with nonoverlapping molecular trait distributions. In the latter case, cryptic species may be described but these mask the coherent evolutionary character of the morphological species. The molecular analysis alone cannot show a clear mechanism for evolution or a molecular basis for recognition of the taxa as separate from classical systematics study.

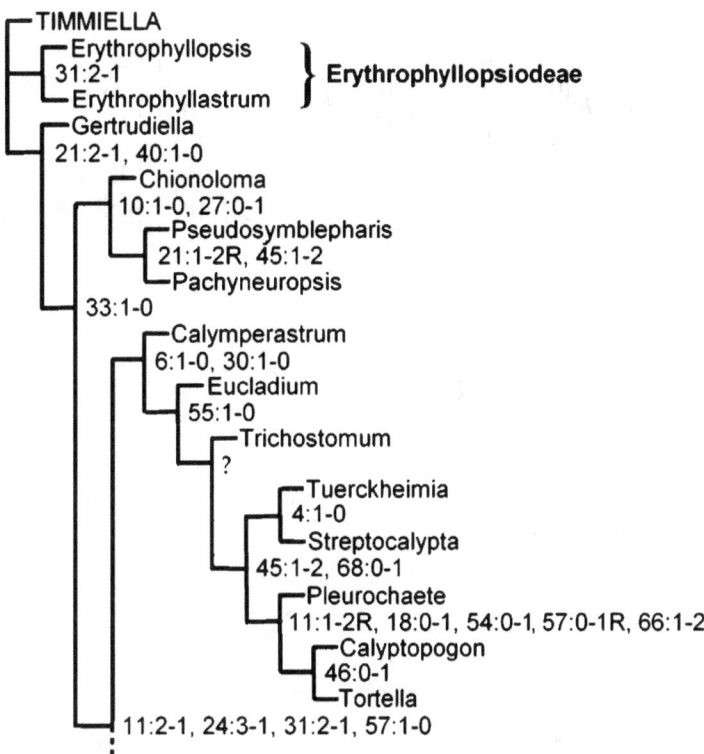

Plate 5.1. — Cladogram equivalent of a natural key from Zander (1993) of basal genera of the moss family Pottiaceae. The code for traits and trait states, available in the original publication, shows trait transformations, R means a reversal. The here basal Erythrophyllopsoideae, represented by two species, is deeply embedded *distally* in another, molecular cladogram, and the morphological cladogram (above, in part) provides a basis for cross-tree heterophyly (Chapter 7) mapping to a molecular cladogram this ancient lineage, a kind of "coelacanth" of the Pottiaceae.

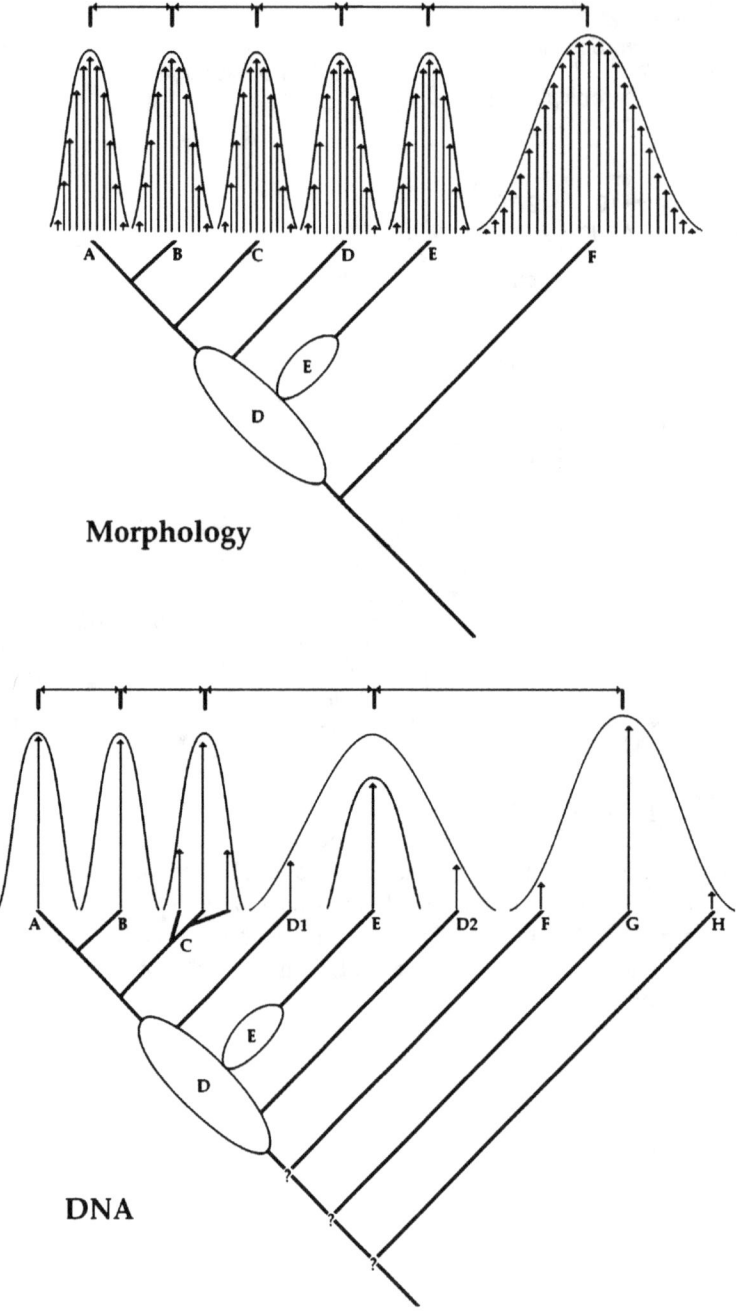

Plate 5.2. — Effect size (difference between distances between population means divided by standard deviation of populations) influences interpretation of hypothetical cladograms. All splits have high Bayesian credibility, i.e., the order of splits is sure. *Morphological clado-gram*: Six taxa are well documented with many samples (vertical arrows) establishing nonoverlapping distributions with large effect sizes. *DNA cladogram*: The molecular sampling is sparse and effect size cannot be determined on the basis of molecular sample distributions. See text for discussion.

Cladistics — Darwin (1859: 420) presented his "descent with modification" concept of an evolutionarily based Natural System as both "arrangement of groups," or genealogy, and "amount of difference," or modification. To Darwin, differences are indicated by formal taxonomic ranks. In a similar vein, Grant (1985) and Mayr and Bock (2002) have pointed out that two modes of evolution have been long recognized in the literature, namely anagenesis (phyletic evolution or differential change) and cladogenesis (evolutionary divergence). What is critical is that cladistic phylogenetics does not investigate, model, or depict in classification anagenesis as macroevolution (speciation and generation of higher taxa), other than presupposition of universal pseudoextinction involving fast anagenesis of the ancestor after cladistic divergence. Divergence in phylogenetics is limited to patristic distance along nodes of Darwinian common descent as branching point phylogeny (Mayr & Bock 2002; Grant 2003; Stuessy 2009). Mayr and Bock (2002) stated: "…in no way is it valid to claim that Hennigian cladograms provide the foundation for understanding the evolution of biological organisms as these cladograms include only branching points (cladogenesis) and not the amount of evolutionary change (anagenesis)." They include both micro- and macroevolution in the term anagenesis. The hypothetical shared-ancestor is thus of unknown nature, and is mostly a point-source place-holder for the analytic process. That anagenesis is a fact for many groups (but usually at supraspecific levels) has been long demonstrated in the fossil record (e.g. Kucera & Malmgren 1998).

Cladistic nesting does not model any process in nature but may be of help in inferring natural evolutionary processes. The phylogenetic system proposed by Hennig (1966), in contradistinction to that of Darwin, focuses solely on sister-group relationships. Basically, of any three taxa, two are clustered as more closely related to each other than to a third in a hierarchy of traits determined by analysis based on a simple model involving maximum shared similarities (e.g., nested hierarchies of exemplars sharing advanced traits, where two shared traits are a stronger indicator of shared ancestry than one shared trait). If a progenitor species, however, gives rise to two descendant species without itself changing or going extinct, then that branching order cannot be modeled in a cladogram. Treated as sister groups, the descendant species would at best be cladistically modeled as a multifurcation, yet such multifurcations are commonly resolved by, in morphology chance reversals or convergences of a few traits, or in molecular

analysis, chance survival of one of many molecular strains. Serial transformations are unmanageable in cladistic modeling, which relies on nesting. Phylogeneticists often switch in discussion (or do not distinguish) a shared ancestor being a trait-characterized node or a diagnosed taxon. If the former, then the ancestor is different from any extant taxon at least by accumulation of molecular changes, if the latter, then the shared ancestor may be an ancestral taxon the same as an extant taxon.

Podani (2013) has discussed diachronic and synchronic trees at length, but remained convinced that a cladogram basically reflects evolutionary history, in that a "cladogram and the species tree are the backbone trees of the evolutionary tree," a notion extensively refuted in the present volume. He explained that a cladogram is comparable with the evolutionary tree when all speciation events are budding. This is not so if the branch order is confounded or masked by randomly generated synapomorphies on the part of multiple daughter species of one core supergenerative species, or by multiple isolated molecular strains.

Plate 8.1 in the present book demonstrates a resolved tree, and a direct caulistic evolutionary interpretation for Plate 8.1 is given in Plate 8.2. There are only one potential true sister group in the moss genus *Didymodon* (Pottiaceae) as presented in Plate 8.2. Given the probability that a wide range of cladograms may have few sister groups (those that truly modeling pseudoextinction), systematics' ship of science is firmly self-anchored to a bottom-ground of phylogenetic relationships and until it casts loose it will sail nowhere.

Because phylogenetics does not usually weight expressed traits, there is no phyletic dimension that may account for linkage of traits through fixation during speciation as a unit by selection, i.e., the traits are not statistically independent and uniquely distributed. Phylogenetics produces a hierarchical arrangement of exemplars that does not model speciation involving selection. There is no explicit naming of ancestors, and thus no explicit ancestor-descendant relationships produced in the phylogenetic analytic process or its resultant classification.

Following Blomberg (1987), there are two approaches to historical criticism: evidentialism and presuppositionalism. The evidentialist applies accepted historical criteria to elucidate reliability. The presuppositionalist first assumes reliability of data and method, then tries to show that the data and method generate a consistent whole, confirming the presuppositions. It is the consistent whole of the cladogram, being the internal simplicity of the pattern isolated from evolutionary theory, that has been substituted for efforts at understanding evolution. The problem is global in that the evolutionary dimension in systematics should allow biodiversity studies to describe and to some extent predict changes in niches and the taxa that fill them. This is not the case with phylogenetic systematics.

Both phylogenetics and phenetics (Stuessy 1990) remove the taxic phyletic (evolutionary change) dimension from systematic analysis, in part because it is "too difficult and subjective" (Stuessy 2009). What is it about macroevolution that two major fields in systematics continue to avoid it these 250 years after Darwin? In phylogenetic analysis, the morphological data set is initially cleansed of evolutionarily unique traits because these do no help determine sister-group relationships. When a maximally parsimonious tree is determined, any additional morphological traits that are distant on the tree and are separately attributed to unrelated exemplars are termed "autapomorphic" and ignored as not contributing to sister-group analysis. Yet these same traits were selected as important in evolution using the same a priori judgmental process as traits later proven phylogenetically informative.

A critique of optimality as a method of hypothesis selection is given by Zander (1998), showing that the empiric Bayesian statistical stance will not tolerate philosophical justifications of cladograms from mere simplicity, maximal posterior probability, or "converging on the truth."

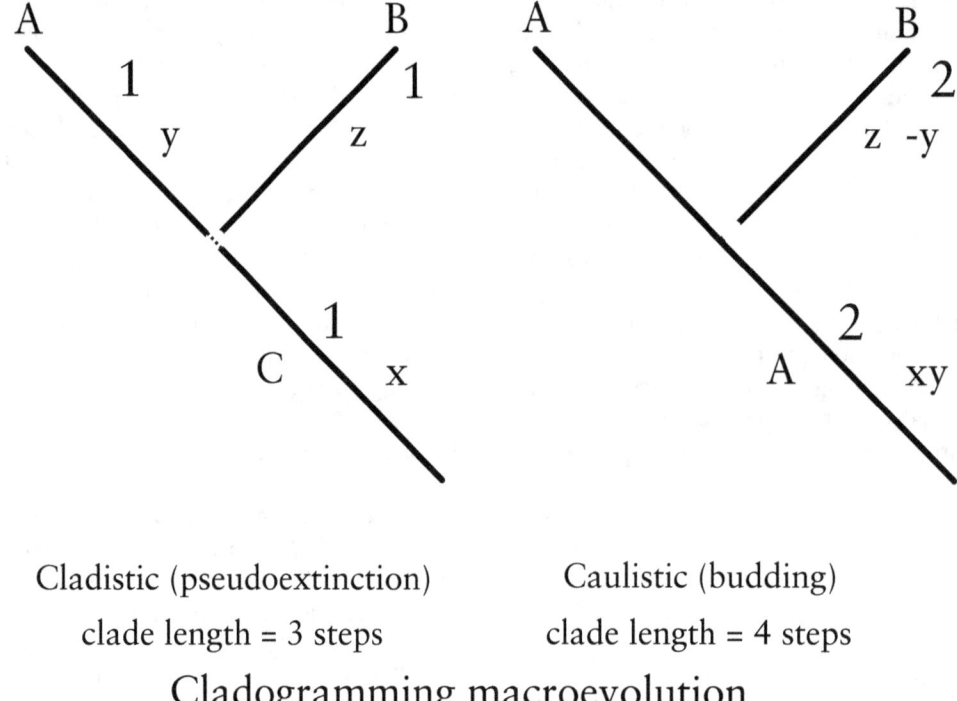

Cladistic (pseudoextinction)
clade length = 3 steps

Caulistic (budding)
clade length = 4 steps

Cladogramming macroevolution

Plate 5.3. — Maximum parsimony in cladistic and caulistic macroevolution. Pseudoextinction is important in cladistics because a shorter tree in maximum parsimony is possible. In the Cladistic (pseudoextinction) diagram, the branch shows taxon C changing anagenetically into taxon A, and B speciated separately. All taxa have plesiomorphic trait x, while A develops trait y and B trait z, yielding three trait transformations or a branch length of 3. Taxon C is assumed as a shared ancestor but is not named. In the Caulistic (budding) diagram, taxon A speciates taxon B but does not change anagenetically and diagnostically remains in stasis. All taxa have plesiomorphic traits xy, but B has two trait changes, transformation to z and reversal of y, yielding four trait transformations or a total branch length of 4 steps. If pseudoextinction cannot be inferred directly from non-phylogenetically informative data, then a longer tree is necessary to reflect a less complex process involving no ad hoc unnamed taxa.

Shared ancestors — Rieppel (2012) has discussed the background of sister-group analysis in a study of the paleobiologist O. Abel's work. Abel found that when two groups each show some complex specialization, neither family can be derived from the other, but that "both must be derived from a hypothetical common ancestor that retained the primitive condition in both organ systems." Many putative ancestral forms were seen to be, then, sister groups on a phylogenetic tree.

Following such theory, phylogenetics commonly states that a shared ancestral node in a cladogram is a hypothesis. A hypothesis, however, has some data or observations of some sort coupled with theory to support it. A shared ancestor is there by cladistic definition, as in "There must be a shared ancestor, since two of any three taxa two must be more closely related" which necessarily lead to a most parsimonious tree whether they evince Abelian specializations or not. Related, in phylogenetics, means distance on a cladogram.

Shared ancestral nodes on a cladogram are contrivances, not hypotheses as is commonly asserted. Even William Hinks, developer of quinarianism (classifying organisms into five-member circles), presented his system as positing hypotheses that could be researched (Coggon 2002: 27). Hypotheses, however, need to be supported by at least some facts. Shared ancestral nodes are an integral part of the cladistic method that allows all cladograms the potential to be fully resolved. If an ancestral taxon has, say, two daughter taxa, then one of those taxa will be more closely "related" in a cladogram to the ancestral taxon (as sister groups) by chance alone

Plate 5.3 demonstrates that a decision based on non-phylogenetically informative data that a cladogram split is due to budding evolution (ancestor-descendant transformation) results in a cladogram longer than that assuming pseudoextinction (see also Plate 8.1). The only way that one might postulate an extinct shared ancestor (and thus pseudoextinction leading to a maximally parsimonious interpretation of that cladogram node) is if there were data and associated theory supporting such; for instance, a group of specialized taxa in isolated recent environments that share some distinctive set of traits that might be ascribed to a more generalized and widespread but now extinct ancestral taxon. In molecular analysis, which one of potentially several molecular races survives for each of three taxa determines the shared ancestor and branch order.

Cladists put their trust in anagenetic change that "disappears" the ancestor. One should, however, look

for stasis first, because if it is there (and fits theory by biogeography, relative age of habitat, relatively generalized morphology, maybe even fossils), then postulating unknown, unnamed, and ad hoc shared taxa is not parsimonious. Deciding that pseudoextinction must have been the evolutionary mechanism for cladogram splits because this allows sister-group generation by maximum parsimony put the cart before the horse. This is discussed elsewhere in this book more extensively.

It has been commented that nature may not be most parsimonious in evolution, but then the question is "Well, how much less parsimonious?" Exactly that amount needed to accommodate both pseudoextinction and budding evolution as shown in Plate 5.3. Podani (2013) argued that "budding manifests itself as a lack of autapomorphy on the edge incident to the mother species (zero-length edge), which has long been known in cladistics...." Yet Plate 5.3 demonstrates that the occasional zero-length branch subtending an OUT is only the tip of the iceberg since many ancestral taxa are assigned a branch length anyway during maximum parsimony analysis. Martin et al. (2010) found that stem-based and node-based trees carried the same information, but their analysis assumed that an ancestor generated two different daughter species and then disappeared. Thus, their "stem-based" trees were like the cladistic tree of Plate 5.3.

The problem addressed by Plate 5.3 is not limited to maximum parsimony analysis. The maximum likelihood method uses a substitution model to assess the probability of particular mutations. A tree requiring more mutations at interior nodes to explain the observed phylogeny will be assessed as having a lower probability. Simplicity, again, is fundamental to the analysis. Bayesian analysis has much the same appeal to simplicity in the face of more mutations clearly needed to explain trees reflecting budding evolution.

The occurrence of polytomies (Korn & Reif 2003: 690) has much the same problem in interpretation, where a "still extant progenitor species gives rise to two or more derivative species, then it manifests itself as a polytomy in the cladogram and one edge with zero length" (Podani 2013: 323). Not necessarily, given chance trait matching or reversals.

A phylogeneticist on the listserver Taxacom (archived for July 21, 2013) asserted that a traditional taxonomic "hunch" may be fully countered by demonstrable phylogenetic relationships. I retorted that "scientific hypotheses" can never be countered by imaginary shared ancestors. The stochastic element generating the shared ancestral node is provided by the false idea that characters themselves evolve (as opposed to an organism evolving). Thus, any two of three taxa that share traits must have "evolved" through those traits away from the third. Circular proof is that the cladogram shows "phylogenetic distance." (I recommend Taxacom for its intellectually delightful scientific exchanges among collegial taxonomists in a warm and supportive environment.)

Why aren't ancestral nodes named by cladists? There are three reasons. (1) The ancestor cannot belong to two clades at once, so a node is more a place holder for an assumed nearly universal process (pseudoextinction—generation of a daughter species and anagenetic change of the ancestor into a different species). (2) If a node were named as the same as a terminal taxon, then the axiom "Every clade is monophyletic" would not be true because one cannot cut between the terminal taxon and the node and declare all distal taxa are monophyletic. (3) Maximum parsimony analysis (Plate 5.3) and methods using Markov chains would be far more uncertain, requiring non-phylogenetically informative data to name nodes when possible.

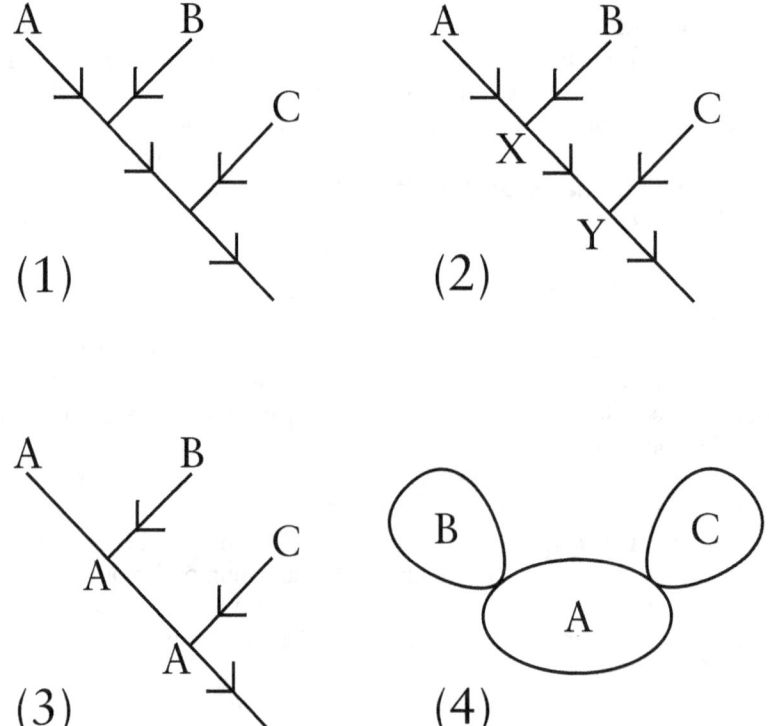

Plate 5.4. — Relevant problems in cladistic diagrams. (1) Cladogram as monophylogram, all clades are monophyletic, but only if no nodes are named. (2) Naming nodes makes inclusion of these taxa (X, Y) in any one clade equivocal, and monophyly is unclear, so nodes are not named even when such inference is possible. (3) Cladogram interpreted as an extant taxon A giving rise to two daughter lineages, B and C. Monophyly is clear but limited. (4) Evolutionary diagram of (3). Such a diagram would require an empty cell if the speciational process were inferred as pseudoextinction as in Plate 8.2. The angle brackets mark areas of monophyly on the cladogram; (1) shows phylogenetic monophyly and (2) evolutionary monophyly, while (1) inadvertently may mask evolutionary paraphyly.

Primitive versus plesiomorphic — The term plesiomorphic is associated with cladistic nesting, meaning a set of traits of basalmost clades. It is a term having to do with nesting relationships of taxa based on transformations of traits. Primitive alternatively means first, or among the first, and refers to linear transformation of both taxa and their sets of traits. A taxon in a clade arising from the base of a cladogram is not necessarily primitive because it may be a highly advanced (derived) survivor of a long, mostly extinct series of evolutionary transformations. If, however, there are other, similar taxa in other clades at the base of the cladogram, i.e., propinquous, then all those taxa might be considered primitive. They probably branch from some similar set of ancestors. Although the basal nodes may not be nameable at the level of exemplars or terminal taxa, they are may be evolutionarily informative. When molecular heterophyly is available, then the nodes may be exactly named, but similarity of taxa in basal clades should give a firm idea of the primitive traits at the base of morphological cladograms.

Comparing morphological and molecular analyses — A molecular apophyletic (descendant) taxon (Plate 6.2 2a), "B", "C", or "D") may arise from a paraphyletic taxon (Plate 6.2 2a) A implied by, "A1" and "A2") of the same or lower rank anywhere as a nested lineage ("apophyletic" in the sense of Carle 1995 or as an "apospecies" sensu Olmstead 1995). The apophyletic taxon will commonly share advanced traits of the direct progenitor taxon. Since molecular apophyletic taxa are apparently common (see below), then the best emphasis in a morphological data set is on central tendencies. The usual model of morphological parsimony is gradualist evolution with pseudoextinction of the ancestor (Rieppel 2011), but this is only one of several possible evolutionary scenarios (Hörandl 2007). This in spite of such declarations as "Groups of species are specifically excluded from being ancestral to other groups of species or to single species. The biological rationale for this distinction is clear; there is an array of processes termed speciation that allow for one species to give rise to another (or two species to give rise to a species of hybrid origin), but there are no known processes that allow for a genus or a family to give rise to other taxa that contain two or more species ("genusation" and "familization" are biologically unknown)" (Wiley et al. 1991: 3). It is held, on the other hand, by the present author that transformations of plesiomorphic traits (representing the shared autapomorphies of the ancestor) of taxa may not be in some cases an appropriate emphasis in a data set in which up to 63 percent of extant taxa may have surviving ancestors (Aldous et al. 2011), and phenetic analysis through overall similarity might be more revealing of closeness of relationship.

Parsimony cladograms should be appropriately weighted (Goldstein et al. 2002: 175) to create results best matching those detailed in classical systematics. This is because the best product of parsimony analysis is a *natural key* or cladogram equivalent (see below) that best reflects the evolutionary weights on individual taxa and on individual traits previously provided by alpha taxonomy and classical systematics. Morphological results may be well supported, through nesting and also through linkage with non-dataset information such as population analysis, growth chamber study, habitat, biogeography, and cytology. In addition to homology analysis, the best traits for morphological cladistics should be those that are quasi-neutral, being (at least for similar taxa) stable across different habitats or at least not associated with specific habitats. Traits and trait combinations deemed liable to selection by being found in very different taxa occurring under the same selection regime, may be good identifiers of the taxon but not of relationships because they are clearly convergent.

Conservative morphological traits are similar to the DNA bases used in molecular systematics to track genetic continuity. One explanation for the demonstrated value of conservative morphological traits in classical taxonomy is that, during speciation, some traits of the progenitor can tag along with new adaptive traits into the new species because these ancestral traits are either valuable generally, or neutral, or not fatally burdensome. Conservative traits are expressed across different habitats in different species, and can be identified readily. Cladistic analyses commonly weight all traits equally, but this allows variable traits to compete equally with conservative traits in determining branch order. Also, conservative and variable traits may not be the same in different portions of a cladogram. These reasons are probably why morphological cladograms have low nonparametric bootstrap support values and are poorly resolved, in that equal-weighted labile traits overwhelm truly conservative traits that track evolution. Morphological cladograms may best be evaluated in terms of the reliability expected from the classical taxonomic descriptions they are based on, using superoptimization and coarse priors (see Chapter 8), rather than on non-parametric bootstrap support values. Given that species exist for hundreds of thousands or millions of years (an average, rule-of-thumb age of a species is five million years), stabilizing selection is quite constant for that particular combination of adaptive traits, while quasi-neutral traits may be fixed in any physiologically tolerable number. What morphological cladistics does is provide a reasonable gross transformation series for a given group, suggesting basal, terminal and intermediate groups. The exact bootstrap support for clades is probably artificial in detail in that fixation of two traits being less probable than that of one trait is locally (on the cladogram) somewhat artificial, and only large scale transformation series are informational.

Morphological traits can be used in aggregate, however, for certain statistical operations. Using the *parametric* bootstrap, Zander (2003a) demonstrated that a morphologically based cladogram can be far more strongly nested than by chance alone. A cladogram (Plate 8.1) of real data (22 species with 20 characters scored) with low nonparametric bootstrap support was examined with constraint trees to compare the total support for each of the three possible branch configurations of each node. The null hypothesis was that the number of steps (unweighted) summing the length of the optimal tree is 1/3 of the sum of the steps of all three alternatives at each node, assuming random distribution of traits. The sum of the number of steps in the most parsimonious cladogram was 34, while the sum of alternative steps (difference between 34 and total all steps) is 16. An exact binomial calculation indicated that this ratio (34:16) or better would occur by chance alone less than one out of 1000 times. The confidence interval is thus 0.999+ that considerable phylogenetic signal is present in the optimal tree (as alternative hypothesis). Using minimum ratios for support at 0.95 probability, 3:0, 4:1, 5:2, 6:3, 7:4, 8:5, etc., it was found that four contiguous internodes were required for this (fairly average) cladogram to demonstrate reliable (0.95 probability) phylogenetic signal of shared ancestry, i.e., the phylogenetic resolution of the morphological cladogram was four internodes. Of course, independence of traits is required, and if these were all conservative traits that varied little in different habitats *for the group*, independence might be assumed.

This was compared (Zander 2003) to support values in molecular analysis. If different gene histories were considered different characters (Doyle 1992) and they were randomly generated at each node, a single gene tree of average 0.95 support per node would require six contiguous nodes for a reliable phylogenetic signal; if two gene trees agree, two internodes are required; if three agree, one is required. This is the case in a particular, unweighted morphological cladogram with no input from superoptimization. Support measures in morphological cladograms may be much increased beyond, say, four internode resolution as above, with proper weighting of conservative traits and identification of progenitor-descendant series. If superoptimized (Plate 8.1), a cladogram collapsed to a caulistic summary (Plate 8.2) has very high support.

Identification of each habitat-neutral, conservative trait is of particular importance in creating morphological cladograms and natural keys. Recognition that certain taxa are more likely to be descendants of certain other, phylogenetically close taxa using non-data set information is good way to weight traits, but because this is not generalizable, differential weighting should be applied only in analysis of small groups of taxa. Two of the ways to discover that differential weighting or other modification (such as removal of taxa) of a morphological cladogram is needed is if (1) non-phylogenetic information (e.g., geography, cytology) indicate that taxa that are apparently derived are more basal on a clade than their apparent ancestors or ancestral morphotype, and (2) if the generalized morphotype of two or more branches is immediately quite different from the generalized morphotype of the next two or more branches contiguous on one clade. This is because one expects the nodes of clades to be similar in morphotype to the nearest exemplars, even short of actual identification of the node as an extant taxon.

At this point, it should be noted that a morphological data set, or in fact any description of a taxon, lacks the "look and feel" element that is commonly a Gestalt of traits that is difficult to measure or describe. This is why illustrations are so important in identification, and why a morphological data set is actually a poor sample of important traits used in taxonomy.

Superficially, a molecular data set may seem to have more traits than morphological data sets, yet the former is commonly of only one or a few specimens per taxon. A molecular data set of, say, 50 species, each represented by one exemplar, and 1,000 phylogenetically significant DNA sites yields 50,000 data points, apparently a robust data set. Yet a morphological data set of 50 species each of 50 traits may involve examination of hundreds or even thousands of specimens as alpha taxonomists build on the work of others over 250 years of study. Even a single revision of 50 species with 50 traits and averaging 50 specimens (each representing a different population) examined for each taxon yields 125,000 data points. If half the traits were too conservative to be informative, i.e., resistant to adaptive selection at least within the taxonomic group of interest, then there are 75,000 data points contributing to both infra- and intrataxon relationships. Although alpha taxonomy is not exact and makes use of informal heuristics, conservative expressed traits varying within a taxon directly track macroevolutionary changes, while molecular traits track genetic continuity and isolation events associated with molecular strains. Thus, the two data sets may be much the same size; they track, however, different aspects of evolution.

A molecular data set, though generating cladograms of high Bayesian credibility (i.e., high phylogenetic valence), may not have the statistical power of what appears to be a smaller morphological data set. One can increase power of discrimination (Aron et al. 2008: 225) in several ways, among them (1) increasing the number of data, (2) decreasing reliability (accept lower credibility scores), (3) use more precise and standardized measures, and (4) use a population with less variation. Molecular analysis uses numbers 1 and 3, while morphological analysis uses numbers 1 and 4 by massive examination of samples and focusing on conservative traits. Following Aron et al. (2008: 226) a study with a small data set that manages to be statistically significant must be due to a large *effect size* (Cohen's d), while a large study that is equally or more significant may have a small effect size. The effect size is a measure of the extent to which distributions do not overlap (Plate 5.2). Given that effect size is the difference between population means divided by population standard deviations, results of morphological study may have great statistical power, particularly as each "trait" represents consolidated observations of tens or hundreds of exemplars exhibiting that conservative trait in particular environmental contexts or selective regimes. Morphological species when they overlap are stated to do so, and (informally) by how much.

Shaw and Small (2005) demonstrated empirically that using only single molecular samples of each taxon can lead to different results in different studies of the same taxa, at least in closely related groups. Omland et al. (1999) emphasized the importance of dense taxon sampling below the species level, but did not go beyond attribution of paraphyly to two oriole species, while "dense" meant to them 2 to 4 samples per species commonly with each sample representing a different subspecies. Price and Lanyon (2992) examined 25 total individuals of 8 species of oropendola birds and asseverated that their study demonstrated the importance of including multiple examples of each taxon. DeSalle et al. (2005) discussed the literature on the problem of too few samples per taxon in mitochondrial DNA barcoding, and suggested that the numbers should reflect a sampling of all species to evaluate variation within the gene se-

quence region used. They examined in their own study 114 individual leeches, up to a maximum of 49 individuals for one species, but considerably fewer for 20 other species. With much molecular analysis there is no or little population sampling (multiple exemplars of one taxon) and thus no statistical power, i.e., the molecular cladograms cluster exemplars which may result in high Bayesian credibility for the exemplar tree but low statistical power to distinguish properties of taxa on a tree by that data.

Viewing morphological parsimony analysis as simply generating a better natural key removes the problems (Zander 1998) of whether the results may be analyzed statistically (Cohen 1994; Wilkinson et al. 1999) or be accepted through a philosophical "simplest solution" argument (Zander 1998). One must deal with the fact that a phylogenetic morphological data set eliminates before analysis all sister-group uninformative traits. Atomizing linked (taxonomically or in selection) traits in a data set (Burleigh 2013) does not render them independent, but it is better hoped that the results are exactly those classical concepts on which the data set is based, simply better organized in large scale. Another problem is that cladograms are dichotomous while nature need not be; Hennig's (1966) principle of dichotomy governing speciation (Rieppel 2011; Rosenberg 2003) is a pervasive bias. Imposing a dichotomous structure on a representation of evolution arbitrarily lowers informational content. It increases the entropy of informational uncertainty with an overburden of arbitrary structure. A natural key allows some degree of reversals in traits when required by polythetic taxon concepts, which is entirely in the spirit of cladistic nested groups involving occasional trait reversals, but a natural key may be multichotomous, or may even have to represent at times a single evolutionary branch (see also natural key to *Didymodon* in Chapter 8)..

Natural keys — Following is a fully detailed cladogram to subfamilies and tribes of the moss family Pottiaceae of Zander (1993), some names updated: This may be used as a precursor to a multichotomous natural key.

Cladogram Equivalent of a Natural Key

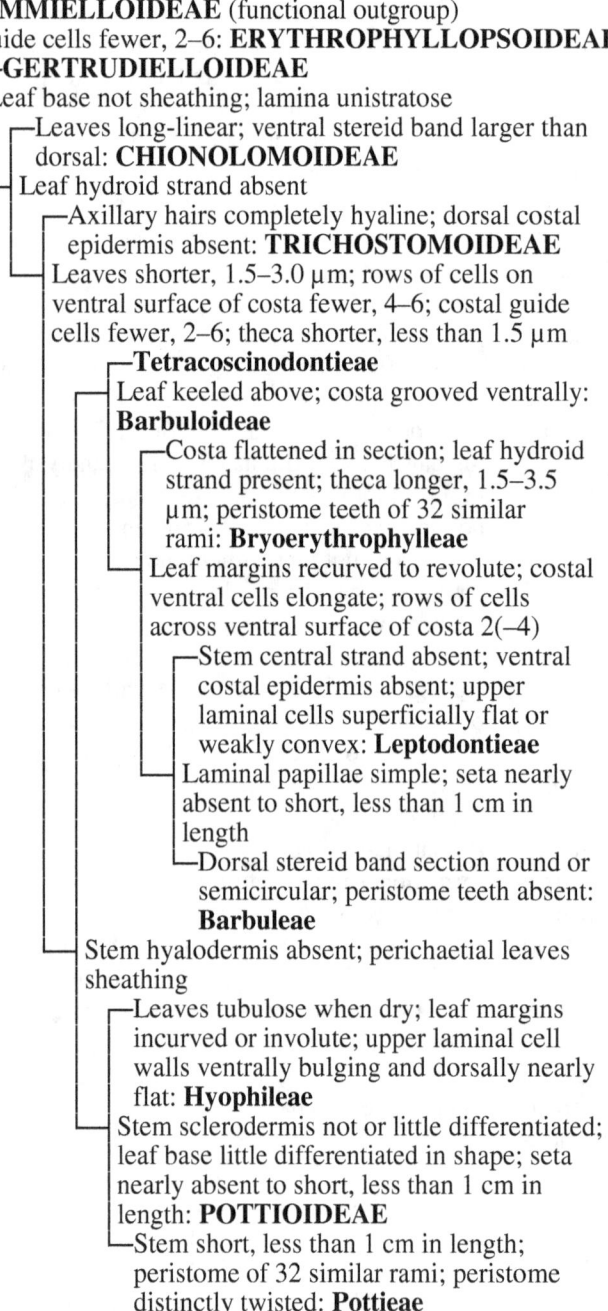

—TIMMIELLOIDEAE (functional outgroup)
—Guide cells fewer, 2–6: ERYTHROPHYLLOPSOIDEAE
—GERTRUDIELLOIDEAE
—Leaf base not sheathing; lamina unistratose
—Leaves long-linear; ventral stereid band larger than dorsal: CHIONOLOMOIDEAE
—Leaf hydroid strand absent
—Axillary hairs completely hyaline; dorsal costal epidermis absent: TRICHOSTOMOIDEAE
—Leaves shorter, 1.5–3.0 µm; rows of cells on ventral surface of costa fewer, 4–6; costal guide cells fewer, 2–6; theca shorter, less than 1.5 µm
—Tetracoscinodontieae
—Leaf keeled above; costa grooved ventrally: Barbuloideae
—Costa flattened in section; leaf hydroid strand present; theca longer, 1.5–3.5 µm; peristome teeth of 32 similar rami: Bryoerythrophylleae
—Leaf margins recurved to revolute; costal ventral cells elongate; rows of cells across ventral surface of costa 2(–4)
—Stem central strand absent; ventral costal epidermis absent; upper laminal cells superficially flat or weakly convex: Leptodontieae
—Laminal papillae simple; seta nearly absent to short, less than 1 cm in length
—Dorsal stereid band section round or semicircular; peristome teeth absent: Barbuleae
—Stem hyalodermis absent; perichaetial leaves sheathing
—Leaves tubulose when dry; leaf margins incurved or involute; upper laminal cell walls ventrally bulging and dorsally nearly flat: Hyophileae
—Stem sclerodermis not or little differentiated; leaf base little differentiated in shape; seta nearly absent to short, less than 1 cm in length: POTTIOIDEAE
—Stem short, less than 1 cm in length; peristome of 32 similar rami; peristome distinctly twisted: Pottieae

The natural key below (here limited by its dichotomous structure) is based on the above cladogram, and assumes that relative degree of nesting is approximately equivalent to relative advancement on a linear, macroevolutionary scale, as a first hypothesis of macroevolution. (This will be modified in discussion of self-nesting ladders, elsewhere.). The couplets include additional information to allow parallel traits in dichotomous branching, eliminates or changes artifacts of the analysis (e.g., peristome absent in Barbuleae when it is mostly present), and makes changes in taxonomic nomenclature following recent research. A multichotomous natural key to a different group is given in Chapter 8.

Natural Key Presented in Dichotomous Form

1. Upper lamina bistratose medially and the cells not vertically aligned (i.e. not directly over each other) near the costa but grading to vertically evenly stacked towards the leaf margin, leaves broadly to linearly lanceolate .... Subfamily **TIMMIELLOIDEAE**
1. Upper lamina unistratose or if bistratose then cells situated directly over one another throughout.
    2. Upper laminal cells ventrally mammillose medially but several rows of cells bulging on both sides marginally, costal guide cells forming a thick-walled, multilayered cylinder.................................... .... Subfamily **GERTRUDIELLOIDEAE**
    2. Upper laminal cells similarly bulging or not throughout leaf, guide cells either not multilayered or if so then thin-walled.
        3. Leaves lanceolate, margins plane to weakly incurved, apex acute, base sheathing, upper lamina KOH red, stereid bands two, guide cells 4–6, rows of cells across ventral surface of costa 10(–16) ....Subfamily **ERYTHROPHYLLOPSOIDEAE**
    3. Not this combination of characters.
        4. Leaves long-linear, margins plane, ventral stereid band larger than the dorsal .... Subfamily **CHIONOLOMOIDEAE**
    4. Leaves lanceolate to spathulate, ventral stereid band absent or generally smaller than the dorsal.
        5. Sclerodermis commonly poorly differentiated, hyalodermis commonly present, leaves lanceolate, margins plane to weakly incurved, upper laminal cells KOH yellow, costa lacking a differentiated dorsal epidermis, clavate axillary propagula rare ....Subfamily **TRICHOSTOMOIDEAE**
        5. Not this combination of characters.
            6. Stem sclerodermis commonly well differentiated from cells of central cylinder, which have abruptly larger lumens, leaves usually broadly lanceolate to narrowly elliptical, usually with two costal stereid bands, leaf base commonly differentiated in shape and ovate or rectangular, upper laminal cells equally convex on both free surfaces, clavate axillary propagula commonly present in some genera . ...Subfamily **BARBULOIDEAE**
                7. Stem black, leaves long-triangular, capsule with a circumstomal ring ....Tribe **Tetracoscinodontieae**
                7. Not this combination of characters.
                    8. Upper lamina usually KOH red, dorsal stereid band usually reniform ....Tribe **Bryoerythrophylleae**
                  8. Not this combination of characters.
                        9. Stem central strand absent.......... ....Tribe **Leptodontieae**
                        9. Stem central strand usually present, or if absent then costa with one stereid band. .... Tribe **Barbuleae**
            6. Stem sclerodermis commonly not or poorly differentiated from cells of central cylinder, which generally grade in size into the cortical cells, leaves usually broadly ligulate to spathulate, usually with one stereid band in the costa, leaf base usually little differentiated in shape, sometimes upper laminal cell free surfaces ventrally bulging and dorsally weakly convex, clavate axillary propagula rare Subfamily **POTTIOIDEAE**
                10. Upper laminal cells usually bulging ventrally and weakly convex dorsally ....Tribe **Hyophileae**
                10. Upper laminal cells equally convex on both free surfaces .... Tribe **Pottieae**

Evolutionary lens — Many taxa may be reduced in expressed character (e.g., size, complexity, range of variation) relative to related taxa (as in mosses, Bateman 1996; Zander 1993). Such multi-character reductions may be parsimoniously misevaluated as shared traits in a morphological cladogram. Dollo evaluation may indicate correctly that morphologically reduced taxa are advanced. If there are reduced taxa, these may be deleted from the initial data set. A cladogram restricted to only taxa with many traits may then serve as a foundation for reanalyzing the reduced taxa one at a time to see where they might fit without confounding morphological "long-branch attraction" in the tree of unreduced character-state transformations.

Taxa that are not reduced, at least among the mosses, are commonly found grouped geographically. Within a species, the most character-rich populations of certain taxa with reduced forms occur in major mountain ranges, hyper-oceanic areas, austral zones, and, perversely in some cases, the margins of ranges. The mountain range phenomenon may be a kind of biotype Massenerhebung effect (widening of floristic zones on mountains surrounded by other ranges). This is also true for genera with species more character-rich than others. Thus, a geographically based evolutionary lens is available that reveals trait combinations difficult to evaluate in small forms or reduced congeneric taxa. These are not necessarily primitive (first of a series) taxa, but are well characterized. This may be due to a kind of reverse Red Queen effect. Instead of a taxon evolving rapidly to keep up with rapid evolution of competing species, taxa in deep stasis, particularly those that occur in microenvironments, are strongly affected by stabilizing selection.

Hard science — Phylogenetic analyses may appear to differ from those of classical taxonomy in better reflecting a "hard science" approach. Cladistics saves tree structure by forcing all relationships into evolutionarily divergent sister-groups. Phylogenetics is on the face of it more rigorous in the Popperian sense (Popper 1959: 71) in approaching an axiomatized system of minimally sufficient and necessary logical terms, and it mathematically generates statistically manipulable masses of data (Avise 2000).

Feynman (1985: 311–312), pronounced "fainman," characterized "cargo cult science" as any science in which the form is perfect, and which "follow all the apparent precepts and forms of scientific investigation, but they're missing something essential ..." because facts that are not reconciled are left out.

That is, powerful methods and elaborate theories must do more than explain that which "gave you the idea for the theory; but that the finished theory makes something else come out right, in addition. In summary, the idea is to try to give *all* [his emphasis] of the information to help others to judge the value of your contribution; not just the information that leads to judgment in one particular direction or the other." Phylogenetics erects a massive edifice of statistical analysis, but the results are not used to develop a causal theory of evolution for a group. Instead, because the raw results are nested like a classification, they are viewed as a classification. Incongruent results are thrown out or relegated. The analysis is not complete.

New theories in science seem like the new overthrowing the old, but they are actually (in most cases) a replacement for the old, and are a complete replacement. Common sense tells us that the Earth is flat, and plane trigonometry suffices. More information tells us the Earth is round and requires spherical trigonometry, yet plane trigonometry suffices locally as an acceptable approximation. Yet more information tells us Einsteinian relativity needs to be accounted for, yet Newtonian physics suffices for regional approximations. This falling back on approximation (Feynman et al. 2011: 2) is a welcome scientific heuristic, and may be done because each new theory encompasses the old. R. Feynman, in a filmed lecture at Cornell University, explained that a new theory is indeed guesswork, but has to completely "fit" the place the old theory filled. His analogy is like the combination of a safe—if five numbers are needed to unlock it, a guess of three numbers will not do. Yes, we must account for new phylogenetic aspects of evolution, yet a new theory must account for the results of classical taxonomy.

The unimaginable is not easily addressed in the context of Popperian hypothetico-deductivism. It is scientific induction that opens theory to the unimaginable, like a spherical earth or modifications of Newtonian physics or genuine transmutation of elements. A pluralistic approach to systematic investigation is not heterodox but is open-ended. "Hard" science is actually the most malleable field of knowledge.

The hard science aspect of phylogenetics is simplistic in spite of non-trivial mathematics and statistics. Because nonparametric bootstrapping (Efron et al. 1996; Felsenstein 1985; Sanderson and Wojciechowski 2000) is at base probabilistic and depends on independence of traits, extreme reductionism (McShea 2005) promotes incorrectly treating traits as independently fixed in speciation. Traits that are basically part of descriptions of evolutionarily coherent organisms may be linked (fixed as an adaptive package during speciation, like the numbers in a combination lock required to open it) and as such may be inappropriate for bootstrapping and other statistics as an indication of accuracy, short of replication of the original classical taxonomic study (Cohen 1994). Zander's (2004) demonstration of a probabilistic equivalence of bootstrap and credibility measures using contrived simulated data sets and exact binomial analysis may be useful for molecular analysis where sufficient independence of data are possibly justifiable, under a set of assumptions (e.g., Cartwright et al. 2011).

Feynman also (1985: 269) tells the story of calculating the length of the Emperor's nose. Since the Emperor would not allow direct measurement, all the people were polled as to their guesses. Then the results were averaged, and that average must be the length of the Emperor's nose because it is a standard, powerful statistical measure. Like phylogenetics, the data were about something other than the process in nature being studied, and the method was understood much better than the natural process or the result.

Taxa may be paraphyletic on a morphological cladogram, but only specimens are paraphyletic on a molecular cladogram where taxa must be inferred from very small taxic samples. One may note here that the mapping of taxa to molecular cladograms either through heterophyly on a single tree (see Element 3) or between a morphological and molecular tree (see Element 4) results in a kind of reciprocal illumination for the morphological tree. Ancestral traits in a morphological cladogram with caulistically mapped taxa from an evolutionary tree may be the same at two or more contiguous nodes, i.e., there is no transformation between them. Therefore, an immediate ingroup is available as functional outgroup for a series of morphological re-analyses of well-segregated subsets, namely those taxa distal to the molecularly mapped progenitor taxon.

Juggling concepts — It is easy to imagine that cladistics is better than phenetics because rather than clustering by just similarity it uses maximum parsimony of trait transformations to nest taxa. Yet, as detailed above, because maximum parsimony in

cases of budding evolution ca be incorrect, cladistics may provide far too many splits to model evolution as a *combination* of pseudoextinction and budding evolution. Many of the splits are randomly generated and provide only imaginary information for other fields of biological science, like conservation. More on this in Chapter 8. Phenetics is informative and requires just as much conceptual analysis to render it into an evolutionary model as does cladistics.

Examples of morphological cladistics — A maximum parsimony cladogram of the Pottiaceae (Zander 1993) was published for 76 genera and 75 morphological traits of this character-rich moss family. An apparent reduction series affecting the sporophyte is present in many genera, with identical modification in species of many genera leading from a operculate and peristomate cylindric capsule on an elongate seta to a simple, irregularly opening, non-peristomate globe with almost absent seta, often associated with increase in spore size. Evolutionary theory equates such reduction with differential r/K selection associated with precinctiveness or local dispersal (Carlquist 1966), or atelochory or nondispersal (Van der Pijl 1972), and data on habitat correlation supports this. Differential weighting of traits was thus justified (Zander 1993) and was used to force reduced sporophyte traits distally on the cladogram, done by increasing character weights on non-sporophyte traits until the cladogram no longer changed when generated.

Zander (2006) used UPGMA (with Dice algorithm to emphasize similarity) cluster analysis to evaluate distribution of unreduced twisted peristomes in the Pottiaceae (generalist structures identifiable with possible progenitor taxa), identifying taxonomically scattered groups with both unreduced and reduced sporophytes based in raw similarity. Given probable irreversibility after Dollo evaluation (see Element 6), reduction series should prove a valuable tool in determining direction of evolution. Although control of trait expression through epigenetic factors (Li 2013; Danchin et al. 2011; Riddihough & Zahn 2010; Turner 2002) may allow reversal of reduction, one must remember that the word "epigenetics" can "explain" any and all things difficult to deal with in terms of standard genetic theory.

CHAPTER 6
Element 3 - Contributions of Molecular Systematics

Précis — The postulation of a caulistic macroevolutionary transformation (as an ancestor-descendent relationship) through naming nodes on a cladogram provides an overarching scientific theory that consiliates different classical, cladistic and molecular studies of evolutionary relationships of the same taxonomic groups. Heterophyly (paraphyly or extended paraphyly as phylogenetic "polyphyly") on a molecular tree implies a deep ancestral taxon identifiable (formally nameable) at the lowest taxonomic level of all exemplar specimens of that taxon. Apophyletic taxa (descendents) are those lineages that derive on the cladogram from a deep ancestral taxon. Self-nesting ladders make classification on the basis of molecular cladograms problematic because molecular branch order may not be the same as the order of evolutionary transformations, particularly for cases of extinct or unsampled extended paraphyly; complex heterophyly involving two self-nesting ladders can be mistaken for evolutionary polyphyly.

Molecular systematic analysis establishes genetic continuity and order of isolation events of exemplar molecular strains (but not necessarily speciation events), within bounds of its many assumptions (Zander 2007a). Molecular analysis seldom ignores morphology, although there are instances of taxa in the literature described only as "DNA sequentia differt" as reported by Bakalin (2011). Of course, molecular phylogenetic study is initially bounded by preselection of exemplars from a previously recognized taxonomic group. Although much information on evolution may be obtained from molecular analysis, evolutionary monophyly is not included. This is because the nesting basis of cladistic analysis does not identify shared ancestors, and an evolutionary deep ancestral taxon does not have to begin generating daughter species at the exact point that phylogenetics expects the unnamed shared ancestor to begin the monophyletic group. Also, modeled trait changes are of microevolution.

Preselection of molecular exemplars (specimens sampled) ensures a general match of clustering because the preselection is from a highly predictive classical classification. The reverse, preselection from a molecular classification, cannot be tested because there is no stand-alone molecular classification. This is because there is not yet an established causal basis of predictive value that applies to molecular analysis alone. Any apparent predictive value is from extrapolation from known cases, similar to regression. Macroevolution through budding evolution or by pseudoextinction (masked by assumption of universal "shared ancestry") has the potential for explaining a causal basis but extinction or non-sampling of molecularly heterophyletic lineages is a major problem. When heterophyletic lineages are available, however, evolutionary information is inferable.

Taxon mapping through heterophyly — Analyses of molecular traits can be done as per standard practice (multiple methods including parsimony and Bayesian analyses, ideally involving multiple sequences, multiple samples of each taxon, and choice among a finite number of models). Curiously, parsimony is too simplistic while Bayesian analyses overspecify the model. The resultant cladogram should be evaluated according to the methods of Zander (2007a) to combine and collapse to 95 per cent credibility all branches, e.g., Plate 7.3. If the cladogram one examines has been created and published by others, the names of exemplars or the taxa they represent should be replaced when necessary with older classical names unless new phylogenetic taxa can actually be supported in the context of possible multiple test problems (Zander 2007a). Wilkinson et al. (1999: 597) point out that "ambiguity in defining variables can give a theory an unfortunate resistance to empirical falsification," in this case, classical names are information carriers while phylogenetic names if used in the enforcement of holophyly are artificial and logically circular. The best molecular cladogram is then subjected to *taxon mapping* (Zander 2010a) using classical names to identify possible deep ancestral taxa as branch nodes bracketed on the cladogram by their heterophyletic (paraphyletic or phylogenetically polyphyletic) surviving exemplars, e.g., Plate 6.2. Of course, in the absence of named taxa at the proper rank, any group of morphologically similar taxa on a molecular tree imply an ancestor of that group having much the same morphological traits. Taxon mapping, however, goes beyond the common practice of mapping morphological traits on a molecular tree, and attempts to assign a taxon name to cladogram nodes, when possible.

In any cladogram, the node or nodes subtending

the exemplars can be named as that taxon that includes all exemplars. Simplistically, several exemplars of one taxon, say, of one species or genus, are derived from an ancestral taxon of the same name. Taxon mapping narrows this through recognition of molecular paraphyly or extended paraphyly (phylogenetic polyphyly) as an indicator of a deep ancestral taxon. Even if a taxon does not exist that is of a rank that fits perfectly for a particular number of fairly uniform exemplars, a generalized ancestor might be inferred for a node or series of nodes. If there is a break in a reasonable progression of generalized ancestors, either a major macroevolutionary event should be considered, or long-branch attraction, choice of a wrong outgroup or effective local (or functional) outgroup, or wrong weighting, or some other bias may be the case. The branch order of a fairly uniform group may be best determined by studying that group alone, with better choice of an outgroup than that presented by global cladistic analysis.

Peripatric and allopatric speciation — Foote (1996) has shown that the longer an ancestral species survives, the more likely two or more descendant species will emerge from it by budding evolution, usually attributed to peripatric speciation (Mayr 1954). Mayr (1982) wrote, "The fundamental fact on which my theory was based is the empirical fact that when in a super-species or species group there is a highly divergent population or taxon, it is invariably found in a peripherally isolated location." He apparently treated peripatric speciation in the allopatric sense. Given that five million years is rule-of-thumb age for a species, and many survive 40 million years, with some remaining in apparent stasis in the hundreds of millions of years, supergenerative species may be or have been common at any one time. According to Batten et al. (2008), "Only forms of balancing and stabilizing selection have been demonstrated in nature..." and that there is evidence that "ecological adaptation occurs only in a minority of speciation events" but their paper emphasized self-organization not habitat-dependent functional transformation. See also Solé and Manrubia (1996). Given evidence from superoptimization (Plate 8.1), there is at least abundant correlation of species form and habitat or range that seems explainable by classical ecological adaptation rather than "self-organization" and "edge of chaos" evolvability.

Although hybridization can contribute to extinctions through genetic swamping or depression by derivatives (Newman & Pilson 1997), according to

Hegde et al. (2006): "Homoploid hybrid derivatives are direct descendents of first- or early-generation hybrids without subsequent introgression, have strong reproductive isolating barriers relative to both of their parents, and generally establish in novel habitats, rarely causing extinction of parents...." Budding evolution is apparently common and does not involve extinction of the ancestral taxon (nor is budding evolution necessarily peripatric if genetic isolation is strong and fast).

Limitations to taxon-mapping by heterophyly — The use of heterophyly to infer deep ancestors is neither mechanical nor sure. Every implied deep ancestral taxon must be subjected to the question "is this reasonable given what one might expect about serial evolutionary transformations at the taxon level?" For example, in plants, a diploid descendant is not expected from a polyploid ancestral taxon unless a real case can be made for diploidization, supported by probably nonreversible, unique morphological changes and specializations in habitat or reproduction. There are two major sources of possible wrong inferences from heterophyly.

(1) The taxa that are heterophyletic must be supported as member of the same taxon by a coarse prior (q.v.) of 0.95 or greater (i.e., sure enough to act on). Many taxa are equivocal in taxonomic position, or simply wrongly placed. (2) Parallelism of descendants (e.g., the same species generated twice by an ancestor) can falsely make an inferred ancestral taxon of a descendant taxon. (3) Apparently well-supported conclusions based on heterophyly may be compromised by unsampled or extinct extended paraphyly. (4) If conclusions are based on internal branches, and two branch support values must be true at once for a conclusion, one should remember that the two 0.95 posterior probability support values must be multiplied, and we get 0.90 or a 9 in 10 chance that both are correct at once, and for conclusions involving three 0.95 support values, multiplication yields about 0.85 or 8.5 out of 10 chance all are correct. Only when working with 0.99 posterior probabilities may up to four clades be used to make acceptable conclusions. (Note that throughout this book this last caveat is largely ignored for demonstration purposes.)

Accuracy and precision — We can define accuracy in the present (non-fossil) context as discovery of congruence between classical taxonomy and cladograms from different data sets including morphological. Precision can be defined similarly as clear-cut, measurable, well-supported differences between nested groups in a cladogram. Molecular cladograms are presently considered precise because of high bootstrap and Bayesian support, and also accurate if they agree with morphological cladograms, which, because of commonly low bootstrap support, are considered much less precise. It is a kind of circular reasoning (Walton 1989) that if molecular analyses commonly support morphological analyses in general, then any additional details supported by molecular data but not the morphological must also be correct and all contradiction must be decided in favor of molecular results because there is more data. Yet even congruence between morphological and molecular cladograms may be suspect (see self-nesting ladders, below).

Circular reasoning can never be completely eliminated from any scientific endeavor (the syllogism is well known to be far too idealistic for actual logical converse), but it can be minimized. Facts, as well-documented observations, always include some element of interpretation, trimming of apparent extraneous information, and some tacked on theory. Theories, incorporating facts, are then to some extent circular. This is why total evidence analysis is valuable, because if all facts are explainable by a theory, then it is less likely to be affected by circularity than if only some, most relevant or critical facts are explainable. This assumes that tendentious elements contributing to circularity of facts are somewhat different for each fact.

Graur and Martin (2004) detail how the accuracy of molecular clock estimates for divergence events have been severely compromised by other researchers by "improper methodology on the basis of a single calibration point that has been unjustly denuded of error." Statistical estimates were converted into errorless numbers in many studies that grossly misrepresented divergence times of ancient groups. Similarly, as argued in the present book, any phylogenetically informative data are rendered uncertain to varying extents by non-phylogenetically informative data stripped from the data set although relevant to macroevolutionary transformations. A certain sophistication in mathematical and statistical methods is necessary to interpret papers using multiple, advanced methods, such as the impressive work of Ding et al. (2006), which are fundamentally based on molecular cladograms.

If there is incongruence between molecular and morphological cladograms (or classifications), then the morphological data are commonly ignored by phylogeneticists as homoplasy or congruence, or a like presumption (as in no evidence) of gross misrepresentation of exemplar nesting. If the morphological data are not thrown out, there commonly can be found some data leading to some corroboration for the molecular tree, by browsing for morphological traits that happen to match in synapomorphy the molecular cladogram; corroboration is declared, then, as found. This is a multiple test (or multiple comparisons) problem in statistics, discussed elsewhere in this book. Also, ignoring data is not in the Bayesian tradition, and an empiric Bayesian analysis (using the results from the Bayes' Formula as prior for another instance with new data) will necessarily lower the support measures for molecular analysis, which cannot be both precise and accurate if there is contrary evidence from other data.

An analogy is commonly helpful in revealing problems with confusing, complex processes. The economy of a nation is, vis-à-vis evolution, similarly complex and difficult to grasp. Let's say a new technique is invented that uses discarded supermarket cash-register receipts to analyze the nation's economy. Statistically analyzing a thousand receipts a month in exemplar cities provides a phenomenal data set with robust results. Analyzing a million a month with more exemplars gives even more precise results. When the question arises as to the exact place of grocery items in analyses of the economy, the answer might be given that groceries are certainly an important dimension of the economy, the results match many other indexes of economic health, and when they do not, the grocery data are far greater than data on, say oil, grain, employment, services, automobiles, and housing, given that each can of beans is weighted the same as a manicure, a barrel of oil, a Buick, or a new split-level ranch-style abode. If pressed, adherents of the new method can simply redefine the economy as supermarket activity.

Self-nesting ladders — The most problematic difference between morphological and molecular cladistics is that morphological cladistics generally clusters evolutionarily similar taxa together. Thus, nodes of a cladogram commonly integrate the traits of taxa within a few nodes of each other, and the cluster is commonly clearly comparable to that implied by classical taxonomic classifications. In molecular systematics, on the other hand, morphological traits may be static over geologic time, yet DNA traits continue mutating. With each speciation event in which the ancestral taxon survives, i.e., resulting in a series of supposed sister-groups with one branch the ancestor, the ancestral taxon is molecularly farther and farther away from its nearest neighbors on the morphological cladogram. This is a "self-nesting ladder" of increasing molecular phylogenetic differentiation, where primitive (taxa at the base of an evolutionary sequence) push themselves into an advanced position in a nested molecular diagram. Self-nesting ladders may be of any length, and the more egregious in molecular cladograms are easily identifiable (see discussion of the moss *Erythrophyllopsis* below). In that a self-nesting ladder when present always resolves evolutionary relationships backwards on a molecular tree, mapping morphological character traits on such a lineage results in backwards inferences. Such reversals or incongruities between morphological and molecular results are commonly attributed to homoplasy and convergence, but an examination of the actual taxa involved in the nesting can often reveal a clear case of self-nesting. *Simple heterophyly* involves one clade with one self-nesting ladder; *complex heterophyly* involves two clades generated by one ancestral taxon and two self-nesting ladders.

Molecular cladograms are not alone in exhibiting self-nesting ladders. There were many reversals in traits in Zander's (1993) maximum parsimony cladogram of Pottiaceae morphology. These involved length of stem; stem sclerodermis and hyalodermis presence; leaf stance when dry or wet; leaf shape and length; conformance of leaf ventral surface and costal groove; leaf base shape; number of rows of cells across ventral surface of costa; transverse section shape of dorsal stereid band; dorsal costal epidermis presence; costal hydroid strand presence; width of medial upper laminal cells; superficial wall width of upper laminal cells; sexual condition; perichaetial leaf shape; seta twisted or not; theca length; annulus type; peristome type; length of calyptra; and spore diameter. Each is a potential source of reversal of branch order on a morphological cladogram, and should be examined with superoptimization.

In all cladograms the order of nesting may not be the same as the order of macroevolutionary transformation. Received phylogenetic wisdom is that an extant ancestral taxon will produce a multifurcation with its daughter taxa in a morphological cladogram. If there are, however, reversals or other modifications of the synapomorphies in the daughter taxa uniting progenitor and ancestors, then maximum parsimony will resolve such multifurcations with the progenitor as a terminal OTU. This is particularly the case when the progenitor is a morphologically generalist taxon and synapomorphies consisting of "not such and so" become "such and so" as rare traits or even autapomorphies (although present elsewhere far away in the cladogram, thus not unique). One modification of the

critical synapomorphic traits may lower a daughter taxon branch one node in the clade. Two modifications may lower a daughter branch by two nodes. This is particularly problematic when traits are binary between "special modification" and "not modified," which lets a generalist ancestor rise higher in a clade. If identical self-nesting ladders occur in the same taxa in both morphological and molecular trees, then the trees will seem congruent (because of lurking variables, as defined by LeBlanc 2004: 303) but both are of wrong branch order and false nesting (Plate 6.1).

A self-nesting ladder reflects the self-nesting information in the exemplars (specimens) sampled. There may well be extinct or unsampled extended paraphyly that makes the ladder false at the taxon level. Self-nesting ladders may be checked with other information, from such fields as biogeography. A possible heuristic that may distinguish evolutionary branch order for two closely related taxa is that for wide-ranging taxa, a taxon found on two continents should be older than one found on one continent, and the latter is probably the descendant. This assumes a distinctive range on each continent, no evidence for confounding long-distance or human-mediated dispersal, and the apparent ability to migrate by short steps to other suitable habitats but no evidence for having done so. For narrowly distributed taxa, it is difficult to distinguish relatively recent local taxa from ancient taxa of reduced range. Cross-tree heterophyly may help in this latter case, see discussion of the Andean moss genus *Erythrophyllopsis* elsewhere in this book, which almost surely represents a deep ancestor for many molecular clades.

In molecular cladograms including one progenitor and two or more daughter species, there is always a self-nesting molecular ladder for that gene tree. One can recognize a self-nesting ladder when, say, three terminal taxa given as ((A, B) C) in a cladogram can be easily interpreted as B and C being derived from A though other data. An example for morphology is the moss *Tortula,* a wide-ranging variable generalist genus, promoted higher in same clade of a morphological cladogram (Zander 1995) than obviously derived, specialized taxa *Stegonia, Crossidium,* and *Pterygoneurum.* The lesson is that both molecular and morphological cladograms are affected by phylogenetic self-nesting ladders and must be examined in light of all information to reveal correct cladistic changes in branch order and in evolutionary sequences of taxic macroevolution. Stevens (2008) suggested that "if hypotheses of phylogeny remain stable, we can have a stable classification based on that phylogeny, and then get on with our work...." This can be considered only if those stable hypotheses were not biased by self-nesting ladders, and were not otherwise inconsistent with analytic results of other data, which necessarily lowers Bayesian support.

Adapting the usual Newick formula for phylogenetic nesting to also show macroevolutionary sequences when known, the budding evolution formula for the basic terminal node on a cladogram with caulistic information is $A > B$, where the progenitor (boldface) gives rise (angle bracket acting as an arrow) to a daughter taxon or clade (lightface). These are otherwise sister groups. When two or more daughter taxa are known the formula would be $A > (B, C)$. If the order of generation is known, as may be revealed in molecular cladograms exhibiting heterophyly, the formula would be $A > (^1B, ^2C)$ or $A > (^1C, ^2B)$, where superscripts before the taxon name or symbol indicate the order. It should be clear, particularly from Chapter 6, Figure 1, that the budding evolution formula may be quite unlike the Newick formula for the same cladogram. The evolutionary formula for a pseudoextinction event is $? > (B, C)$.

The Newick formula of a molecular cladogram, e.g., the terminal group $((A, B) C$, as seen in Plate 6.2, may be simply modified such that if progenitor A is sister with daughter B and daughter C is lower in the cladogram, then the composite parenthetical formula $((A > B) > C)$, which gives the same information as $A > (B, C)$, that is, descendants not ordered, but puts it in the context of the particular molecular cladogram by not losing information on phylogenetic nesting (however misleading) in that particular cladogram. Plate 6.1 illustrates various views, evolutionary, morphological cladistics, and molecular cladistics, of a progenitor with three daughter descendents ordered B, C, and D but for which nesting may or may not represent actual order of generation of daughter taxa.

When there are more than one descendant taxa arising from a single progenitor taxon, a natural key, such as might be developed from a morphological cladogram, should not be restricted to dichotomous branching, which is an artificial imposition on evolutionary models. In Plate 6.1, mechanically generated cladograms involving taxa A, B, C, and D might be the source of a dichotomous key that implies great evolutionary nesting of ancestral Taxon A to reflect a self-nesting ladder:

1. Specialization X	B
1. Without specialization X	2
2. Specialization Y	C
2. Without specialization Y	3
3. Specialization Z	D
3. Without specialization Z	A

But with a trichotomous natural key, Taxon A can be correctly represented as theoretically primitive (first in a series), while B, C. and D are not ordered:

1. Without specialization, generalist	A
1. With specialization, highly adapted	2
2. Specialization X	B
2. Specialization Y	C
2. Specialization Z	D

Macroevolutionary series

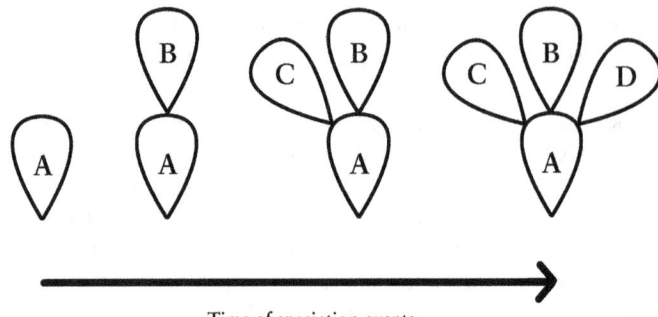

Time of speciation events

Morphological cladograms

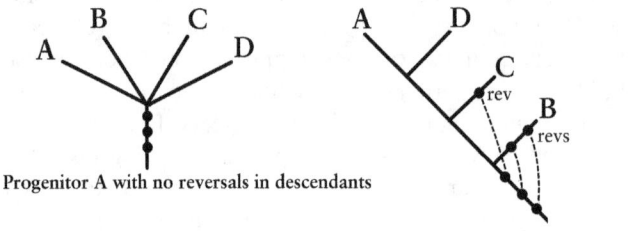

Progenitor A with no reversals in descendants

Progenitor A with reversals in descendants

Molecular cladograms

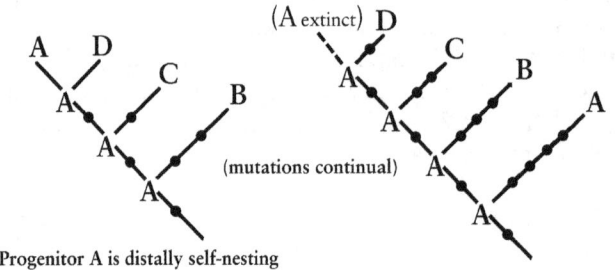

(mutations continual)

Progenitor A is distally self-nesting

Progenitor A is proximally self-nesting

Macroevolutionary formula for all diagrams

$$A > (^1B, {}^2C, {}^3D)$$

Plate 6.1. — Self-nesting ladders of serial budding evolution demonstrated with contrived clades terminal on a much larger cladogram (that is not given because A, B, C and D form a group not related to remainder of taxa). Note that in the two molecular cladograms, the ancestor A may be also entirely extinct or extinct both proximally and distally but survives as an extant molecular lineage between C and B. Branch order may be easily compromised by extinction or poor sampling. See also Plate 10.1.

Distinguishing macroevolutionary transformation order from nested order — Plate 6.1 is relevant here. If one does not know the order in which a progenitor generates daughter taxa but remains in mor-

phological stasis, and one has no heterophyly as guidance, then the order is not necessarily that of nesting. In Plate 6.1, the **macroevolutionary series** indicates true order of local geographic speciation of daughter taxa B, C, and D from morphologically static surviving progenitor A. This is what evolutionary analysis is supposed to recover.

In the **morphological cladograms,** the first is of **progenitor A with no reversals in descendants** shows a multifurcation, which is the ideal representation of a lack of nesting of the progenitor and three daughter taxa. The second morphological cladogram of **progenitor A with reversals in descendants** shows modifications occurring in some of the daughter taxa in synapomorphic traits (dots) that would have held the multifurcation together. In this case the daughter taxa are lowered in the clade, which occurs randomly. Here the nesting order happens by chance to be the transformation order, i.e., by chance alone a match is forced with the molecular cladogram below, and a *false congruence* in branch order results from the matching self-nesting ladders. A *false incongruence*, though, is more probable between the molecular and morphological cladograms.

The morphological cladogram may also (not shown) have descendants as sister groups distal to the progenitor if by chance the traits changes in such descendants by chance alone match. Reversals and matches are to be expected because morphological traits are generated by only a few traits that in combination may be adaptive to the environment, physiology and developmental constraints of that combination of traits that are conservative. Each species is here conceived as a core of conservative traits and a not large number of labile traits that may be integrated with the conservative traits when fixed heuristically as a small set in evolutionary adaptation.

In the **molecular cladograms** of Plate 6.1 the order of generation of the daughter taxa (as exemplars) is shown, but in the first cladogram the progenitor taxon is distal on the cladogram along with the latest daughter taxon, D, as sister groups, because the tracking sequence continuously mutates (dots) but A is morphologically static and are all molecular strains but the last are extinct. In the second cladogram, the progenitor has a surviving line basal to the daughter taxa and the line giving rise to the daughter taxa is presently extinct, thus the self-nesting ladder leaves the progenitor proximal on the cladogram. The second cladogram may be relevant to the problem of the position of the orangutan in the hominid cladistic tree. According to Grehan and Swartz (2009), based on morphological data, humans and orangutans are sister groups, with chimps and gorillas less closely related. This is contrary to well-documented molecular results that chimps are the sister group of humans, with gorillas less closely related, and orangutans

most distant. Given that patterns are not explanations, it is possible to explain the contradictory patterns by suggesting that orangutans survived as an isolated molecular lineages in Asia, while a presently extinct lineage of orangutans give rise to gorillas and chimps, then humans. This matches the Plate 6.1 molecular cladogram of progenitor A proximally self-nesting. This explanation was given by Zander on the Listserver Taxacom (October 14, 2012) and was recouched in cladistic terms by Curtis Clark (California Polytechnic State University, Pomona):

"... that orangutans are a grade, that was at one time widespread, and that gave rise to chimps, bonobos, mountain and lowland gorillas, and humans, without itself being transformed by anagenesis. Those orangs that gave rise to humans had previously given rise to chimps, and so there would be expected to be strong molecular similarities between humans, chimps, and the extinct orang subgroup that gave rise to them. But those orangs still share morphological features with the SE Asian orangs, and those features were less changed when humans speciated than when chimps, bonobos, or gorillas speciated. So the orang-human morphological similarities can appear to be symplesiomorphies relative to chimps, bonobos, and gorillas, while at the same time being evidence of close relationship" (C. Clark, October 14, 2012, see Taxacom Listserver archives, http://taxacom.markmail.org).

In Plate 6.1, the **macroevolutionary formula** representing the structure common to all analyses is $A > ({}^{1}B, {}^{2}C, {}^{3}D)$ where the superscripts show the order of daughter taxon (or exemplar) generation left to right. The molecular cladogram-specific equivalent is $(((A > D) > C) > B)$. The generalist formula would be simply $A > (B, C, D)$, when order of generation of the taxa (or exemplars) not known or inferable.

It should be clear that all self-nesting ladders deal with gaps in evidence for lineages that are implied by nesting patterns explainable by no other process but macroevolutionary intermediate ancestral taxa in morphological stasis. This is quite like the "ghost lineages" discussed in paleontological literature (e.g. Cantalapiedra et al. 2012), which are postulated for extant organisms with very ancient fossil evidence and a long gap between then and now. The coelacanth has a gap are of around 80 million years, and the taxa filling this gap termed a ghost lineage (Cavin & Forey 2007; Sidor & Hopson 1998).

Because the ancestral taxon may generate two or

more descendant lineages, a phylogenetic analysis of changes in adaptational traits associated with budding evolution, like that of Baum and Larson (1991), is confounded in that there are no adaptational or conservative trait changes in the line of the ancestral lineage, just between the ancestral lineage and the descendant taxa. Without analysis of non-phylogenetically informative data, the problem is not revealed.

Stasis — That morphological stasis does happen is incontrovertible. There are many "living fossils," notably *Triops cancriformis*, the 300-million-year survivor European tadpole shrimp. Both mutations in *cis*-regulatory sequences and in gene-associated tandem repeats (Frondon & Gardner 2004) have been associated with rapid evolution of phenotypic traits. The conservation of such gene-associated orthologous tandem repeats across mammalian orders despite high mutation rates have been shown to be indicative of strong stabilizing (non-neutral) selection. Thus, we have the theoretical potential of an abundance of pre-adapted, pre-speciation phenotypic traits that may persist in an ancestral, potentially supergenerative ancestor, and confound expectations of pseudoextinction (elimination of ancestral taxon during speciation).

In morphological cladograms, daughter taxa may be in a multifurcation or situated below or above the progenitor taxon in a clade. Progenitors and daughter taxa must be distinguished as a transformational group of very similar taxa, then a theory of that transformation must be advanced that is better, preferably much better than any alternative theory. If one can find a generalist, wide-ranging taxon with very similar taxa that are specialist in habitat and anatomy, found in recent local habitats, one has a fairly good theory. See also Plate 8.1.

Given arguments about self-nesting ladders in molecular cladograms, one might expect daughter taxa to be sister to or below the progenitor taxon. Given that the basis for the analytic process proposed here is that the true macroevolutionary historic structure of the taxa involved is the same whether analyzed classically, or cladistically with morphology or molecular data, differences in results need explanation. A single macroevolutionary transformation series (usually branching as an evolutionary tree) occasionally with breaks due to pseudoextinction is that explanation no matter how difficult it is to obtain from biased cladograms.

Binomial confidence interval for multiple nodes —

Critical to discussions of support for clades is the fact that the joint probability of any set of hypotheses is the product of their individual probabilities. The chance that three fair coins will come up heads when tossed individually is 0.5^3 or 0.125. The chance of three dice, a tetrahedron, a cube, and a dodecahedron, each coming up with a single pip when tossed individually is $0.25 \times 0.17 \times 0.08$, or 0.0034. The chance of three clades each supported at 0.95 probability being all correct at once is 0.95^3, or 0.86. The chance of six clades each supported at 0.99 being all correct at the same time is 0.99^6, or 0.94. Conclusions based on sets of hypotheses must always be judged under this stricture.

Whole cladograms are seldom provided with confidence intervals (here including posterior probabilities) that reflect their perceived chance of being correct. In the literature, however, many cladograms are used in their entirety to model broad conclusions, e.g., many genera grouped into multiple families. These cladograms are commonly viewed as "mostly correct." But what does "mostly correct" mean? The binomial confidence interval (BCI) is here advanced to provide a measure of confidence in whole cladograms that are used for broad conclusions. It provides the proportion of nodes (or internodes) with Bayesian support measures that one can expect to be correct all at once than total nodes being correct at once, *defining "correct" as joint probability of at least 0.99.*

Although the chance of all nodes in a cladogram being correct is simply the product of the Bayesian posterior probabilities of each of them, one must use a binomial calculator to deal with combinations when less than all are considered. Using a binomial calculator (e.g., Stat Trek 2012), simply enter the average posterior probability of the nodes, enter the number of nodes as number of trials, and increase the number of successes until the cumulative probability for all nodes *is equal to or greater than 0.99.*

The minimum number of nodes each at 0.95 probability in a 40-node cladogram to be *expected to be all correct at joint probability 0.99* is at least 35, or 9/10 of the nodes. In other words, a minimum of any combination of 35 0.95 nodes in a 40-node cladogram will have a confidence interval of 0.95 probability. In a 100-node cladogram all of 0.95-supported nodes, the fraction is approximately the same—a minimum of 9/10 can *be expected to be correct at a joint probability of 0.99*. So we might expect any cladogram with all 0.95 nodes to be at least 0.90 totally correct, giving a binomial confidence interval of 0.90 for the whole cladogram. Likewise, the binomial confidence interval for a cladogram of all 0.99-

supported nodes is about 19/20 or 0.95 for both 40-node and 100-node cladograms.

Thus, for cladograms with nodes variously supported at any combination of 0.95 to 0.99 nodes, the number of nodes that may be expected to be correct at 0.99 is between 18 and 19 out of 20. This is the extent of "mostly correct," which seems robust. Translated into probabilities, however, the full cladograms are between 0.90 and 0.95 correct. Therefore, for most cladograms that are published and used for broad conclusions, the confidence in those cladograms, each considered as a whole, seldom reaches 0.95, a standard for confidence in statistics.

The exact binomial confidence interval for a cladogram with nodes of many different support values is calculable from an average of the Bayesian posterior probabilities for all nodes in the cladogram. The average support for the 28-node molecular cladogram (each node of 0.95 or better support) of the Pottiaceae presented in Zander (2007) based on that of Werner et al. (2004) is 0.988 probability, and binomial analysis yields a BCI of 26/28, or 0.93 for the full cladogram. In a molecular cladogram of 64 nodes (La Farge et al. (2002) used to separate groups of genera of mosses belonging to different families, the average Bayesian posterior probability was 0.76 (multifurcations were ignored, unsupported nodes were assigned 0.33 probability); the BCI for the whole cladogram (that minimum percentage of nodes we can expect to be correct) is 40/63, or 0.64. This may seem a low value, yet those clusters that match the groupings of classical taxonomy gain corroboration for those groupings. When classical taxonomy conflicts, on the other hand, then the cladogram cannot support broad conclusions that involve those conflicts. Of course, the BCI can be also be used for any multitaxon clade of interest in a cladogram.

This discussion does not assume that no major rearrangements of any high support values will occur as second-best alternative to the optimal cladogram. It deals only with the implications of sequential Bayesian probabilities as support values.

It is possible to collapse any cladogram with support measures on the nodes that are lower than 0.95 posterior probability (or equivalent) into smaller cladograms with nodes of only 0.95 or higher probability (Zander 2007). Two or more contiguous branch arrangements can be combined into one node (or internode) using a simple probabilistic calculation: the chance that at least one of two or more events will happen is one minus the product of the chance they will not happen. Chained internodes, each internode with posterior probabilities lower than 0.95, are combined into one implied reliable internode using a formula that determines the chance that at least one internode among two or more is correct, by calculating an implied reliable credible interval (IRCI). The formula for the IRCI is simply one minus the product of the chances of each of all concatenated arrangements being wrong (where the chance of being wrong is one minus their Bayesian posterior probability). See also discussion of the formula for implied reliable internodes by Zander (2007). A cladogram with all nodes reduced to those at 0.95 or above is more easily comprehended.

Whole cladograms, even those with nodes supported at 0.95 and 0.99 posterior probabilities, are to some extent only optimal, or "best hypotheses." This brings up the old argument that optimality (simplicity, maximum parsimony, maximum posterior probability) alone is sufficient for most scientific purposes. Such philosophical justifications for optimal or shortest trees such as "simplicity," "converging on the truth," "most parsimonious," or "least falsified" as a single criterion of satisfactory results in phylogenetic analysis have been replaced with various methods of gauging statistical support (with reference to the Central Limit Theorem, which basically comes from physics) for trees or branches.

In addition, as noted above, resolution alone is insufficient to demonstrate reliability because random data usually produce resolved trees, and length of branches alone is insufficient because (1) large random data sets generate long branches, and agreement alone between two cladograms is insufficient because if one or both arrangements are at less than 0.50 probability, then, by Bayes' Formula, the BPP must be reduced, not increased, and (2) disagreement or agreement of cladograms is rendered problematic by self-nesting ladders and extinct or unsampled extended paraphyly (see Chapter 7).

Phylogenetists' use of Occam's Razor (Posada & Buckley 2004) continues to be abused, however, in that the difficulty of evaluating the relative importance of suboptimal solutions has been slid across to optimal sequence alignment and model selection, and simply assuming that all the other many assumptions associated with methods of phylogenetic reconstruction are correct or correct enough. An expectation of success in post hoc testing is the psychological justification of the otherwise illogical idea of preferring simplicity. A corollary to Occam's Razor pertaining especially to historical reconstruction is that explanations must remain multiple when no one of them is probabilistically adequate (Zander 1998).

Problems in molecular analysis — Molecular analysis successfully nests sets of specimens by their present (synchronic) phylogenetic relationships but the reverse, the revelation of past (diachronic) evolutionary relationships, is not modeled in cladistics (Zander 2008b). A set of nested parentheses constitute the essential cladogram, and "tree" lines connecting the nested exemplar specimens are actually little more than visual aides (Plate 6.2). That molecular systematics matches to a significant extent past morphological study and evolutionary classification is the surprise, not the reverse, given the lack of reasoned diachronic inference in phylogenetics, the prevalence of paraphyly, and of self-nesting ladders (Chapter 7).

By reasoned, I mean that although molecular analysis is strongly based on good molecular data and powerful statistical techniques, the application of such techniques in inferring ancestor-descendant relationships is rudimentary, being restricted to assumption of universal pseudoextinction in clustering (cladification) of exemplar specimens and their associated taxonomic names, something of a Barmecides feast. One can interpret (Zander 2008b), however, molecular heterophyly (paraphyly and phylogenetic polyphyly) as an indicator of progenitor-descendant diachronic (caulistic) evolution and infer an evolutionary tree, i.e., a "Besseyan cactus" (Bessey 1915) or commagram.

(1) Molecular cladogram with Newick formula

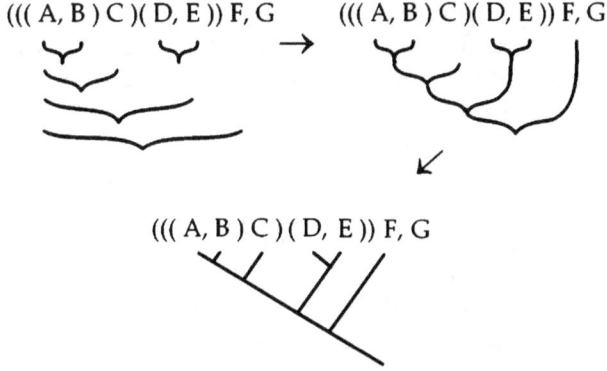

(((A, B) C)(D, E)) F, G   (((A, B) C)(D, E)) F, G   →

(((A, B) C)(D, E)) F, G

(2) Evolutionary tree and macroevolutionary formula

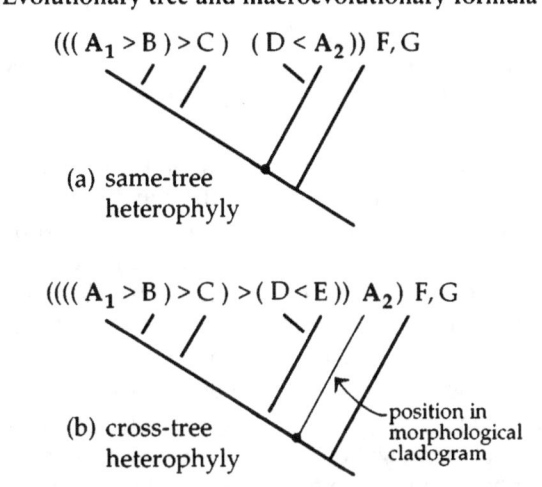

(((A₁ > B) > C) (D < A₂)) F, G

(a) same-tree heterophyly

((((A₁ > B) > C) > (D < E)) A₂) F, G

(b) cross-tree heterophyly

position in morphological cladogram

Plate 6.2. — Contrived comparison of phylogenetic and evolutionary trees and formulae. (1) Molecular cladogram as represented by Newick Formula, showing that the dendrogram is largely a visual aid to seeing (parenthetical) nesting of the taxa. In the phylogenetic formula, "E" is a new phylogenetic species molecularly distinguishable but identical in expressed traits to "A". (2) Newick Formula here modified to reflect sequential caulistic evolutionary relationships, showing evolutionary trees and macroevolutionary formulae without modification of tree into Bessyan cactus: (2a) Same-tree heterophyly based on taxon A occurring in non-contiguous parts of a molecular tree, i.e., non-monophyletic, represented by A and E (recognized as A2) in molecular cladogram. (2b) Cross-tree heterophyly of same taxon A occurring basally in a morphological cladogram but added to the molecular cladogram (as A2); this is a different cladogram from 2a and here E is not A, but a different species. Basally broken lines indicate progenitor-descendant pairs determined through a separate superoptimization; the black dot indicates an inferred molecular isolation event without change in expressed traits in one branch; bold-faced letters indicate surviving ancestral lineages or molecular strains; thin line shows position of a taxon on a morphological cladogram of same taxa; simple split (e.g., involving F) indicates either possible pseudoextinction event or equivocal support for ancestor-descendant distinctions. In all evolutionary trees, A is the major surviving ancestral taxon for most other taxa. G is outgroup or the remainder of a rooted cladogram. When heterophyly occurs, then branch order of exemplars is informative of inferable deep ancestors, but when heterophyly is not apparent (as when either A1 or A2 is extinct or unsampled, then branch order of taxa is uncertain.

Exemplars, particularly the usual single exemplar per taxon, in a molecular analysis (the specimens used as OTUs) represent species and higher taxonomic groups only with uncertainty because, among other reasons (Syring et al, 2007; Ramdhani et al. 2009: 2011), (1) DNA mutates gradually while morphology of the same taxon may stay in stasis through stabilizing selection, and (2) the larger group supposedly represented may have been phylogenetically paraphyletic in the past (Cavalier-Smith 2010; Zander 2010a), but only one lineage survived, or, because insufficient sampling was done to determine sequence variability in the taxon represented. Exemplar specimens have more meaning when there are two or more samples of the same taxon on a molecular cladogram in that paraphyly implies a mapped taxon in a caulogram; but even then, extinct or otherwise unsampled heterophyletic molecular strains are to be expected, which contributes uncertainties.

Examination of the Werner et al. (2004) moss data set, graciously provided by Olaf Werner, shows that *Anoectangium aestivum* and *Gymnostomum viridulum* differ by seven sites (2 first position, 2 second, 3 third), but the two specimens of *Splachnobryum obtusum* (the only species of which two specimens were analyzed) differ by 21 sites (4 first position, 7 second, 10 third). Although the rates of change in sites, especially in the codon, surely differ, it is doubtful, given the taxonomic specialization of the analytic team, that the former two samples are not in the same genus, or the latter involve misidentifications. Given that the word "exemplar" implies example representation of the molecular sequence characteristic of other specimens in the species and genus, one might expect in view of the (unusually?) large internal variation in *Splachnobryum obtusum* that the specimens studied are samples from a more heterogeneous molecular assemblage at the species level than expected. What this means is that although many exemplars of the same taxon, even a species, all nest together on a molecular cladogram, they probably do not have the exact same DNA sequence, and are isolated different strains. If there are commonly multiple strains, then this may explain many cases of molecular paraphyly and polyphyly.

Species and higher taxa are not uncommonly heterophyletic, that is, occurring in two or more branches isolated by at least one branch of a different taxon of the same rank or higher. Reiseberg and Brouillet (1994) estimated that at least 50% of all plant species are products of local geographic speciation and are therefore paraphyletic, while Funk and

Omland (2003) found actual species-level paraphyly or polyphyly in 23% of more than 2000 species sampled. Aldous et al. (2011) estimated, using simulations, that more than half of extant species have extant ancestral taxa. "Sister-group" is doubtless often a misnomer because many nodes on a molecular cladogram clearly may be inferred as mother-daughter groups. Thus, "putative isolation events of molecular strains" might be a better term for nodes on a molecular cladogram.

Given the apparent commonness of molecular paraphyly, then two nodes on a cladogram are commonly of the same taxon. This being so, the fundamental analytic premise of cladistics—that two of any three taxa must be closer in relationship—is void. Given the arguments in superoptimization (Chapter 8), this is also true, to a large extent but for different reasons, in morphological cladistics.

Theoretically, isolated lineages of a taxon (in my opinion, at any taxonomic rank, and see Barraclough 2010) may be in morphological stasis through stabilizing selection (Koonin 2009; Mallet 1995), habitat tracking (Eldredge 1989: 206), or simple "phylogenetic inertia" (Griffiths 1996; Shanahan 2011) in the absence of selection, though in the latter case drift might be expected to operate in populations (Brandon 1990; Griffiths 1996). This goes quite against the phylogenetic expectation that any isolated subgroup will necessarily become a new species, probably sooner than later through "reciprocal monophyly" (not a process but a term for a desideratum). In fact, there is now software (Ence & Carstens 2011: 473) that identifies, on the basis of sampled molecular data within a species, which intraspecies molecular lineages "can be validated as distinct" in that they have the "potential to form new species before these lineages acquire secondary characteristics such as reproductive isolation or morphological differentiation that are commonly used to define species." This is presumptuous.

In this book I often cite supportive statements in works of other fields than systematics—philosophy, evolution, physics, cosmology, and mathematics. Alert readers may wonder if I might be "cherry picking" particular statements out of context that happen to support my contentions. It is, in fact, impossible to not run across, quite regularly, statements in other fields that support a non-axiomatic, non-theoremic, multi-methodological pluralistic science that aims to explain all relevant facts, not just as best possible with Procrustean data sets, but with one explanation much better than other explanations; an explanation that involves natural processes (e.g., macroevolution) rather than hierarchical classification-like descriptions (e.g., phylogenetic trees).

Given a certain admittedly tendentious sensitivity, I commonly find apropos quotations concerning quite technical aspects of a critical nature. For instance, just the other day I opened a used copy of the Portable Darwin (Porter & Graham 1993) and there, highlighted in fluorescent ink by a previous reader sprung forth the following comment by Darwin on pseudo-extinction:

"It must not, however, be supposed that groups of organic beings are always supplanted and disappear as soon as they have given birth to other and more perfect groups. The latter, though victorious over their predecessors, may not become better adapted for all places in the economy of nature. Some old forms appear to have survived from inhabiting protected sites, where they have not been exposed to very severe competition; and these often aid us in construction of our genealogies, by giving us a fair idea of former and lost populations. But we must not fall into the error of looking at the existing members of any lowly-organized group as perfect representatives of their ancient predecessors" (C. Darwin, Descent of Man, and Section in Relation to Sex, in Porter & Graham 1993: 332).

Problems — Taxic stasis is associated with punctuated equilibrium, which may be valid for a large percentage of taxa involved in an analysis (Gould 2002: 606; Gould & Eldredge 1993; Stanley 1981). Gurushidze et al. (2010), in addition, have reviewed evidence for the persistence of ancestral molecular sequences through speciation events as extended incomplete lineage sorting. This last may, in fact, help the identification of ancestral taxa. The amount of present-day paraphyly is a measure of present morphological stasis, and in a cladogram is also an indicator of the degree of past levels of paraphyly for a particular group. It signals an inherent uncertainty in recovering a molecular tree.

Molecular data, when analyzed with statistical methods, is assumed in phylogenetics to be generated by Markovian processes, and only present-day information is therefore needed to reconstruct past gene histories. It has been shown by simulations, however, that molecular data may to a significant extent be non-Markovian (Cartwright et al. 2011). Markov processes are also ideal mathematical solutions, and in practice transmit information limited by noise and

made inefficient by redundancy (Littlejohn 1978: 153). In addition, there is evidence for convergent molecular adaptive evolution involving rapid evolution of regulatory mechanisms in response to new environmental and genetic situations (Amorós-Moya et al. 2010).

Breen et al. (2012) found that there is profound selection on protein coding genes such that genes determining some proteins are dependent on other genes (epistasis), and "amino acid substitutions that were beneficial or neutral in one species should often be deleterious in another, " i.e., "...epistasis is pervasive throughout protein evolution: about 90 per cent of all amino-acid substitutions have a neutral or beneficial impact only in the genetic backgrounds in which they occur, and must therefore be deleterious in a different background of other species." Robust epistasis explains why "...the vast majority of amino-acid substitutions that occur in one species cannot occur in another regardless of whether or not positive selection plays the dominant role in the course of fixation of amino-acid substitutions in specific genetic contexts." This limitation on molecular evolution is important in that analysis of DNA in phylogenetics commonly includes both non-coding and coding sequences.

There are many problems and assumptions associated with molecular analysis (Amorós-Moya et al. 2010; Avise 1994; Doyle 1992; Hudson 1992; Lyons-Weiler & Milinkovitch 1997; Maddison 1996; Marshall 1997; Mooi & Gill 2010; Pamilo & Nei 1988; Templeton 1986; Zander 2007a), including the effect of unsampled extended paraphyly on branch order introduced here. Recently, Stegemann et al. (2012) found horizontal transfer of entire chloroplast genomes are possible between naturally grafted plants of tobacco species, and they are hereditable. It was suggested that this is a possible reason for inconsistency between phylogenetic analyses of chloroplast and nuclear sequences.

Molecular lineages at best represent (as nested parentheses) only the genetic continuity and isolation events associated with the "exemplar" specimens studied (Zander 2007c, 2010a). Extension by deduction or analogy of exemplar clustering to other specimens of the same taxa represented by the exemplars or to higher taxa ignores the needed adequate sampling to determine homogeneity of the molecular traits through the law of large numbers, and ignores any process affecting homogeneity involving the central limit theorem (as discussed by Aron et al. (2008). Vanderpoorten and Shaw (2010) have pointed out that lack of molecular support for a morphologically distinguished taxon is only negative evidence. Coherence in the systematic sense requires recourse to the intensive sampling of classical systematics, not phylogenetic axioms (e.g., Farris et al. 1970).

Examples of problematic axiomatic assumptions in phylogenetics in addition to those discussed by Brower (2000) are: exemplars represent their higher categories, ancestral taxa do not survive speciation events, analysis of changes in non-coding DNA are sufficient to classify evolutionary changes in expressed traits, and ancestor-descendant relationships cannot be added to sister-group relationships without confusing phylogenetic analyses and thus classification. Any assumption becomes axiomatic in the structuralist sense and problematic when it is acted on (for example) by changing classifications, as opposed to simply being an element in abduction (hypothesis generation). Although classical systematics is used by phylogeneticists to distinguish taxa when they are sister groups or sets of sister groups (Knox 1998: 37; Mallet 1995: 298), classical decisions are considered by phylogeneticists to be insufficient, apparently, to distinguish at the same rank the same kind of groups if they are paraphyletic or apophyletic; this is a consistency problem for the principle of holophyly, this in addition to the fact that holophyly is not a refutable scientific theory (Knox 1998; and see Bock 2004).

Paraphyly and polyphyly in phylogenetics are much the same phenomena. There is little to distinguish phylogenetic paraphyly:

$$(((\mathbf{A}, B)\, \mathbf{A})\, C)\, D, E$$

from phylogenetic polyphyly, which is actually just extended paraphyly:

$$(((\mathbf{A}, B)\, C)\, \mathbf{A})\, D, E$$

because heterophyly of taxon A implies a deep ancestral taxon A in both (see Plate 6.2). Evolutionary polyphyly, quite a different thing, requires *demonstration* of two or more caulistic ancestral taxa as different ancestors for two or more exemplars of A. Illustration of homophyly, phylogenetic paraphyly and polyphyly, and the implied ancestral taxa involved in each is given by Zander (2010a).

Classical taxonomy derives a *taxon* from a multidimensional tensor-like data set of specimens, morphological traits, geography, ecology, habitat, chromosome number, and other taxonomic dimensions, but molecular analysis is limited to deriving a *tree* from a 2-dimensional vector data space of exemplar specimens and sequence data. The exemplar specimen can be assumed to be a sample of the taxon but the assumption that the molecular data, and resultant optimized tree, applies to all or most specimens of that taxon is nonsense in light of poor sampling of the taxon.

In using molecular barcoding (using one sequence alone to distinguish many taxa) to detect new or cryptic species, one may have a few molecular samples corroborating (i.e., "not incompatible with") morphological differences (instead of supporting them) but the morphological differences have to stand alone. That is, they have to be taxonomically (or evolutionarily) distinct. They have to not just be correlative of molecular differences, but have to be much better than other combinations of morphological traits (i.e., have to avoid the problem of finding by chance alone a distinct combination of traits). In Bayesian terms there is no increase in credibility with results that do not stand alone. Refutation, beyond contradiction, requires a clear causal explanation (even if to some extent probabilistic) that may be investigated, with a clear definition of what is evolution for a particular group. In a like manner, attempting to falsify molecular results (Schwartz & Maresca 2006) does not directly support morphological conclusions.

Pseudoconvergence — Molecular phylogenetics then has two largely unrecognized major problems, self-nesting ladders and unsampled extended paraphyly. Together these can lead to apparent convergence in a molecular cladogram, here termed pseudoconvergence because it is forced by the faulty method, and is not real.

Cladistics deals only with similarity, not dissimilarly (aside from distance on a cladogram). Mistakes in estimating branch order necessarily result, then, in false convergence, this being lineages that are out of order or wrongly tagged as sister groups. There are two kinds of pseudoconvergence, explicit and implicit.

A simple example of a self-nesting ladder has the ancestral taxon terminal on a clade and sister with the latest descendant lineage, e.g. $((((\mathbf{A}, B)\, C)\, D)\, E)$, F where B, C, D and E are descendant taxa, and F is the rooted remainder of the larger cladogram.

Explicit pseudoconvergence occurs when macroevolutionary stem-thinking (as opposed to cladistic tree-thinking), using all available information, finds A the ancestral taxon inferable on the cladogram $((((B, C)\, D)\, \mathbf{A})\, E)$, F. The pair B and C are only sister groups in the phylogenetic (nesting) sense. A separate, isolated **A** lineage as sister to B is assumed extinct or otherwise unsampled. That is, the correct clade is ((((**A1**, B) C) D) **A2**) E), F, as inferred from all data. In fact, because deep ancestor **A** is present all along the clade, there is a potential for an extant molecular strain of taxon **A** to appear between any two descendant taxa in a molecular cladogram.

A second explicit pseudoconvergence is created with the ancestral taxon used as an outgroup or is otherwise forced lower in the rooted cladogram than its descendants. The descendants then are wrongly ordered or paired.

Implicit pseudoconvergence occurs when the ancestral taxon is entirely extinct or unsampled. Such unsampled ancestral taxa will contribute to problems with branch order and sister-group pairing in the same manner as in the case with explicit pseudoconvergence. Consider the clade ((((B, C) D) F) G), E. Suppose the ordering and pairing do not match classical expectations of evolutionary descent of taxa or a morphological cladogram. One can then postulate an extinct or unsampled ancestor **A** as explanatory. This may seem similar to the unnamed shared ancestors at nodes in cladograms, but this is not so. Inferred ancestral taxa can be named, perhaps to the genus level, and assigned a hypothetical ancestral morphology based on all data.

It may be possible to infer much of the morphology of an extinct or unsampled ancestral core taxon from footprints in the evolution of extant, more specialized descendants. This is possible for groups in which the core generalist species is absent, the descendant species are not so disparate in morphology as to suggest more than one core species, and the group is amenable to the speciational burst concept (dissilience) at the genus level (as judged by related genera having clearly distinguishable core and radiative species). It is doubtful that one can take this too far, such as naming cryptic species, although the idea of virtual fossil hunting is, of course, attractive.

Examples of molecular systematics —Numerous examples appear in the literature of well-supported molecular analyses that match morphological trees and classically generated classifications in large part. But there are many studies with poor congruence. An extensive study (Wilson et al. 2011) of the accuracy of using DNA barcodes alone to identify Sphingidae moths found 83% of unknowns identifiable to genus but with many false positives, but with a more strict criterion 75% were assignable to genus with less than 1% false positive. This is not impressive. Heterophyly, however, is a fruitful source of evolutionary information. Zander (2007a, 2008a,b, 2010a) reviewed a sampling of published phylogenetic papers, and used heterophyly of classical taxonomic groups represented by exemplars in these molecular trees to infer (map on a molecular tree) ancestral taxa as new caulistic elements. The analysis of the Werner et al. (2004) of Zander (2008b) is summarized in Plate 7.3 showing how both same tree and cross-tree heterophyly imply theoretic progenitor-descendant relationships.

The future of molecular systematics — Given that budding evolution must be modeled when ancestor-descendant relationships are clearly present, future molecular analysis must take into account superoptimization based on expressed traits, e.g., morphology, through natural keys based on well-interpreted morphological cladograms. That is, the most parsimonious or maximum likelihood solution to a caulistic model. Such a model is not simply a low probability alternative to a solution for synapomorphy transformations or Markov chains of traits as with present-day cladistic or phylogenetic analysis. The shortest or most likely tree assuming universal pseudoextinction is not the end of the analysis, but the shortest or most likely tree given identified instances of budding evolution would be.

CHAPTER 7
Element 4 - **Contributions from Cross-Tree Heterophyly**

Précis — Morphological cladograms that are well founded by recourse to Dollo evaluation at the taxon level and relevant non-phylogenetic information may aid in uncovering taxa clearly more basal on the morphological tree than on the molecular tree. The implication is that the exemplar specimen distal on the molecular tree is a surviving extant representative of a deep ancestral taxon. This may be explained by the mechanism of a self-nesting ladder. All lineages dependent on that deep ancestor between the distal and proximal positions are theoretic descendant lineages. Heterophyly between morphological and molecular cladograms may be termed "cross-tree heterophyly."

Morphological and molecular data reflect two rather different aspects of the same phenomenon, evolutionary transformation. We can analyze these separately (Plate 7.1). Morphological data are in part well supported because truly conservative traits (those at higher taxonomic levels, refractory to changes in selective regimes) are tested every time there is an isolation event of gene history. With each gene history isolation event, one or more mutations in (mostly) non-coding sites are introduced in the population but the conservative morphological traits remain fixed either through developmental constraint or well-buffered neutrality. Conservative morphological traits are appropriately well-weighted given this gauntlet, and should not be compared with molecular traits on a one-to-one basis.

Cross-tree heterophyly may be a difficult concept for those used to copious literature in which taxon trees are interpreted as clusters of taxonomic groups, these presented as evolutionarily coherent by hierarchic shared ancestry. Each clade is monophyletic in this interpretation. Cladograms, however, particularly molecular cladograms, cannot be so easily translated to evolutionary and therefore systematic groups. Imagine a molecular cladogram in which a basal taxon, because it has generated many major lineages, turns up to be terminal. A dark black fat line drawn on the cladogram from its base to the cladistically terminal but evolutionarily basal taxon then represents the basal taxon as ancestral to all lineages that come off the line. This molecular cladogram cannot be divided into clusters of monophyletic taxa in the usual way. This chapter deals with the implications.

Although it may be argued that one should not compare an exemplar tree with a taxon tree, if patristic distance is rather large, then cross-tree heterophyly is evolutionarily informative.

One might expect that the conservative morphological traits lag behind changes in DNA sequences used to track phylogeny. Every time a taxon of primitive (meaning first, being basal in a caulistic tree) but static traits pups (or buds) a new taxon but survives the speciation event, its molecular tracking DNA bases continue to change but not its expressed traits. Thus, a primitive taxon can, through a series of speciation events, "climb" a molecular tree. It climbs out of its true cluster of evolutionarily similar taxa into a terminal molecular cluster of advanced taxa. The mechanism of the self-nesting ladder is simply the record of molecular changes left in daughter lineages. There may be long and short ladders. An example of a long ladder is the moss genus *Erythrophyllopsis* (Pottiaceae), discussed elsewhere in this book (see index and Plate 7.3).

If we assume both morphological and molecular data reflect the same evolutionary phenomenon, suppose molecular analysis results in the rooted cladogram ((A, B) C)... with a terminal (A, B) supported at 0.95, but morphological results assert ((A, C) B)... is correct. Can we compare the two statistically?

Suppose there are two binary morphological traits that C has in common with A that would have required convergence of A and C to get the morphological clade different ((A, C) B) ... from the molecular clade ((A, B) C) .... What is the chance of that convergence? Let us assume that the morphological traits were not conservative but rather labile traits that change at 50% probability. Both taxon B and taxon C would have to each change in two traits to get this difference in cladograms, so we have four trait changes involved. The chance of those four traits transforming to their alternate form is then 0.50^4, or 0.06. Because this is a low probability, the complement, 1 minus 0.06, or 94%, is the chance that the morphological cladogram is correct. It would be even higher if the traits were demonstrably and informatively conservative.

Comparing the two via the Bayesian formula where a small bit of morphological support (1 minus 0.96) is made to support, as a prior, the molecular cladogram, versus when a small bit of molecular support (1 minus 0.95) is made to support, as a prior, morphology:

> Molecular probability of 0.95 with prior support of 0.06 from morphological support of the molecular cladogram yields a posterior credibility of 0.55 for ((A, B) C).

> Morphological probability of 0.94 with prior of 0.05 from molecular support yields a posterior credibility of 0.45 for ((A, C) B. (Note these posteriors add to 100%.)

Thus, we have an impasse, with almost equivocal support for conflicting molecular and morphological cladograms. The impasse certainly cannot be ignored if Bayesian analysis is stated as the analytic method used, and should not be ignored in any case. The usual phylogenetic solution is to declare the morphological results unscientifically intuitive and simply map the morphological traits to the molecular cladogram. Another, more complex fix is to add the morphological data set to the molecular data set in what is called a "total evidence" analysis, in which case differences between the morphological and molecular results are commonly decided in favor of the data set with much more unitary, unweighted evidence per specimen exemplar. There is, also, little information on molecular variation between exemplars of the same taxon that might reduce certainty.

Morphological data are reflective of a different view of evolution than are molecular data. Morphological data are present-day results of past macroevolutionary events and involve diachronic transformational relationships traced by proven long-stable conservative traits and apply to *taxa*. Molecular data are present-day results of past microevolutionary events reflecting genetic continuity and isolation events of molecular strains but not necessarily macroevolutionary transformations, and apply to *single specimens*. Although some morphological traits can be strongly convergent in being sensitive to selective regimes, the conservative traits can lag behind constantly mutating molecular traits, and the difference between molecular and morphological cladograms may be caulistically informative.

Morphological results that differ from molecular results are therefore not necessarily contrary or falsificatory. If molecular analysis results in ((A, B) C) ... and morphological analysis results in ((A, C) B) ...,

then a scientific theory can be presented that reconciles them, namely that an ancestral taxon the same as exemplar **B** is progenitor of both A and C. Traits on the morphological tree need not be merely mapped (relegated) to the molecular tree or buried in an lump-all-data-together total evidence analysis. This pluralist solution reminds of Haack's (1993: 81) analogy of the crossword puzzle that is solved by mutual support from two rather different belief or experiential systems. One should note that to the extent that a taxon is more basal on morphological cladograms, that taxon should not be used for classification by clustering on the molecular cladogram.

Heterophyletic isomorphism — There is a growing discrepancy between the results of morphological and molecular analyses of the same taxa (Assis & Rieppel 2010) as more exemplars are sampled. Cross-tree heterophyly, i.e., the superimposition of morphological and molecular cladograms, may provide acceptable scientific theories about evolutionary relationships, e.g. Plate 7.2. Taxa low in the morphological tree but high in the molecular tree are theoretically ancestral taxa of all lineages in between. A morphologically strongly plesiomorphic taxon occurring basally on a morphological cladogram but distally in a molecular cladogram is best representative of conciliation through mapping taxa across two cladograms. It is explainable as either simply a surviving taxon, or the lone surviving branch of an extinct (or not yet sampled) taxon of extended paraphyly.

In structuralist terms, cross-tree heterophyly reveals an isomorphism (Zander 2010b) inherent in both data sets. Differential aspects of morphological and molecular cladograms are simply different modifications of the cladistic dimension by the unrepresented caulistic dimension. The implied shared hidden structure is retrieved in both structuralist and empiricist terms by postulating a mapped taxon (ancestral bridging group at a particular rank) between a position on a morphological cladogram and the taxon's possibly quite different position on a molecular cladogram. Empirically, a deep ancestral taxon is far more likely as a scientific explanation than theories of massive homoplasy giving rise to massive convergence, expressed as a newly evolved separate molecularly distant taxon identical in all morphologically diagnosable respects (Cain 1944: 290; Crisp & Cook 2005: 122). Note that morphological cladistic analysis has placed a great premium on homology analysis, yet the morphological evidence (of sister groups) is thrown out in molecular systematics or buried by joining data sets. A deep ancestral taxon, or evolutionary isomorphism, is a good explanation for differing morphological and molecular cladogram positions of what may be ancient, surviving taxa.

The chance of morphological convergence must be evaluated at the level of taxa, not traits alone. What is the chance through chance alone that conservative traits will combine in different parts of an evolutionary tree to reconstitute a taxon exactly (i.e., as the sum of all its important traits)? Consider a contrived morphological data set of 30 binary traits, and 20 exemplars each representing a different species. We eliminate, say, half (15) of the traits as strongly liable to selection in certain environments implying a higher probability of convergence, or which are developmentally linked in some way. Fifteen conservative binary traits can be combined 2^{15} or 32768 ways (i.e., two states per trait). For a genus of 20 species the probability of two of them totally convergent through random, independent combination is very small. This can be demonstrated using a variation of the birthday problem in combinatorics. At times, as now, when $p = 1 - q$, the direct calculation of p may be complex, and we can calculate q more easily. There are 32767 / 32768 ways two taxa can *not* have the same conservative traits. For 20 taxa there are (n (n − 1)) / 2 or 190 pairs possible. The probability of any two pairs *having* the same combination of conservative binary trait states is then 1 minus (32767 / 32768)[190], or 0.006. This does not, in addition, consider traits of more than two states or new traits occurring. Although the set of conservative traits at the genus level and above is smaller, there seems to be no paucity of traits distinguishing genera in families with large numbers of genera, thus small chance of total convergence at any taxon level.

Millstein (2000) has reviewed theoretical pluralism in paleobiology, comparing stochastic (theory-free) and deterministic (theory-based) methods of addressing macroevolution. Rieppel and Grande (1994: 239) addressed the problem of incongruency between the results of morphological and molecular analyses, suggesting that methodological pluralism might provide an answer, keeping analyses separate but treating the separate results as contributing to total evidence, in their case to help choose one cladogram over another. Reiners and Lockwood (2010) have argued the value of such pluralism (controlled by "constrained perspectivism") in ecology, while Santos and Faria (2011) suggest pluralism as an end to what they see as a cold war between researchers that are strictly molecular and others that are strictly morphological. Assis and Rieppel (2010) suggested that an important issue is to "make empiric evidence scientifically relevant by trying to find out *why* [their emphasis] such contrastive signals are obtained in the first place." Past work (Zander 2006, 2007a, 2008b, 2010, and other authors on the subject of paraphyly) and the present study suggest that deeply buried ancestry is a proper and testable explanation.

Examples of morphological and molecular heterophyly — Plate 7.1 compares a (1) synchronic (same time, cladistic) cladogram, and a diachronic (through time, macroevolutionary) caulogram with (2a) progenitor-descendant relationships implied by same-tree molecular heterophyly, and by (2b) cross-tree, molecular/morphological heterophyly. The greater than and less than signs are arrows inserted into the standard Newick Formula indicating sequential macroevolution. Bold-faced letters are progenitors. G in the cladograms, which are all rooted, represents continuation basally of a larger tree, and there is no information on macroevolution for the first split, which is either due to pseudoextinction or is simply unknown.

(1) Synchronic phylogenetic pattern

(((A, B) C) (D, E)) F, G

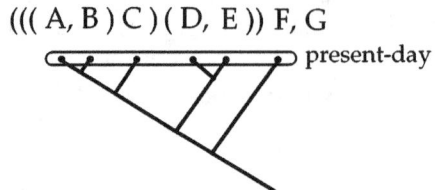

(2) Diachronic evolutionary pattern

(((A$_1$ > B) > C) (D < A$_2$)) F, G

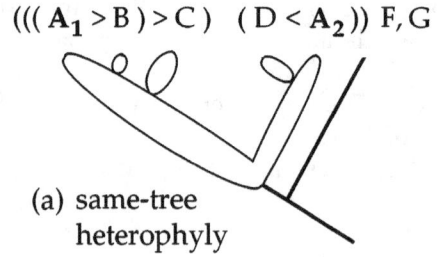

(a) same-tree
heterophyly

((((A$_1$ > B) > C) > (D < E)) A$_2$) F, G

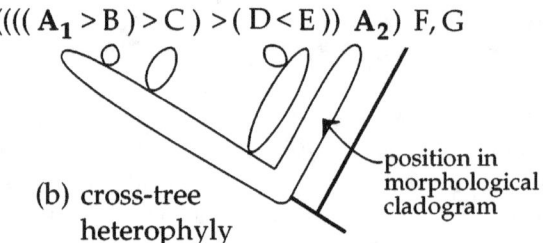

(b) cross-tree
heterophyly

Plate 7.1. — Comparison of (1) synchronic cladogram, and a diachronic caulogram with (2a) progenitor-descendant relationships inferred from same-tree molecular heterophyly, and (2b) from cross-tree, molecular-morphological heterophyly. The greater than and less than signs indicate sequential macroevolution. Bold-faced letter are progenitors. Parentheses are used to indicate cladogram branch order as in the Newick formula.

Schneider et al. (2009) presented a chart comparing molecular and morphological trees for 34 genera of vascular plants emphasizing the ferns. Cross-tree heterophyly can be observed in a modification (Plate 7.2) of their two-part diagram and inferences made about ancestry, in this case indirect ancestry, of several branching lineages: *Cycas* (Cycadales) and *Ginkgo* (Ginkgoales) (1) provide ancestral taxa (as a pair contributing to some as yet unresolved set of the same taxa) for *Pinus, Gnetum, Chloranthus,* and *Austrobaileya. Angiopteris, Marattia,* and *Danaea* imply that Marattiopsida (2) is the ancestral taxon (pres-

ently resolvable only at that rank) of all other ferns and *Equisetum. Gleichenia* and *Phanerosorus* imply that Gleicheniales (3) is the ancestral taxon for *Hymenophyllum* and indirectly, by extension, other ferns more distal on the molecular cladogram. *Dicksonia, Cyathea* and *Plagiogyria* imply that Cyatheales (4) is ancestral to a number of molecularly crown fern genera including *Lygodium*. Taxa selected here as ancestral are those that are lower in the morphological tree than in the molecular tree, and thus provide a clear choice for mapping taxa on the molecular tree.

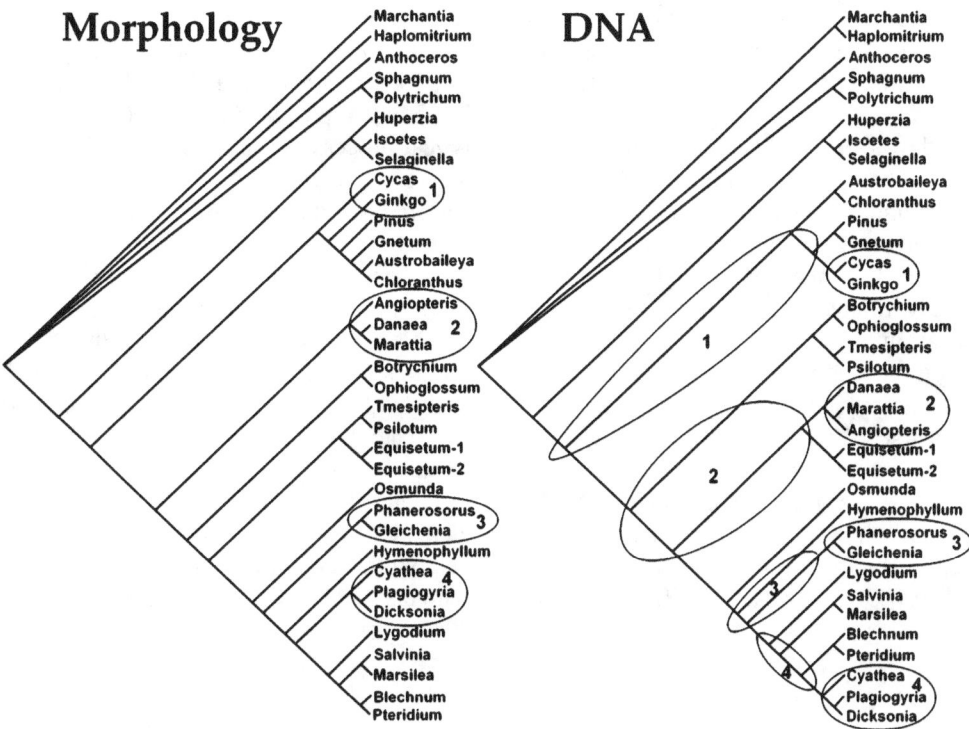

Plate 7.2. — Exemplary chart of cross-tree heterophyly modified from Schneider et al. (2009) comparing morphological *(left)* and DNA *(right)* cladograms of selected vascular plants, particularly the ferns. Taxa that are significantly more basal in the former than in the latter imply structuralist isomorphic ancestral taxa, which are mapped as ellipses over relevant areas of the tree on the molecular cladogram: 1 = Cycadales and Ginkgoales; 2 = Marattiopsida; 3 = Gleicheniales; 4 = Cyatheales; all of which are apparently surviving ancestral taxa. These contribute to an incomplete "Besseyan cactus" (Bessey 1915) of lines and ellipses on the molecular tree, showing both sister-groups and some ancestor-descendant evolutionary relationships. The diagrams were made in part with Treeview (Page 1996).

Alert readers will note that in the molecular cladogram (Plate 7.3) of the moss family Pottiaceae, subfamily Erythrophyllopsoideae is placed as a caulistic entity at the base of the evolutionary tree to represent the extremely plesiomorphic genus *Erythrophyllopsis* in the morphological cladogram (Plate 5.1), yet two species of *Erythrophyllopsis* are nested high in the molecular cladogram apparently in the Pottioideae. This is an artifact of the molecular cladogram, and *Erythrophyllopsis* is indeed in the Erythrophyllopsoideae. Classification does not need to follow caulistic mapping of taxa on a molecular cladistic tree when there are complications, and no evolutionary theory should follow any strictures of classification. The latter confounds the expectation of evolutionary monophyly (all organisms in a group plus any dependent group of equal or greater rank derive from a shared ancestor), but that may be as axiomatic as phylogenetic monophyly (all organisms derive from a shared ancestor must be of the same taxonomic rank and name). Evolutionary parallelism (e.g., Rajakumar et al. 2012) is here considered an expected feature when studying macroevolutionary transformations revealed as paraphyly from a demonstrated caulistic taxon, and may be non-artefactual (Arendt & Reznick 2007; Cronquist 1975; Gould 2002; Shanahan 2011). If so, then the Darwinian maxim that evolutionary monophyly must begin with a single shared taxon may be somewhat challenged in that several individuals or lineages of the same taxon of any rank may derive from one different taxon at different points on a molecular cladogram. Darwin asserted (discussed by Dayrat 2005) that classification should reflect both genealogy and similarity, but if caulistic study demonstrates prevalent parallelism as a minor challenge, it must be accounted for in non-cladistic classification. Apparent determinative parallelism has been demonstrated at the highest levels of evolutionary community structure (Ricklefs & Renner 2012).

Plate 7.3 shows a split between *Barbula* sect. *Convoluta* (represented by exemplars of *B. bolleana* and *B. indica*) and *Barbula* sect. *Barbula* (represented by an exemplar of *B. unguiculata*). The former section has plane leaf margins as in the clustering subfamily Trichostomoideae, and the latter recurved leaf margins as in the clustering Pottioiceae. Both clearly belong to the genus *Barbula* by the distinctive stem section anatomy, gemmae type, blunt leaf apex, and areolation (see Zander 1993). Kučera et al. (2013), based on splitting in a molecular analysis of many *Barbula* species along with species of other Pottiaceae genera, split *Barbula* into *Barbula, Hydrogonium* and *Streblotrichum,* reflecting the information in Plate 7.3 in formal classification. Their

work, however, in my opinion, simply supports the idea of *Barbula* and *Pseudocrossidium* as basal taxa in an evolutionary tree (Plate 8.8). Cladistic clustering is not a process in nature but a result of an analytical process that is unfinished. That is, mere nesting in a foreign subfamily requires insight into why. *Barbula* species should or should not all be in the tribe Barbuleae of the Pottioideae unless there is more reason to modify this than cladistic clustering. There is indeed a scientific process-based explanation for this cladistic nesting, namely that the two *Barbula* sections are engaged in *parallel self-nesting ladders,* each ladder generative of taxa in a different subfamily. The genus *Barbula* in the tribe Barbuleae is therefore primitive and coherent as a group of species, and is not represented or even representable by cladistic nesting or clustering on this molecular cladogram. This deals with both cladistic nesting and with classical taxonomic evaluation.

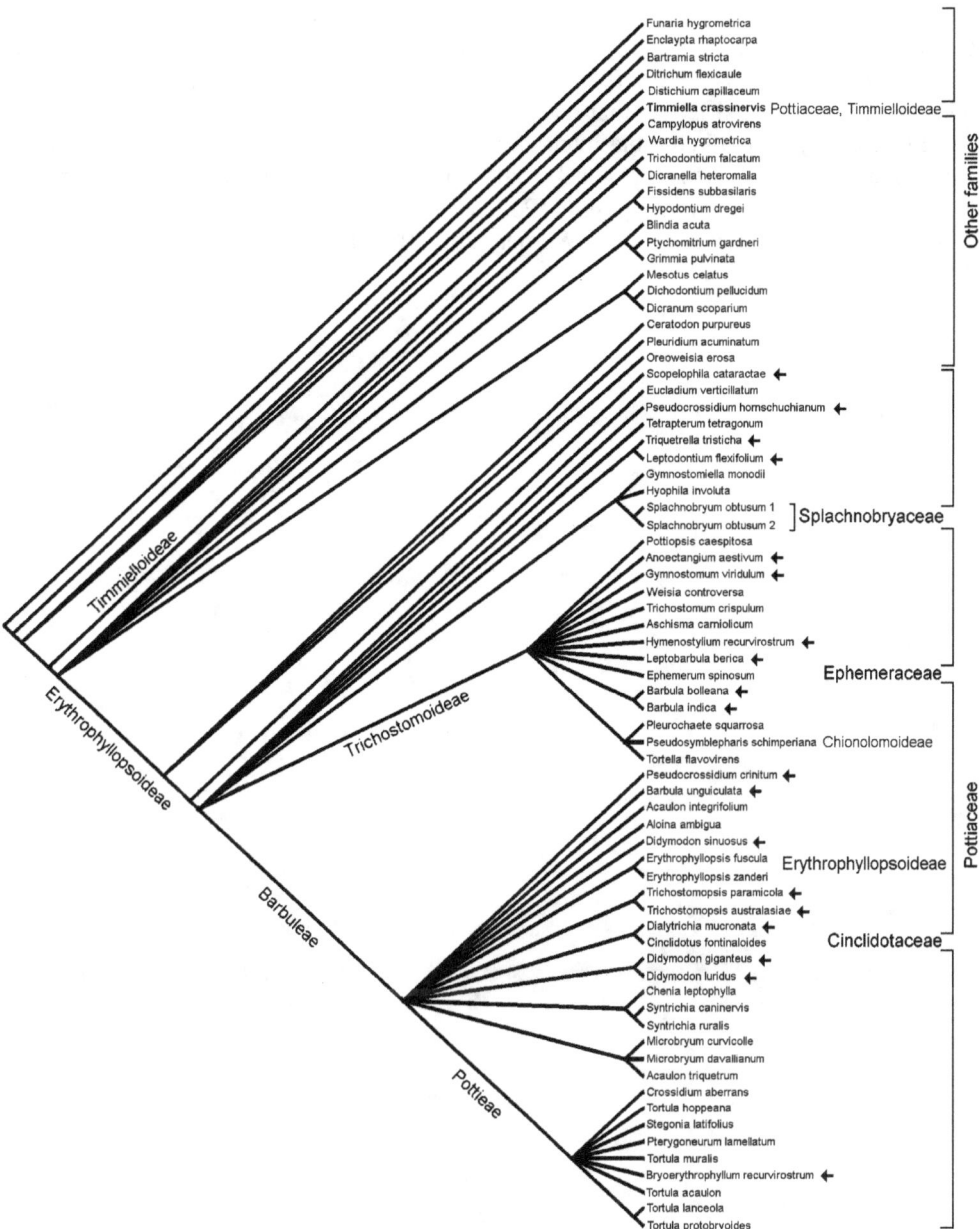

Plate 7.3. — Combined same-tree and cross-tree heterophyly. The cladogram, adapted from that of Zander (2008b), summarizes at 0.95 Bayesian credibility the Werner et al. (2004) molecular analysis of the Pottiaceae (Musci) and related families. The outgroup is *Funaria* in the Funariaceae. Three families are macroevolutionarily derived from (i.e., deeply nested in) the Pottiaceae, these are Ephemeraceae, Cinclidotaceae and Splachnobryaceae. *Timmiella* has been confirmed (Cox et al. 2010) to be molecularly heterophyletic with the remainder of the family implying it is directly derived from a pottiaceous progenitor that also budded several additional intercalated families of mosses. A basal position in a morphological cladogram (Plate 5.1) implies that *Erythrophyllopsis* is a deep ancestral taxon with a clear self-nesting ladder. The proper position of the strongly self-nesting *Erythrophyllopsoideae* is basal to Barbuleae in an evolutionary (not cladistic) tree. Different species of each of *Barbula* and *Pseudocrossidium* are strongly heterophyletic on the same molecular cladogram implying a deep ancestral taxon including these genera. In fact the Barbuleae is clearly scattered (and

heterophyletic) among Trichostomoideae and Pottieae (**arrows**). That Barbuleae is indeed heterophyletic on the genus level (*Barbula, Didymodon, Pseudocrossidium*) seems clear. *Pseudocrossidium* is apparently heterophyletic as supported by recent treatments of the genus (Cano 2011; Jiménez et al. 2012) that recognize or at least segregate together the two species in spite of the present 0.95 credible interval involved in heterophyly. The possibility of self-nesting ladders means that higher ranks should not be clustered solely on the basis of a molecular cladogram. *Acaulon integrifolium* (Pottiaceae) is the correct name (Werner et al. 2005) for the exemplar originally published (Werner et al. 2004) as *Goniomitrium acuminatum* (Funariaceae).

CHAPTER 8
Element 5 - **Superoptimization and Consolidation**

Précis — Superoptimization is an attempted formalization of scientific insight. Cladograms may be made more parsimonious in general (not stepwise) by naming the internal nodes as possible ancestral taxa. This is done by examination of sister clades and deciding, if possible, which is the probable ancestral taxon given non-phylogenetic information from morphology, biogeography, ecology, cytology, and other information associated with classical systematics and biosystematics. A preliminary evolutionary tree can be constructed. Molecular cladograms are inspected for heterophyly (Elements 3 and 4), then collapsed to cladograms with only clades of 0.95 credible intervals using the Implied Reliable Credible Interval method. Classical taxonomy, morphological cladistics and molecular phylogenetics are consolidated through a Bayes Solution using "coarse priors." If a cladogram is not available or does not reflect accepted classical concepts, superoptimization (distinguishing macroevolutionary change by pseudoextinction and budding evolution) may be done as best possible with supraspecific categories. An example is given with the moss genus *Didymodon* leading to name changes at the genus and species levels with the basic transformational evolutionary unit being the genus as represented by a core generalist species chosen to best fit Dollo transformation at the taxon level. Although morphological and molecular cladistics are uncertain, they help determine primitive versus advanced macroevolutionary positions of taxa.

One can or should see the utility and logic of intratree and cross-tree heterophyly in implying diachronic, sequential changes in taxa. Although a well-supported identity of the gene tree with the species tree is ideal for support of sister groups, heterophyly can be distinguished from simple incomplete lineage sorting (where a taxon appears twice in sequence on a molecular clade) by careful evaluation of conflicting cases of heterophyly and Dollo evaluation at the taxon level. That is, one asks if this heterophyly (as apparent evidence of an ancestral taxon) is reasonable in light of expected serial macroevolutionary transformation for this group and any other evidence. Incomplete lineage sorting, in any case, *is* evidence of a shared ancestral taxon of a name including both exemplars that exhibit this, but the implied ancestral taxon cannot be expected to be a particularly "deep" ancestral taxon.

Naming nodes in cladograms increases the parsimony by decreasing the number of postulated but unnamed shared ancestors. Some may argue that parsimony is decreased when the same traits are forced into multiple origins yet the constraint of two or more nodes being the same ancestral taxon comes from different, non-phylogenetically informative data or from molecular heterophyly, and such constraint is decisive.

Some nodes cannot be named because sister-group taxa are clearly equally derivative in specialization and recent environments, or which have very different complex traits for which evolution from a shared ancestor is more reasonable than from each other. This potential pseudoextinction event "breaks" a caulogram. In the example in Plate 8.1, only one node is posited that could be evidence of shared ancestors different from one or the other of the sister taxa. It is expected that a paucity of potential shared ancestors in the sense of pseudoextinction is true for many other taxa.

Maximum parsimony of posited taxa would require that all shared ancestors be named to the extent possible. Paraphyly involving taxa surviving at least one speciation event or even more has been estimated as widespread in extant taxa by Funk and Omland (2003), who indicated that species level paraphyly or polyphyly occurred in about 23% of assayed species. Rieseberg and Brouillet (1994) suggested that at least 50 percent of all plant species and possibly much more are products of geographically local speciation, of which half are likely to be not monophyletic, and that in plants "...a species classification based on the criterion of monophyly is unlikely to be an effective tool for describing and ordering biological diversity." According to Levin (1993), "...local speciation by geographically marginal or disjunctive isolates [resulting in paraphyly] is the rule instead of the exception and may match patterns of geographic subdivision." Based on simulations, Aldous et al. (2011: 322) asserted, "...for about 63% of extant species, some ancestral species should be itself extant..."

They further remarked that biologists anecdotally regard as small the number of extant species with an extant ancestor, "though we have been unable to find useful data, perhaps in part because cladistics dogma discourages asking this question." Thus, based on the above studies, there are empirical and theoretical reasons to minimize the number of unnamed and unobservable nodes in a cladistic tree that act as hidden causes (Zander 2010b).

This suggests an apparent rarity of speciation events involving Hennigian pseudoextinction (generation of two daughter species with disappearance of the progenitor), although this is the basis of the molecular coalescent model (Rosenberg 2003). Evidence for pseudoextinction, or at least for a minimum of influence on macroevolution by a supergenerative ancestor, would be the absence of clear subgenera or sections in a genus. Distinct subgenera or sections would be evidence for a supergenerative ancestor, even one that is extinct or unsampled.

Minimization of unobservable entities is important but complete macroevolutionary understanding doubtless cannot be achieved—as Einstein (Gilder 2008: 86) said, every theory includes unobservable quantities. This additional parsimony maximization, or *superoptimization*, of a cladogram is accomplished by designating one sister lineage at each node, when possible, as the progenitor of the other. Superoptimization using expressed traits (including biogeography, ecology, etc.) complements molecular analysis. This is because the latter alone cannot determine branch order of taxa with confidence, even with dense sampling, because of the probable commonness of extinct or unsampled molecular extended paraphyletic clades.

Although some nodes may be clarified caulistically by intratree and cross-tree heterophyly, for other ancestor-descendant tree splits information outside the phylogenetic data set must be used. Any information that makes it likely that one lineage gave rise to the other may be used to maximize parsimony of the caulogram, including geographic distributions, cytology, and Dollo evaluations of total morphological change at the taxon level. The cladogram nodes are assigned the taxonomic name of a terminal exemplar or, as is done in heterophyly evaluations, a taxon name high enough in rank to include all distal terminal exemplars in that lineage. There is no reason that some nodes (perhaps a third as suggested by the work of Aldous et al. 2011) cannot be left as pseudoextinction events, but these are indistinguishable from nodes simply having equivocal information on ancestor-descendant relationships.

In many cases, a molecular cladogram contradicts accepted or near-certain classical groupings of taxa. In such an event, superoptimization may be done within supraspecific groups, or else the data generating the cladogram should be (if reasonable) reweighted until it does reflect near-certain classical groupings. Or if no cladogram at all is available, the superoptimization may proceed with the guide of a natural key. The idea is to determine which taxon is an ancestral taxon, and which are direct descendants. There may be, for instance, many species as descendants of a supergenerative species, see example with *Didymodon* below.

A morphological or molecular cladogram is ideally meant to illuminate evolution among classical taxa. Morphological cladograms should help reveal macroevolutionary aspects of primitive and advanced groups, not nested results of presumed and unproven pseudoextinction events. Only if a molecular cladogram is based on adequate statistically robust sampling and if mechanisms of molecular evolution are understood (e.g., self-nesting ladders) should changes be made in classical classifications on the basis of DNA data.

Classical systematics is often concerned with the detection of gaps between groups of organisms. There is software that is intended to help identify such gaps, but classical systematics also weights other traits, including non-gap information such as relative importance of autapomorphic characters, and salient conservative characters that are important in group coherence.

There is apparent agreement among phylogeneticists that phylogenetic methods involve distinguishing which two of three taxa are more closely related (Nelson 2004: 128), and also that phylogenetic analyses should incorporate Hennig's postulation that a shared ancestral taxon disappears (pseudoextinction) after generation of two daughter taxa (Avise 2000). There is no accounting, however, for surviving ancestral taxa or ancient paraphyly with all but one branch unsampled (e.g., extinct) that would affect the order of branching and confound mapping of traits on cladograms.

Assis and Rieppel (2010) asserted that mapping of *traits* on molecular trees is empirically empty because these are not refutable by synapomorphies. Laurin's (2010) evaluation of evolutionary trend detection used only simulations of character change "using known evolutionary models" of character change. Mapping *taxa*, however, through inference from heterophyly on molecular trees results in hypotheses of ancestry that may be tested against fossil evidence, biogeography, and other information not used in generating a molecular cladogram, and maximizes parsimony of postulated shared ancestors. Although one particular macroevolutionary scenario may not seem well supported, if there are no reasonable alternatives, one can invoke Cohen's (1994) arguments against relentless search for statistical support in such unambiguous Sherlockian eventualities.

Dollo's Rule — Dollo's Rule (Dollo 1893; Goldberg & Icić 2008; Gould 1970; Lönnig et al. 2007) is the generalization that an organism can never return exactly to a previous evolutionary state, though often nowadays mistaken as meaning that individual traits are doubtfully reversible, as noted by Cavalier-Smith (2010), Gould (1970), Hall (2003), and Jackman and Stock (2006). This is important at any step in generation of a natural classification, but may be of particular analytic value when a taxon is basal in a morphological study but terminal in a molecular study (see Element 4). In the present paper, Dollo's Rule is considered in the original sense as applicable at the whole organism level through a developmentally or selectively unified combination of traits—as opposed to occasional homoplasy of portions of the genome or expressed traits atomized in a data set—and full convergence at the taxon level is considered rare or improbable (Gould 1970).

Two morphologically complex and different taxa that do converge cannot, by Dollo's Rule, converge completely. There are always some telling traits from a previous separate lineage that are dragged along during convergence. Identification of such traits is then decisive when identifying the same taxon with exemplars distant on a molecular tree probabilistically because the telling traits occur only in that one taxon, and very rarely elsewhere. That rarity of occurring elsewhere is a measure of probability, and this can be gauged as a proportion of all taxa that conceivably tolerate developmentally those dragged-along traits. This probability is small.

Thus, any cladogram topologies that are incredible or improbable in light of standard evolutionary theory need re-examination for constraint on direction of evolution. Typical features that may allow a successful Dollo evaluation are convergence of organisms indistinguishable at some taxonomic level, polyploidy, hybridy, unique trait complexes, wide and ancient distributions, recent habitats or pollinators or parasites or predators, paleontology, vicariance events, consilience (see Glossary) of morphological and molecular derivations, and developmental pathways that are essentially one-way (e.g.,

Bridgham et al. 2009). There are, of course, well-known developmentally based violations of Dollo's Law in certain complex *traits*.

Examples include apparent re-evolution of shell coiling in snails; reactivation of wings in wingless walking sticks; eye atavisms in cyclopean brine shrimp; modes of vulva formation in nematodes; ancestral traits of the lateral lines, muscles, and gill rakers of cichlid fishes; eye reactivation in eyeless copepods; teeth in chickens; and re-occurrence of a second molar in lynx, as reviewed by Zander (2010a). But these are confined to complex organs (e.g., bats and birds have wings). Examples of apparent total convergence at the taxon level (as discussed by Collin & Miglietta 2008; Jardine & Sibson 1971: 144) may be better explained (Zander 2010a) as morphological stasis plus molecular heterophyly due to temporal or geographically isolated populations of ancestral taxa. Not all reversals of complex traits are strictly homologous, as vestigial hips in snakes and whales functionally depend on different developmental pathways (Bejder & Hall 2002). Thus, although some complex traits can be deemed reversible, others are irreversible, and judgment based on biosystematic and developmental facts about the total organism beyond phylogenetic analysis of morphological and molecular traits is required.

Crawford (2010) has published a thorough review of phylogenetic and other methods for determining at least recent progenitor-derivative species pairs in plants. Biosystematic study provides biological evidence other than that of descriptive morphology that may support or require modification of alpha taxonomy; it is often experimental or quasi-experimental (Cook & Campbell 1979) or statistically analytic (Tobias et al. 2010).

Traits outside those commonly used in phylogenetics — With attention to the details involved in superoptimality, the evolutionary tree fills out as a complete theoretical description of gross aspects of macroevolution of the groups involved. Required, of course, is information on geography, ecology, ethology, chemistry, genetics, and many aspects of expressed traits affected by evolution that may necessarily be represented only in a monograph or flora or faunistic study of a large region, combined with a willingness to engage in the Dollo evaluation.

Microevolutionary changes in single isolated traits (including ecology and aspects of adaptive morphology) are commonly reversible, as is assumed in the usual non-Dollo-enforced phylogenetic analyses, but macroevolutionary traits involving anatomy and general bauplan are not or rarely so (Grant 1991: 329). Macroevolutionary traits are built up as a complex, interdependent edifice over time, constraining reversals. The evolutionary ratchet of Levinton (1988: 217) involves epigenetic, genetic, and selectional features not easily lost or reversed. This is, theoretically, due to accumulative functional integration that is epigenetically buffered by regulator genes and promoter sequences. The identification or at least inference of such constraints at the taxon level helps decide direction of descent with modification of taxa. J. Glime (Bryonet, June 22, 2012) found that the water moss *Fontinalis,* which almost always has no costa (midrib) in the leaf, produced short costae when grown in an artificial stream with much air exposure. She suggested that a suppressive regulator gene or promoter sequence acts on costa development under usual conditions, and that expression is an ancestral state in this case.

Superoptimization of *Didymodon* — Zander (1998, 2001) did a most-parsimonious cladistic analysis of 22 New World species of the moss genus *Didymodon* (Pottiaceae) with a data set of 23 morphological characters, and *Barbula unguiculata* as outgroup. The North American specimen of "*Didymodon sinuosus*" has since been reidentified as a new species, *D. murrayae* Otnyukova. The data were treated as non-additive (non-ordered) and equal (no) weighting was used. Three trees differing only in placement of *D. nichol-*

sonii and *D. murrayae* were obtained, and an optimal tree (Plate 8.1) was selected for analysis because it was closest to the (Zander 1998) UPGMA result with the same species. Tree length is 63 steps, consistency index is 0.44.

Authorities for botanical scientific names in this book, if not given, may be found in the treatment of the Pottiaceae by Zander (1993) or on the Web site Tropicos of the Missouri Botanical Garden.

The cladogram was subjected to superoptimization, namely the identification by name of all possible ancestral taxa to reduce the numbers of nodes identified only as unnamed, unobservable, superfluous postulated "shared ancestors" as hidden causes. Each branch in Plate 8.1 was drawn as broken at the juncture of an inferred descendant and progenitor to make an evolutionary tree. This tree is only an interpretation of the cladogram, however, and is limited by dichotomous structure.

A thorough analysis would require following a complex method based a review of the literature on determining (inferring, informedly guessing) which of two sister-group taxa or clades are progenitor and which descendant, a study beyond the scope of the present work. For purposes of this example, however, the results are unequivocal. The following numeric codes in the cladogram indicate inferred aspects of macroevolutionary transformation of the budding type:

1. A major morphological differentiation that signals the first groupings above species level.
2. A species generalist in morphology that might easily generate specialized descendants.
3. A widely distributed species that is found in many habitats and may be relatively old.
4. A habitat specialist.
5. A species with asexual reproduction common and sexual reproductive organs rare or absent.
6. A species of local distribution, often of recent habitats.
7. A species that after superoptimization is reasonably considered an ancestral taxon of many species.
8. A species of multiple subspecies or varieties.

Examination of the modified cladogram (Plate 8.1) shows only one node (except the most basal, of course) that could not be named at the species level (that for *Didymodon asperifolius* and *D. vinealis*). *Didymodon asperifolius* doubtfully belongs with *D. vinealis* and has a morphology that is similar to *D. fallax* but more primitive (having quadrate adaxial costal cells). It probably shares an immediate but extinct (or unsampled) ancestral taxon with *D. fallax,* and is an instance of possible pseudoextinction. This is contrary to what the actual cladogram reveals, but

construction of the database is not infallible or sufficiently inclusive of difficult to describe traits, nor is identical weighting necessarily representative of conservative traits. Mechanical analysis must be interpreted.

Two species (codes 1 and 7), *Didymodon fallax* and *D. vinealis,* are the implied progenitors of many descendant species. These may be called supergenerative core taxa. The groups associated with all inferred deep ancestral species have been recognized as sections of *Didymodon* (Zander 1993). One group

(Code 1), with *D. australiasiae* as progenitor, has been considered a different genus, *Trichostomopsis,* thus this genus was generated by another because it is well-nested in that other genus. *Didymodon rigidulus* (Code 1) s. lat. is progenitor of two asexually reproducing species, and these are in a third section of *Didymodon,* thus one section is generated from another. *Didymodon nigrescens* (Code 1) is a generalist species that probably generated one propaguliferous species (Code 5), and another as habitat specialist of local distribution (codes 4 and 6).

Given that all immediate descendants of one core species have the same extant shared ancestral species and a term is not available for these important features of evolution, a special name for such daughter species might be the English word "stirp" (plural stirps) as a lineage descending from a single ancestor. This English word is not spelled exactly like the Latin "stirps" (plural stirpes), which is more commonly used in the legal sense of distribution of a legacy equally to all branches of a family (per stirpes), but the sense is similar.

Vrba (1985) proposed a bias in the way that species are generated and go extinct. Generalist species (*eurytopes*) apparently survive longer—when one food source or habitat changes, they can make use of other food or survive in another habitat. Specialist species (*stenotopes*), on the other hand, are far more sensitive. Specialists, however, apparently speciate more frequently, even if they go extinct more frequently, too, as they too adapt to degradation of necessities. The core species of the present book are not just eurytopes but are also fully charactered. The stirps are not just stenotopes, but are usually also highly modified in traits such that adaptation to new conditions may be difficult or impossible because of the physiological burden of specialization. An extreme form of "dead-endedness" is associated with "evolutionary suicide," in which there is, for a short but significant time, extreme selection against a trait that is otherwise evolutionarily advantageous, such as in wildlife harvesting (Sasaki et al. 2008).

Thus, certain species may be identified as intergenerically primitive, and generative in many cases of a number of specialized derivative species—a set of stirps. An area of species diversity or biotype multiplication may provide an evolutionary lens effect that serves to illuminate a genus' ability and direction to evolve, given phyletic constraint and developmental restrictions. Progenitors are inferred to give rise to other progenitors, thus conserving (contributing to Dollo parsimony at taxon level) the more complex traits less easily acceptable as reversible. An example is re-evolution of reduced or absent peristomes here considered improbable though theoretically possible.

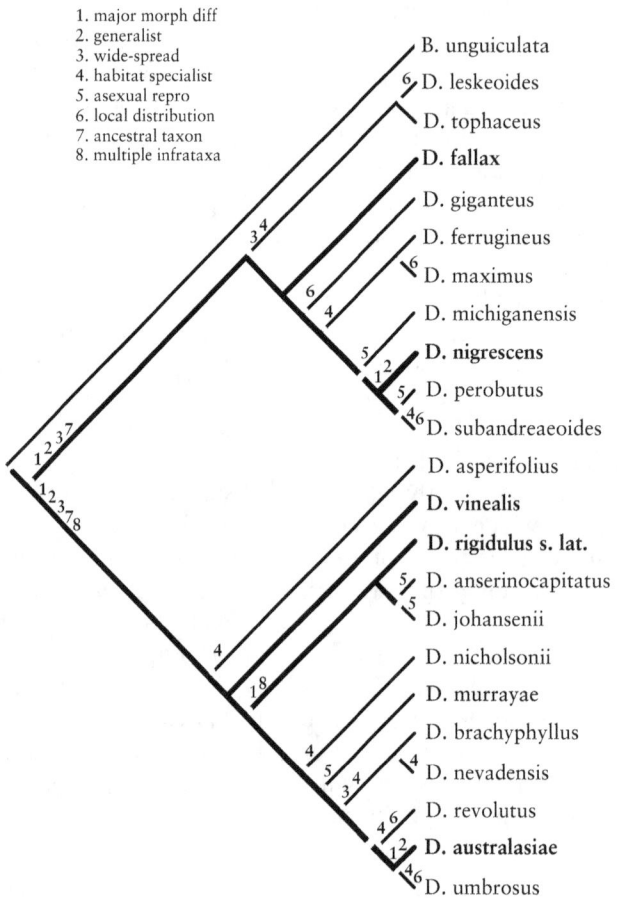

1. major morph diff
2. generalist
3. wide-spread
4. habitat specialist
5. asexual repro
6. local distribution
7. ancestral taxon
8. multiple infrataxa

Plate 8.1. — Superoptimized evolutionary tree of 22 species of the moss genus *Didymodon* (Pottiaceae), with *Barbula unguiculata* as outgroup. The breaks in lines signify inferred descendant lineages (stirps), here forming descendant clouds from two supergenerative species of the four (bold faced) core species (or main ancestral taxa). Branching integral lines show genetic continuity at the species level. Codes refer to justifications or other features of inferred macroevolutionary transformations. *Didymodon fallax* and *D. vinealis* (bold) are the most prolific ancestral taxa in terms of apparent descendants, but the two are not well distinguished as to which is the ancestor of which in this cladogram but the latter is somewhat more specialized. *Didymodon nigrescens* and *D. australasiae* are, however, by their deep nesting, clearly descendants of *D. vinealis* and *D. fallax*, respectively, although they themselves are ancestral to specialized daughter species. Most nodes in this cladogram can be seen to be superfluous as shared ancestral taxa different from the sister groups, that is, there are no true sister groups in this cladogram. But see discussion of *D. asperifolius* below.

Following is a polychotomous natural key derived from the superoptimized morphological cladogram of *Didymodon* (Plate 8.1). This key emphasizes serial macroevolutionary transformations, which compromises the usual key faculty of sequentially segregating trait combinations until each taxon is well distinguished. Thus, natural keys are not merely difficult for users because technical characters may be used but also because only an artificial key can deal well with trait reversals. In any case this key matches the evolutionary tree in Plate 8.2, but also details important distinguishing traits of the taxa in a macroevolutionary context. Note that a natural key may have more than two "couplets," sometimes less.

Prim. means "primitive" as in first of a macroevolutionary taxic series; **Deriv. Prim.** means "derived primitive," that is, primitive in terms of its descendants but immediately derived from another taxon; **Adv.** means "advanced," particularly, derived and specialized in some way and not inferred as the progenitor of a significant new evolutionary series, possibly a dead end in evolution or simply most recent.

Didymodon rigidulus s. lat. is now considered three separate species, and is represented in the key by the evolutionary formula that indicates a theory that the generalist species *Didymodon acutus* generated *D. icmadophilus,* a species of higher elevations with unusual undifferentiated basal leaf cells, and also probably generated *D. rigidulus* s. str., a species of specialized wet habitats with small gemmae in leaf axils. Readers will note that this natural key also uses autapomorphies as distinctions.

A natural key is difficult to comprehend if one is used to dichotomous keys. A natural key is most simplistically constructed by listing major generative species (or a blank when no extant generative species is known). These are indented themselves if one is seen as generated by another (as done in Natural Key to *Didymodon* below). Then equally indented under each are the daughter species (even if there is only one or if there are several). Then indented under the daughter species are their own daughter species and so on. Blanks (no species given) represent unknown ancestral taxa or inferred pseudoextinction events (as, for instance, the ancestral taxon of *D. asperifolius* and *D. fallax*). Then, for each line, add before the species (or blank) a description of the inferred "evolutionary trajectory" or unique adaptive solution or evolutionary neutral advanced trait of that species. Specialists can make informed decisions on this. The numbering is simply for the macroevolutionary level of taxa from the tree base, with basal taxon being number one. *The number of indentations are equal to macroevolutionary patristic distance from the base, that is, the level minus one.* Readers might try creating a natural key for well-understood taxa in their area of specialization.

Natural Key to *Didymodon*

1a. **Prim.** Leaves lanceolate, reddish brown in nature, with a small oval window ventrally on costa near apex; costa bulging dorsally, with quadrate to short-rectangular adaxial cells; laminal papillae usually multiple; peristome long and twisted, absent in a variety .... *Didymodon vinealis*

    2a. **Adv.** Leaves shorter, leaf base squared; more arid habitats; peristome short and twisted or rudimentary .... *Didymodon brachyphyllus*

        3a. **Adv.** Leaves with multilayered photosynthetic cells on ventral surface of mid-costa; sporophytes absent .... *Didymodon nevadensis*

    2b. **Adv.** Leaves with bistratose cells medially, often across leaf; peristome long and twisted . ... *Didymodon nicholsonii*

    2c. **Adv.** Leaf apex sinuose, bi-tri-stratose, deciduous as a propagule; sporophytes unknown .... *Didymodon murrayae*

    2d. **Deriv. Prim.** Costa much flattened, ventral stereid band absent, upper laminal cells bistratose; peristome long and weakly twisted .... *Didymodon australasiae*

        3b. **Adv.** Leaves very long lanceolate, basal laminal cells with slits; peristome long and weakly twisted .... *Didymodon umbrosus*

        3c. **Adv.** Leaves short-ovate, unicellular propagula in leaf axils; peristome absent to short, straight .... *Didymodon revolutus*

    2e. **Deriv. Prim.** Leaves green or reddish in nature, costa not bulging dorsally, distal laminal cells only weakly papillose or smooth; peristome short and straight to long and twisted .... *Didymodon rigidulus* s. lat., or ***Didymodon acutus*** **>** (*D. icmadophilus*, *D. rigidulus* s. str.)

        3d. **Adv.** Leaf apex cylindric, fragile in pieces as a propagule; peristome straight, to long and weakly twisted .... *Didymodon johansenii*

        3e. **Adv.** Leaf apex turbinate, deciduous as a propagule; sporophytes absent *Didymodon anserinocapitatus*

    2f. **Deriv. Prim.** Moist areas; leaves narrowly channeled, papillae simple, recurved, carinate.. .... Unknown ancestral taxon.

        3f. **Adv.** Mountainous areas; deep red plant coloration; stem central strand often absent, peristome short and straight .... *Didymodon asperifolius*

        3g. **Deriv. Prim.** Moist sites; adaxial cells of costa elongate, papillae usually simple; peristome long and twisted .... *Didymodon fallax*

            4a. **Adv.** Leaves ovate-lanceolate, usually without papillae, with small auricles or long decurrencies, wet habitats; peristome short and straight, occasionally rudimentary or absent .... *Didymodon tophaceus*

                5a. **Adv.** Leaves long-acuminate lanceolate, with large auricles; sporophytes absent .... *Didymodon leskeoides*

            4b. **Adv.** Leaves usually without papillae, very wet habitats; peristome nearly straight to long and twisted .... *Didymodon ferrugineus*

                5b. **Adv.** Leaves much enlarged; sporophytes absent ...... *Didymodon maximus*

            4c. **Adv.** Leaves and plants much enlarged, laminal cells with trigones; sporophyte absent .... *Didymodon giganteus*

            4d. **Adv.** Leaved catenulate when dry, small spherical gemmae in leaf axils; sporophytes absent .... *Didymodon michiganensis*

            4e. **Deriv. Prim.** Leaves dark brown to black in nature, distal marginal cells crenulate; peristomes twisted to straight .... *Didymodon nigrescens*

                5c. **Adv.** Leaves ovate, clusters of unicellular gemmae in leaf axils; sporophytes absent .... *Didymodon perobtusus*

                5d. **Adv.** Leaves dimorphic, the smaller strongly concave in series in some parts of the plant; sporophytes absent *Didymodon subandreaoides*

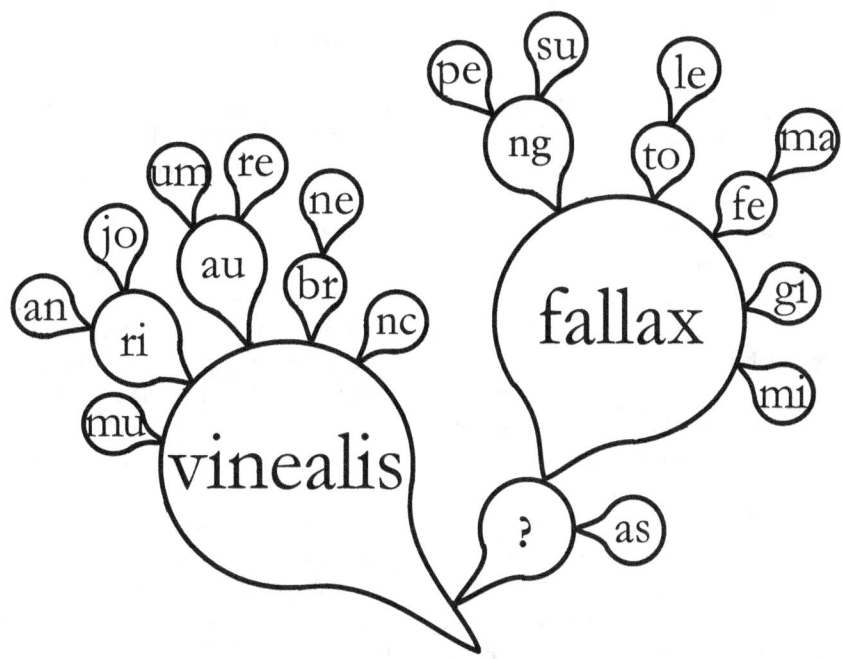

Plate 8.2. — An evolutionary tree as a commagram (Besseyan cactus) derived from the natural key above, which itself was in part derived from the superoptimized morphological cladogram of 22 species of the moss genus *Didymodon* (Pottiaceae), plus *Barbula unguiculata* as outgroup, not shown here. Species epithets are indicated by the first two letters that distinguish them, other than the two major supergenerative species, *D. vinealis* and *D. fallax.* The genus *Didymodon,* based on this evolutionary tree, is split (later in this book) into six inferred genera, represented by the two large and three somewhat smaller commas above (pads of the Besseyan cactus). The question mark represents an inferred unknown shared ancestor.

One may find in Plate 8.1 that certain derived species are nested more terminally on the tree than their supergenerative ancestral taxon, even though a multifurcation is expected because superoptimization indicates they are all derived from that one ancestral taxon. For example, *D. giganteus, D. ferrugineus, D. maximus,* and *D. michiganensis* are nested beyond deep ancestor *D. fallax.* This may be explained as a kind of self-nesting ladder in which random trait changes that match those of other daughter species will force nesting of some derived taxa, in addition to forcing some derived taxa lower in the clade. *Didymodon asperifolius* is associated with the *D. vinealis* lineage in the morphological cladogram of Plate 8.1, but this species is similar to *D. ferrugineus* except for the primitive and conservative (for the entire group) trait of quadrate adaxial costal cells.

Though there are 23 traits in the data set, certain of those traits are more likely to change and be fixed in a population because they have changed recently and other traits have not changed. Thus, changes and reversals that are recent or local on the cladogram are more likely to be tolerated by the same set of other traits, a kind of phyletic constraint. The potential traits that are most liable to change upon speciation are therefore not the entire set of 23 traits, but those recently changed and compatible with the organism. Traits unlikely to change are core traits of the group. Thus, self-nesting ladders below and beyond the supergenerative ancestral taxon should be expected, generated randomly among fewer than 23 traits.

An exception to the pure dissilient genus concept (one core species with several descendent species) is the existence of several highly modified stirps that *are more similar to each other than to the core generative species.* Such a case is with species similar to *D. murrayae* occurring elsewhere in the world and not included in this study. These are reviewed by Otnyukova (2002), and could well have speciated among each other rather than in parallel from the core species. A focused study may be able to elucidate directions of taxic transformation.

This evolutionary tree as modified from a cladogram shows clustering along the lines of the recognized sections of the genus (Zander 1993), though these subgroups were neither confirmed nor rejected by Jiménez (2006) in his revision of Old World *Didymodon*. The tree thus matches in my opinion to considerable extent the conclusions of classical taxonomy. Of some interest are the large numbers of clearly derived species as judged by geographic and anatomical criteria, such that each ancestral taxon is surrounded by a small or large cloud of descendants.

In any case, in the original study, parametric bootstrapping (Zander 1998) found good resolution at a minimum of four nodes (see section on comparing morphological and molecular analyses, above). The evolutionary tree is, of course, ultimately based on morphological descriptions from classical taxonomy, though limited by dichotomous structure and no character weighting. This tree needs to be compared with the results of a taxically densely sampled molecular analysis to see if heterophyly of the species might confirm or reject the inferred deep ancestors obtained through omnispective superoptimization. See discussion of the ITS study by Werner et al. (2005a), below. The aim is to determine an evolutionary caulistic explanatory structure shared by classical taxonomy, morphological cladistics and molecular analysis.

An important feature of this example of superoptimization is that although the cladistic tree is not particularly well supported (Zander 1998), the additional information used in superoptimization analysis distinguishes two clear caulistic lineages, and two well differentiated smaller groups. The exact order of branch nesting from the ancestral taxon of the several daughter species may be addressed, possibly, with molecular information. The distinguishing of the ancestral taxa is enough to begin a comparison with molecular analyses that will further try the evolutionary tree.

Consolidation of classical taxonomy, morphological cladistics, and molecular phylogenetics — There are four stages for consolidating the results of all known sources of on macroevolutionary transformation at the taxon level, exemplified here with *Didymodon*.

(1) Morphological cladistic analysis based on classical descriptions with proper weighting creates a cladogram best reflecting classical information. It is superoptimized, as above, and a preliminary evolu-

tionary tree is constructed from a natural key.

(2) A molecular analysis is done (preferably with many sequences and many exemplars per taxon) and the resultant cladogram is inspected for macroevolutionarily informative heterophyly, both same-tree and cross-tree. The molecular cladogram is collapsed to one of only 0.95 or better credible intervals using the Implied Reliable Credible Interval method (page 59).

(3) The molecular cladogram is evaluated against the morphological cladogram or natural key in the context of a Bayes' Solution using "coarse priors" (see below) that represent evolutionary monophyletic reliability of classical and morphological cladistic work.

(4) A final evolutionary tree is generated as a best representation of taxic serial macroevolutionary transformations. Of course, hybridy when known is also taken into account.

A molecular tree is considered corroborated if it matches the results of classical taxonomy or morphological analysis. If this is the case then, logically, it is not only not corroborated, it is refuted to some extent by the incongruence. What extent? In the Bayesian context, a credible interval of more than 0.50 from a supportive study will increase the credible interval of both combined. The complement of this fact is that any study that establishes a result that conflicts with another study and has support for that other study less than 0.50 must decrease the credibility of both studies combined. Thus, if one analysis shows a credible interval of a particular clade of 0.95 and another study shows support for the same clade of only 0.30, then the combined credibility (or posterior probability) using the Bayes' Formula is 0.89 for that clade. It goes down, not up. Essentially, this means that contrary information, unless it can be shown to be irrelevant, cannot be ignored or merely explained away as an example of some general concept (homoplasy, convergence, hybridization, etc.) that may be true in some examples but not demonstrable in the present instance.

One must remember that comparing a tree of nested transformations to a tree of serial transformations can introduce false expectations. A molecular tree should always be reduced to serial macroevolutionary transformations, even if only the assumption that a number of contiguous nodes of the same taxon (at whatever level) implies that taxon as a single shared ancestor of all evolutionarily monophyletic exemplars. That which splits in a clade may be an extinct shared ancestor as in the Hennigian scenario, or a long-surviving ancestral taxon pupping off daughter taxa, but both are conflated in phylogenetic nesting.

If we are going to combine classical taxonomy with phylogenetics, then a common measure is needed. In the Bayes' Solution this means we must use the credible interval (CI) or the equivalent Bayesian posterior probability (BPP). When dealing with classical taxonomy, trying to apply the exact measures of clade support seen in molecular analysis is seemingly arbitrary or hyperexact. But it is possible to translate the kind of reliability viewed in classical systematics into Bayesian terms. One should note that a molecular tree gives credible intervals for nesting of specimens, which are only *inferred* to represent all specimens of a taxon.

The genus as basic element of evolution — Nature teaches us taxon concepts. We later may develop rules or ontogenies to guide taxonomic study (O'Leary & Kaufman 2011). The superoptimization of the moss genus *Didymodon* s. lat. indicates that it should be split into smaller genera in which the genus is the basic element of evolution for this large group, because, operationally, a core generalized species is progenitor of specialized descendants in the same genus and generalized progenitors in other genera. This is a *genus-level speciational burst or dissilience associated with a named core species*. It is an exact definition for a genus or other supraspecific taxon, at least for some groups. Most evolutionary speciational bursts described in the literature are along a time gradient, this is taxonomic. The whole genus is an evolutionary unit (a core-and-radiation group) though specialized descendants may be dead-ends. Simpson (1953: 392) discussed "explosive" adaptive radiation, but did not associate it with an ancestor-descendant burst as above, but emphasized the rapid occupation of adaptive zones.

In addition to specialization the descendent species of a widespread, multiplex core species may be dead-ends because small groups that experience size fluctuations are more apt to go extinct over time. Raup (1981) applied this to genera and higher groups in his simulations, but local, stenotypic species should have the same problem.

Doubtless, in *Didymodon* s. lat. as split here, most evolution of genus from genus occurs via modifications of the supergenerative taxa so that the twisted peristome is preserved from one genus to another. Each section of *Didymodon* s. lat. consists of a large, core group of refractory, conservative *traits*, and a small group of traits that change among the radiative descendant species in response to selection. This small group of traits is the reason that apparent nest-ing in each section in the morphological cladogram is smeared out, because the few traits in selective combinations in related taxa easily reverse or parallel. There is also no reason parallelism cannot generate the same descendant more than once, given the few traits active for (or tolerated by) any particular set of core traits. Thus the macroevolutionary genus concept is paramount in *Didymodon* taxonomy and classification. Of course, other taxonomic groups may have different concepts as basic to their evolution and classification.

Coarse priors and the Bayes' Solution — There is no reason that probabilistic support as Bayesian priors need to be on a scale of 100 probabilistic intervals. Such precision may be impossible to assess. Here it is suggested that 10 levels of probabilistic support may in practice be estimated by informed scientific intuition for any taxonomic hypothesis, not including 1.00 or zero. We can assign Bayesian credible intervals to each. These levels of support are coarse (or stepped) priors, which may be used in systematics for estimates of evolutionary monophyly. They are easy to use, and powerful.

(1) "Five Sigma" (0.998 or better) super-certainty (i.e., "quite certain," "damn sure,"). Statistical certainty is a real feature of some analyses, see Cohen (1994)

(2) Almost certain. Say, only once wrong out of a hundred times would the hypothesis be wrong. Expected level of correctness in critical research. Assigned credible interval is 0.99.

(3) Just acceptable as a working hypothesis; just at the lower limit of supporting some action, like a nomenclatural decision. Expected to be correct for non-critical, easily reversible decisions 19 out of 20 times. Credible interval is 0.95.

(4) "Some support" is not alone decisive for action. It can be narrowed down to half-way between certain (1.00) and totally equivocal (0.50), or 0.75. Using 0.75 as prior and 0.75 as probability yields 0.90, then using that as prior and 0.75 again as probability yields 0.96 as posterior. Thus empirical use of Bayes Formula with the answer to the first use being the prior for the second and so on indicates that perhaps three occurrences of "some support" with no contrary evidence is sufficient for action. Credible interval is then 0.75.

(5) A "hint" of support is certainly not actionable alone, nor are even several hints impressive. Using 0.60 probability as representative of a hint, being just beyond totally equivocal, requires 0.60 to be used as a prior seven times in successive empirical analyses

with Bayes' Formula, with no contrary information, to reach 0.96. The credible interval for very minor support is 0.60.

(6) Totally equivocal support probability (assuming only two reasonable alternatives, yes or no, support or refutation) is 0.50. Using 0.50 as prior in Bayes' Formula does not change the probability. An example of an equivocal coarse prior is when one has a multifurcation, e.g. ((ABC)D)E. Monophyly of A and B is 0.50 probability, with neither support for nor against, in this cladogram. Support for monophyly at 0.95 from another cladogram, say, (((AB)C)D)E, allows acceptance of monophyly of AB.

(7 to 10) Support against a hypothesis, is the reverse of the above, that is, 0.40, 0.25, 0.05, 0.01 in support "for" the hypothesis (the remainder for any opposing hypotheses). See Table 8.1.

Bayesian credible intervals may be non-intuitive. For instance moderate support might be expected to be 0.25 probability that a hypothesis is right, since, indeed, 25 percent of the range is support. Yet a figure of 0.25 also includes, necessarily, the baggage of 0.75 probability that the hypothesis is wrong. So in comparing two conflicting hypotheses, a coarse prior of 0.95 in support of a hypothesis means 0.05 in support of the other hypothesis. And molecular clade support of 0.90 means 0.10 in support of the alternative classical or morphological cladistic hypothesis. So when the Bayes' Formula is used to calculate support for a single hypothesis, one uses support only for that hypothesis, both above and below 0.50. Support of 0.80 for one hypothesis implies support of a maximum of 0.20 for the alternative hypothesis. In addition, a very high level of support, such as 0.99 may be acceptable for action in some situations but not others (critical medical decisions), so a Bayes' Solution must include an estimate of "risk if wrong" before a decision is made. The Bayes' Formula itself is only part of the Solution.

There are situations in which a hypothesis with high, say, 0.90 support is apparently confounded by another analysis with only 0.25 support for the first hypothesis. But that second analysis of low support, which would otherwise lower the joint probability, has no one alternative hypothesis of more than 0.25. Does that hypothesis at 0.25 probability then support or reject the first? There is extensive discussion about this (Salmon 1971), including arguments invoking Bayes' factors and maximum likelihood. Since the problem must occur often, statitisicians need to deal with it. I tend (as doubtless do others) to reject the second analysis as not relevant or helpful because too easy to be the result of randomized data.

In addition, support for non-monophyly can be calculated with the Implied Reliable Credible Interval (see page 59, also Zander 2003) from phyletic distance on a cladogram. Probabilistically there exists at least a single node at 0.998 probability between two clades distant by two nodes each with 0.95 support.

Given that the Bayes' Formula, if one does not use a calculator (the Silk Purse Spreadsheet is available online, Zander 2003b), is tedious, Table 8.1 allows rapid estimate of Bayesian support for a particular hypothesis of monophyly given levels of support from an agreeing or conflicting hypothesis. The rows represent coarse priors as may be estimated from classical and morphological cladistic study, while the columns are Bayesian support from molecular cladograms, and the table gives posterior probabilities in the grid.

Coarse priors	Clade support probabilities										
	0.999	0.99	0.95	0.90	0.85	0.80	0.75	0.70	0.65	0.60	0.55
0.999	**0.999**	**0.999**	**0.999**	**0.999**	**0.999**	**0.999**	**0.999**	**0.999**	**0.999**	**0.999**	**0.999**
0.99	**0.999**	**0.99**	**0.99**	**0.99**	**0.99**	**0.99**	**0.99**	**0.99**	**0.99**	**0.99**	**0.99**
0.95	**0.999**	**0.99**	**0.99**	**0.99**	**0.99**	**0.99**	**0.98**	**0.98**	**0.97**	**0.97**	**0.96**
0.75	**0.999**	**0.99**	**0.98**	**0.96**	0.94	0.92	0.90	0.88	0.85	0.82	0.79
0.60	**0.999**	**0.99**	**0.97**	0.93	0.90	0.86	0.82	0.78	0.74	0.69	0.65
0.50	**0.999**	**0.99**	**0.95**	0.90	0.85	0.80	0.75	0.70	0.65	0.60	0.55
0.40	**0.999**	**0.99**	0.93	0.86	0.79	0.73	0.67	0.61	0.55	0.50	0.45
0.25	**0.997**	**0.97**	0.86	0.75	0.65	0.57	0.50	0.44	0.38	0.33	0.29
0.05	**0.981**	0.84	0.50	0.32	0.23	0.17	0.14	0.11	0.09	0.07	0.06
0.01	0.910	0.50	0.16	0.08	0.05	0.04	0.03	0.02	0.02	0.02	0.01

Table 8.1 — Bayesian posterior probabilities using coarse priors (left column) and molecular branch support probabilities (top row) in Bayes' Formula. Bold faced probabilities are posteriors at 0.95 or above. Coarse priors given are: 0.999 (quite certain), 0.99 (rather certain), 0.95 (support just acceptable to act on), 0.75 (moderate but not decisive support), 60 (hint of support), 0.50 (equivocal, yea or nay), 0.40 (hint of support against), 0.25 (moderate support against), 0.05 (sufficient support against to stand alone), and 0.01 (rather certain support against). Coarse priors less than 0.50 are some support for but also imply more support against a hypothesis, thus all coarse priors less than 0.50 reduce the posteriors. Note that a scientifically intuitive coarse prior of 0.999 (quite certain) increases (via the Bayes' Formula) any software-generated clade support level over 0.50 to 0.999.

Table 8.1 demonstrates that coarse priors for a hypothesis of (evolutionary) monophyly developed from classical taxonomy or morphological cladistics when considered high (0.999, 0.99 or 0.95) in light of all available information will be supported by molecular agreement at any level of clade support 0.55 or higher. Quite certain (0.999) and rather certain (0.99) credible intervals in molecular cladograms are little lowered by coarse priors. For example, only a contrary classical hypothesis supported by 0.99 (that is, implying a prior of support for its contrary hypothesis of 0.01) will refute molecular clade support of 0.999, while only one of at least 0.95 (meaning a coarse prior of 0.05) will refute clade support of 0.99. The 0.999 molecular clade support probabilities are simply examples of notional almost absolute certainty.

In practice, molecular support of 0.999 would be reduced to 0.99 by a standard penalty of one percent for unaccounted assumptions (Zander 2007a). With evolutionary morphologically based relationships, this level of support is common and both cladistically and classically acceptable, see Cohen's 1994 paper "The world is round (p < .05)," such as two foxes are more closely related to each other than either is to a cat (or in the context of expertise a third fox, clearly unrelated by some entirely reasonable criteria). An implied very high morphological coarse prior may be the unstated reason cladistics was long defended by arguments of parsimony, simplicity, and "converging on the truth."

There is a hidden limit to the ability of coarse priors to deal with differences in classical, morphological cladistic, and molecular analyses. It may well be that a coarse prior of 0.99 supports the clade (AB)C,D, as does a molecular posterior probability of 0.99. All this means is that nesting of A and B versus C is very well supported given the data and evolutionary model. But if A is the ancestral taxon of both B and C, e.g., $A > (^1B, {}^2C)$, then the nesting is biased by an inappropriate evolutionary model (pseudoextinction), and the statistical analysis is misapplied.

Bayesian and classical likelihood analyses differ by the former allowing prior distributions to be included in calculation, but the latter does not. Both use the "likelihood principle," that the likelihood function contains all the information from the sample that is relevant for inferential and decision-making purposes

(Winkler 1972: 390). This is somewhat circular as far as phylogenetics in concerned because the information is restricted to phylogenetically informative data and the inferences are about phylogenetics, i.e., sister groups. It is a sufficient statistic for phylogenetics analysis, yet must not be interpreted as analytically sufficient for macroevolutionary analysis. The use of coarse priors from classical and morphological cladistic analysis is to help make a Bayesian Solution a sufficient statistic for macroevolutionary analysis.

Preselection and coarse priors — Molecular analysis is flawed by preselection of exemplars. Since molecular clade support is entirely dependent on preselected exemplars, agreement is expected unless there are biases (self-nesting ladders, pseudoconvergence). If exemplars were selected randomly and an adequate molecular sampling was made (dense sampling) of each taxon, then agreement of classical taxonomy (including morphological cladistics) and molecular analysis would constitute support for whatever evolutionary inferences are discernable. If they disagree, then equal support for two well-supported but contrary results based on dense sampling yields a true equivocal statement unless the molecular analysis is clearly compromised by a probable extinction (or unsampled heterophyly) of multiple molecular lineages of the same taxon.

Unfortunately, molecular exemplars are selected from a cluster of taxa already determined by classical taxonomy and informally superoptimized into a natural key or classification. The results of the sparsely sampled molecular analysis should reflect the natural key, if we assume no or little extinction or non-sampling of supergenerative core taxa. If the molecular study agrees with classical results, this is expected and is not support for the classical tree or vice versa. If it disagrees, and the morphological relationships are re-examined and found good, the statistical primacy of the morphological relationships must be respected and the coarse priors remain in effect.

In two cases the molecular tree is informational. (1) If the coarse priors are low in support for the classical tree or equivocal, then information may be judged from general position on the molecular tree (e.g., wrong family), avoiding biases like self-nesting ladders. (2) When classical analysis cannot determine an ancestral taxon by clear macroevolutionary transformations associated with environment and habitat, then heterophyly is informative.

The use of Bayes' Formula is well-justified in the case of independent data, say, supporting some one result. In cases when molecular analysis involves dense sampling of a taxon, e.g., comparing several large families, the Bayes' Formula with coarse priors should work well to evaluate combined results. In addition, if one believes that the statistical properties of particular molecular data are such that a small sample can represent what a large sample will reveal, then, again, the Bayes' Formula will work with coarse priors. Molecular analysis is, however, most valuable for the ability to infer deep ancestors from heterophyly of taxa represented by exemplars, which is a separate means of inference than superoptimization of a morphological cladogram or natural key.

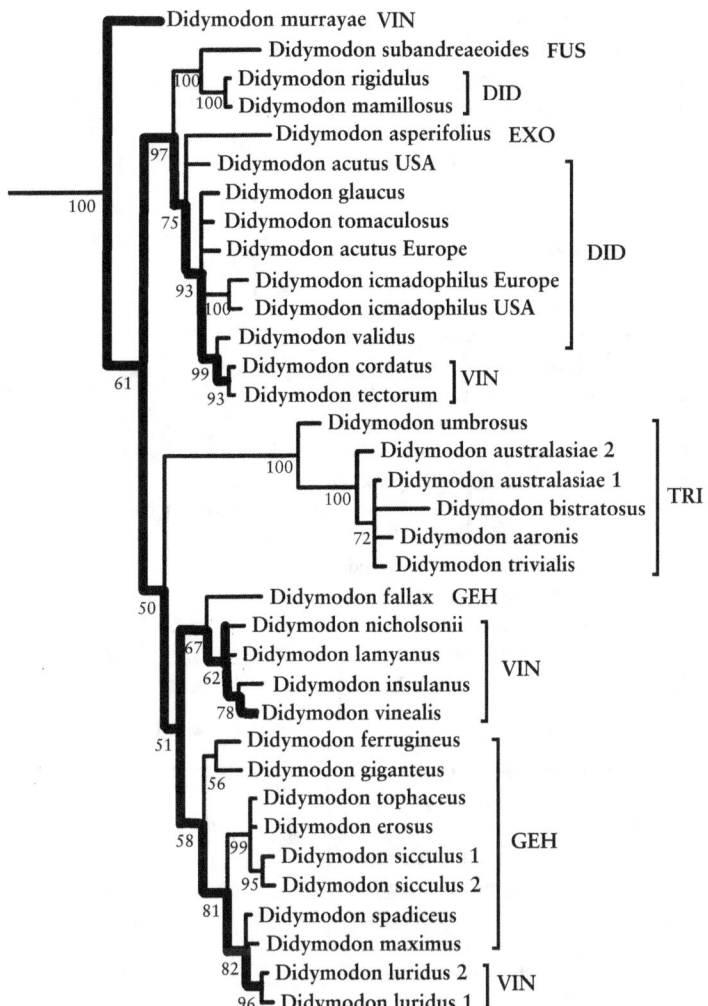

Plate 8.3 — Molecular (1TS) analysis of *Didymodon* (Pottiaceae, Bryophyta) modified from Werner et al. (2005a). Species are grouped according to new segregate genera proposed in this book. Genus *Vinealobryum,* as a deep ancestor, is marked with a bold line. Bayesian posterior probabilities from the Werner et al. (2005a) study are given for clades. *Didymodon* s.lat. is segregated here into six genera: DID = *Didymodon* (s.str.), EXO = *Exobryum,* FUS = *Fuscobryum,* GEH = *Geheebia,* TRI = *Trichostomopsis,* VIN = *Vinealobryum.*

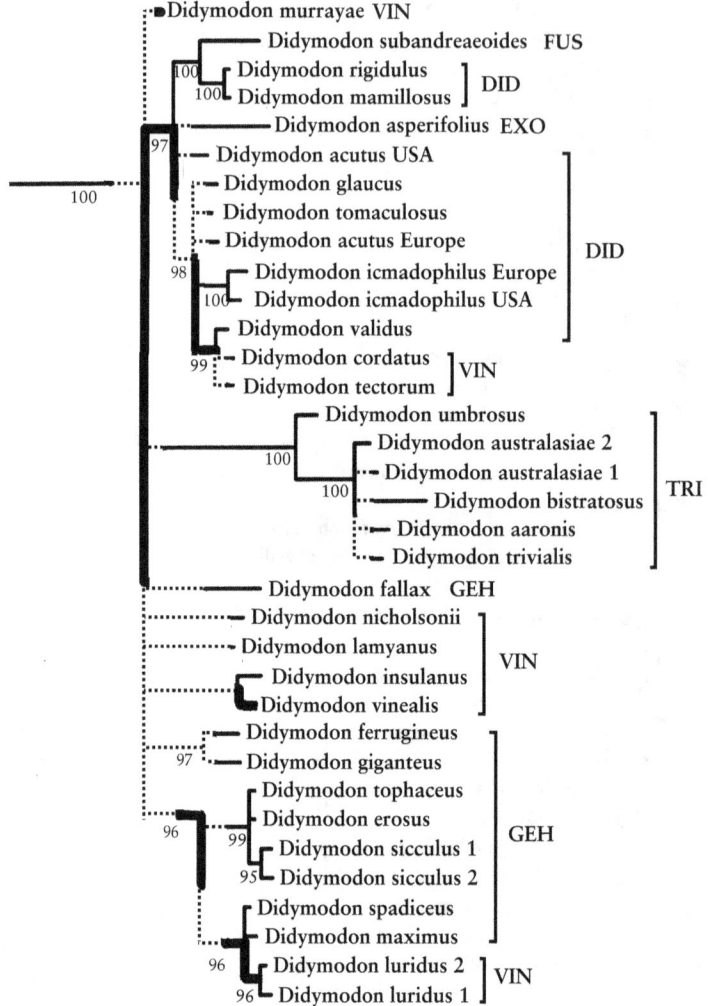

Plate 8.4 — Molecular (1TS) analysis of *Didymodon* (Pottiaceae, Bryophyta) modified from Werner et al. (2005a). Cladogram collapsed to all clades of at least 0.95 posterior probability using the implied reliable credible interval (IRCI) formula. Dotted lines are clades corrected to 0.95 support.

Didymodon *and molecular analysis* — The molecular (ITS) analysis of *Didymodon* (Pottiaceae, Bryophyta) (Werner et al. 2005a) did not show heterophyly at the species level because most species were represented by single exemplars. The sections of the genus are fairly well clustered, expected, of course, because of preselection of taxa.

The Werner et al. (2005a) cladogram of the molecular analysis of *Didymodon* is presented in a modified form in Plate 8.3. Species are grouped according to the new segregate genera proposed in this book. Genus *Vinealobryum* is considered (from superoptimization of morphological cladogram) the ancestor

of the remaining genera, and as a deep ancestor is marked with a bold line. Bayesian posterior probabilities from the Werner et al. (2005a) study are given for clades. Using the newly proposed genera (below), *Vinealobryum* is terminal on all clades but that of *Trichostomopsis*, which may indicate that *V. vinealis* is leaving behind a trail of descendant species. Plate 8.4, on the other hand, indicates that such a conclusion is far too early because of lack of reliability. There are signs of self-nesting ladders but these are represented differently in the maximum parsimony, minimum evolution and Bayesian cladograms (see Werner et al. 2005a) made with the same data. Given that the support values were rather low, more work is needed. Probably necessary for better resolution is a study with many examples of each species and with multiple sequences, and/or more rapidly mutating sequences or proteins, possibly resulting in informative heterophyly.

Plate 8.4 is the same cladogram as 8.3 collapsed to all clades of at least 0.95 posterior probability using the implied reliable credible interval (IRCI) formula. This formula is, again, simply one minus the product of the chances of each of all concatenated arrangements being wrong (where the chance of being wrong is one minus their Bayesian posterior probability), see Zander (2007). This cladogram seems reliable even if much collapsed, but one must remember that branch order of three clades requires that the clades be scrutinized for multiple test problems (Chapter 15).

The functional effect of natural selection on *cis*-acting regulators on single genes have been recently demonstrated in human evolution (Rockman et al. 2005). A silenced gene cluster is thought to degrade over the passage of time, and if so, it may be that the trait complex may degrade in stages. If so, then one might expect to see a central group of taxa with the intact trait cluster, surrounded by a group of taxa with gradually a reduced trait complex.

In fact, we do see this in published cladograms of *Didymodon* and the other genera of Trichostomoideae (Werner et al. 2005b), where taxa or groups of taxa with long, filamentous, twisted peristomes (generalist structures associated as primitive in reduction series) are deeply embedded in the cladogram among related taxa of short, long-triangular, straight peristomes, rudimentary peristomes, or none at all. Thus, it may be that the dissilient or burst genus arrangement of a core species with a halo of reduced forms is the evolutionary group in the Pottiaceae (a spray or a series of multifurcations). In the superoptimized morphological evolutionary tree of *Didymodon* this same assumption is applied, that of a generalist, wide-ranging taxon as ancestral to morphologically and environmentally specialized descendants. This needs confirmation from studies that may detect heterophyly in the same (molecular) tree or cross-tree (molecular and morphological) pair.

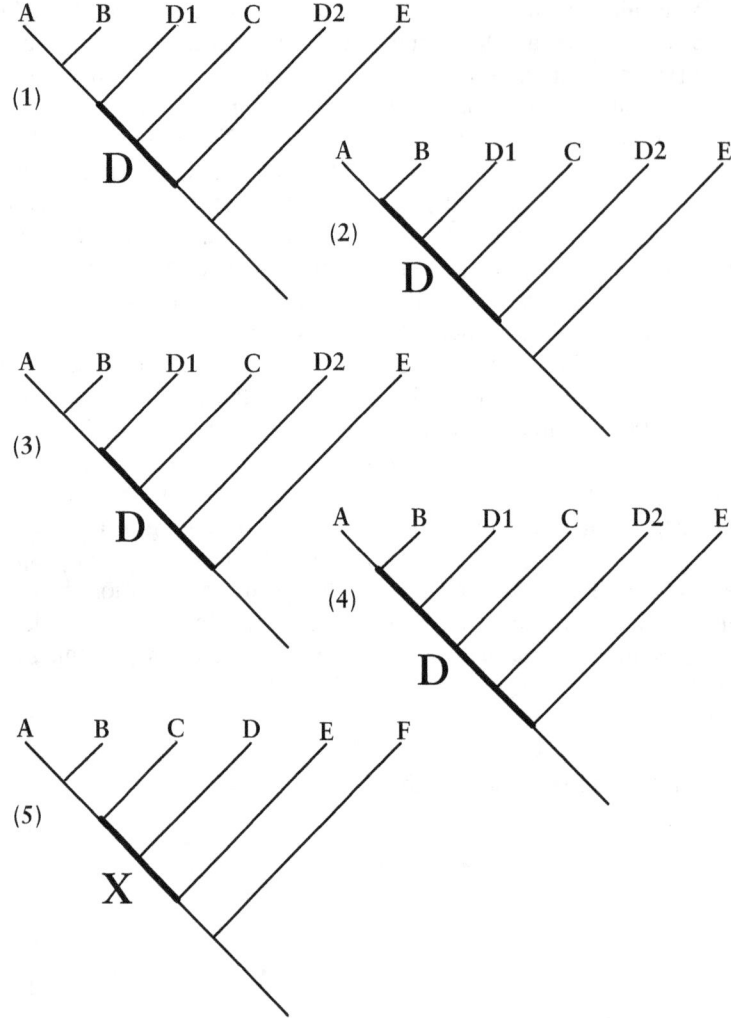

Plate 8.5. — Superoptimization and heterophyly in molecular analysis diagrammed. The bold line indicates a deep ancestral taxon. (1) Typical inference of a deep ancestral taxon "D" inferred from D1 and D2 extant exemplars. (2) Deep ancestral taxon "D" may, however, extend to (AB). (3) Or to E. (4) Or "D" may be direct ancestral taxon of all exemplars. (5) In the absence of heterophyly, there may still be a deep ancestral taxon "X" of unknown identity that may be perhaps inferable on the basis of data other than phylogenetic. Demonstration of two or more nodes in sequence attributable to the same named taxon obviates the fundamental analytic assumption of cladistics—that two of any three lineages or exemplars must be more closely related. Superoptimization with information about geographic distribution, apparent adaptive specializations, and other traits can supplement heterophyletic inferences of taxic macroevolutionary transformations and aid adoption of a new fundamental analytic assumption—macroevolutionary transformations at any taxon level. Reconciliation with a superoptimized *morphological* cladogram is possible through postulations of self-nesting ladders in one or the other of the cladograms.

Genera based on a dissilient genus concept — The presence in this cladogram of four clear major ancestral species, *D. fallax, D. nigrescens, D. vinealis* and *D. rigidulus,* each with a cloud of derived species gives opportunity to recognize genera based, not on holophyletic clades, but on caulistic clustering. Thus, theory is put into practice. These findings match well the more recent classical classifications at the infrageneric level. Recognition at the genus level salutes the clear importance of these groups as distinctive linear (as opposed to nested) macroevolutionary transformations. *Didymodon asperifolius,* on the other hand, is here recognized at the genus level by its unusual morphology. An evolutionary tree summarizing the position of these segregate genera in the family Pottiaceae is given in Plate 8.2. I have previously suggested the splitting of *Didymodon* along these lines (Zander 1993: 158). The combinations given here are largely those for the North American representation of these genera, with which I am most familiar. Given the large size of *Didymodon* s.lat., more than 120 worldwide, additional combinations must be considered by other specialists if this rearrangement is accepted.

—

DIDYMODON

***Didymodon* Hedw.** — The generitype of *Didymodon* sensu stricto is *D. rigidulus* Hedw. *Didymodon anserinocapitatus* and *D. johansenii* (in Plate 8.1) need no name changes.

—

TRICHOSTOMOPSIS

***Trichostomopsis* recognized (again)** — Of the evolutionary tree species, *Didymodon australasiae* is the type of the genus *Trichostomopsis*. Both this species and *D. umbrosus* are already in combination in *Trichostomopsis* at the species level.

Trichostomopsis Card., Rev. Bryol. 36: 73. 1909. Type: *Trichostomopsis crispifolia* Card.
  Asteriscium (Müll. Hal.) Hilp., Beih. Bot. Centralbl. 50(2): 618. 1933, *hom. illeg. non* Cham. & Schlecht., 1826.
  Barbula sect. *Asteriscium* Müll. Hal., Linnaea 42: 342. 1879. Type: *Barbula umbrosa* Müll. Hal.
  Didymodon sect. *Asteriscium* (Müll. Hal.) R. H. Zander, Cryptogamie, Bryol. Lichénol. 2: 383. 1981 [1982]. Type: *Barbula umbrosa* Müll. Hal.
  Didymodon sect. *Craspedophyllon* Card., Rev. Bryol. 36: 81. 1909.
  Husnotiella Card., Rev. Bryol. 36: 71. 1909. Type: *Husnotiella revoluta* Card.
  Kingiobryum H. Rob., Bryologist 70: 9. 1967. Type: *Kingiobryum paramicola* H. Rob.

Trichostomopsis revoluta (Card.) R. H. Zander, comb. nov.
  Husnotiella revoluta Card., Rev. Bryol. 36: 71. 1909, basionym.
Trichostomopsis angustifolia (Warnst.) R. H. Zander, comb. nov.
  Didymodon angustifolius Warnst., Beih. Bot. Centralbl. 16: 289. 1904, basionym. (Homotypic synonym: *Didymodon bartramii* R. H. Zander)
Trichostomopsis bistratosa (Hebr. & R. B. Pierrot) R. H. Zander, comb. nov.
  Didymodon bistratosus Hebr. & R. B. Pierrot, Nova Hedw. 59: 354. 1994, basionym.
Trichostomopsis challaensis (Broth.) R. H. Zander, comb. nov.
  Trichostomum challaense Broth., Biblioth. Bot. 87: 30. 1916.
Trichostomopsis marginatum (H. Rob.) R. H. Zander, comb. nov.
  Trichostomum marginatum H. Rob., Phytologia 21: 389. 1971, basionym.
Trichostomopsis paramicola (H. Rob.) R. H. Zander, comb. nov.
  Basionym: *Kingiobryum paramicola* H. Rob., Bryologist 70: 9. 1967, basionym.

In the moss family Pottiaceae, the genus *Trichostomopsis* Card., of 10 species, has been lumped into *Didymodon* for several years following Zander (1993). Further study (Jiménez et al. 2005) indicated that *Trichostomopsis* is indeed distinct through advanced traits of bistratose laminal margins, lack of adaxial stereid band, flattened costa, and tendency to the unique trait of transversely slashed or perforated basal cells of the leaf. This is due in part to the discovery (Werner et al. 2005a, 2005b) that the robust, Andean taxon *Kingiobryum paramicola,* previously placed in Dicranaceae, was in fact deeply embedded molecularly in the genus *Didymodon* of the Pottiaceae among two species of similar morphology. Given that this distinctive species adds character by adding to the number of species in which characteristically stabile, conservative traits occur, *Trichostomopsis* should be recognized as macroevolutionarily generated from *Didymodon* (segregate *Vinealobryum*). Molecular phylogenetic analysis of this relationship (Werner et al. 2005b) sank the species into *Didymodon* through strict phylogenetic monophyly because it was deeply nested in that genus, but, if recognized macroevolutionarily at the genus level to include relatives, then the earliest name is *Trichostomopsis*. In that *Trichostomopsis* is characterized by a highly reduced adaxial stereid band in the leaf costa, and *T. paramicola* has such unreduced, then one might assume the latter is primitive in the clade though appearing terminal in the Werner (2005b) cladogram. A cladistic position for *Didymodon bistratosus* was not given in the original morphological cladogram, but its evolutionary position with *Trichostomopsis* was clearly evaluated by Zander et al. (2005).

Trichostomopsis paramicola has been demonstrated (Werner et al. 2004b) to be molecularly close to *Trichostomopsis australasiae* (Hook. & Grev.) R. H. Zander, which itself is a very close relative (intergrading in parts of the range) with *T. umbrosa* (Müll. Hal.) R. H. Zander. Given that *T. australasiae* is widespread in the world and has several closely related but somewhat reduced apparent derived species in Asia and elsewhere, and has a variably developed peristome, it can also be postulated as progenitor. *Trichostomopsis paramicola*, which lacks a peristome, does have the unique transversely split basal cells of *T. umbrosa*, a species that grades into *T. australasiae*. The gametophyte is robust, but although many primitive taxa of the Pottiaceae occur in South America and are robust, size alone is not primitive. The complete loss of the peristome in *T. paramicola* is perhaps associated with long isolation and different rates of evolution of gametophyte and sporophyte. Several genera of the Pottiaceae have distinctive gametophytes but variably reduced sporophytes (Zander 1993), implying that modification of the sporophyte is generally more rapid, possibly through the greater selection pressure of being structurally more exposed to environmental changes. Thus, *T. australasiae* has more claim to being progenitor. It is possible that molecular heterophyly in a densely sampled study might clarify this.

Trichostomopsis umbrosa is clearly derived with unique traits in very linear leaf shape, and much modified leaf base, and is now worldwide as a human-distributed weed in cities and botanical gardens. A superoptimized evolutionary tree may be postulated as: (*T. umbrosa* < (*T. paramicola* < **T. australasiae**)) < **Vinealobryum**. A Besseyan cactus evolutionary tree showing the derivation of *Trichostomopsis* from *Vinealobryum n* in the context of the family Pottiaceae is given in Plate 8.8. Superoptimized parenthetical representations of caulograms as above can be made by describing nesting in the usual way with the Newick formula, then adding greater than or less than signs as arrows to show direction of generation of daughter species. Critical ancestors may be boldfaced.

—

GEHEEBIA

Geheebia, a genus for *Didymodon* sect. *Fallaces* — The genus *Geheebia* has been used in the past only for the species *Geheebia gigantea* (Funck) Boulay (synonym *G. cataractarum*), a physically relatively large taxon of *Didymodon* sect. *Fallaces* with distinctive trigones (knots, or collenchymatous thickenings) in the corners of the laminal cell areolation. Although distinctive, the trigones appear weakly in other robust species of the section and otherwise this one species cannot be set apart at the genus level.

Geheebia Schimp., Syn. ed. 2: 233. 1876. Type: *Geheebia cataractarum* Schimp.
 Barbula sect. *Fallaces* (De Not.) Steere in Grout, Moss Fl. N. Amer. 1: 174. 1938.
 Barbula subsect. *Fallaciformes* Kindb., Eur. N. Amer. Bryin. 2: 246. 1897. Type: *Barbula fallax* Hedw.
 Barbula subg. *Geheebia* (Schimp.) Szafr., Fl. Polska Mchy 1: 213. 1957 [1958].
 Barbula sect. *Graciles* Milde, Bryol. Siles. 117. 1869. Lectotype: *Barbula rigidicaulis* C. Müll. fide Saito, J. Hattori Bot. Lab. 39: 601. 1975.
 Barbula sect. *Pseudodidymodon* Kindb., Eur. N. Amer. Bryin. 2: 246. 1897, *nom. illeg.*

incl. sect. prior.

Barbula sect. *Reflexae* Mönk., Laubm. Eur. 280. 1927, *nom. illeg. incl. sect. prior.*

Barbula subsect. *Reflexae* (Mönk.) Chen, Hedwigia 80: 203. 1941, *nom. illeg. incl. sect. prior.*

Didymdon sect. *Fallaces* (De Not.) R. H. Zander, Phytologia 44: 209. 1979. Type: *Barbula fallax* Hedw.

Didymodon sect. *Graciles* (Milde) Saito, J. Hattori Bot. Lab. 39: 501. 1975, see Zander, Phytologia 41: 24. 1978.

Dactylhymenium Card., Rev. Bryol. 36: 72. 1909. Type: *Dactylhymenium pringlei* Card.

Limneria Stirt., Trans. Bot. Soc. Edinburgh 26: 428. 1915. Type: *Limneria viridula* Stirt.

Prionidium Hilp., Beih. Bot. Centralbl. 50(2): 640. 1933. Type: *Prionidium setschwanicum* (Broth.) Hilp.

Tortula sect. *Fallaces* De Not., Mem. Roy. Acc. Sci. Torino 40: 287. 1838. Type: *Tortula fallax* (Hedw.) Turn.

Trichostomum subg. *Zygotrichodon* Schimp., Syn. ed. 2: 169. 1876. Type: *Trichostomum tophaceum* Brid.

Geheebia fallax (Hedw.) R. H. Zander, comb. nov.
  Barbula fallax Hedw., Sp. Musc. Frond. 120. 1801, basionym.
Geheebia ferruginea (Schimp. ex Besch.) R. H. Zander, comb. nov.
  Barbula ferruginea Schimp. ex Besch., Mém. Soc. Sci. Nat. Math. Cherbourg 16: 181. 1872, basionym.
Geheebia laevigata (Mitt.) R. H. Zander, comb. nov.
  Tortula laevigata Mitt., J. Linn. Soc., Bot. 12: 160. 1869, basionym.
Geheebia maxima (Syed & Crundw.) R. H. Zander, comb. nov.
  Barbula maxima Syed & Crundw., J. Bryology 7: 527. 1973 [1974], basionym.
Geheebia maschalogena (Ren. & Card.) R. H. Zander, comb. nov.
  Barbula maschalogena Ren. & Card., Bull. Soc. Roy. Bot. Belgique 41(1): 53. 1905, basionym.
Geheebia tophacea (Brid.) R. H. Zander, comb. nov.
  Trichostomum tophaceum Brid., Muscol. Recent. Suppl. 4: 84. 1819 [1818], basionym.
Geheebia leskeoides (K. Saito) R. H. Zander, comb. nov.
  Didymodon leskeoides K. Saito, J. Hattori Bot. Lab. 39: 508. 1975, basionym.
Geheebia spadicea (Mitt.) R. H. Zander, comb. nov.
  Tortula spadicea Mitt., J. Bot. 5: 316. 1867, basionym.
Geheebia waymouthii (R. Br. bis) R. H. Zander, comb. nov.
  Weissia waymouthii R. Br. bis, Trans. & Proc. New Zealand Inst. 31: 439. 1899, basionym.

—

EXOBRYUM

***Exobryum*, a genus for *Didymodon asperifolius*. —** This is a genus evolutionarily midway between *Vinealobryum* and *Geheebia*, but with significant autapomorphic traits of its own. *Exobryum* is not recognized on account of an extant core and radiation structure, but by an inference that it is a fairly specialized remnant of a mostly extinct core and radiation group, with distinctive conservative traits. *Didymodon asperifolius* can sometimes be immediately recognized by a red-yellow translucency in leaves of dry plants, like oiled paper. The adaxial surface of the costa may have either quadrate or short-rectangular cells. The distal laminal cells are also rather large compared to other taxa in *Didymodon* s. lat. The KOH reaction, as well as the natural color of the lamina are sometimes light orange, but usually quite red. Some plants may appear green but the laminal cell walls are red under high magnification.

Exobryum R. H. Zander, gen. nov. Type: *Barbula asperifolia* Mitt.

Plantae aurantico-virides, rubro- v. flavo-brunneae; folia quum sicca translucentia; caules filo centrali carentes vel filo debili, hyalodermide nulla. Folia triangularia vel ovato-lanceolata, infragilia, quum madida valde reflexa, apice anguste vel late acuta, carinata atque per superficiem adaxialem costalem anguste canaliculata. Costa cellulis superficialibus adaxialibus in dimidio distali folii quadratis vel breviter rectangularibus, ad apicem summam folii cellulis adaxialibus costalibus quadratis praesentibus, sulco brevi naviculato fenestrelliformi carens; cellulae ducum uniseriatae; stratum stereidarum adaxiale plerumque praesens e stereidis paucis parvis compositum. Cellulae distales laminales unistratosae, parietibus vulgo flavo-virentibus vel rubrescentibus, maxime incrassatae, papillis saepe nullis, interdum simplicibus, una supra quidque lumen praeditae. Peristomium erectum, breve. Plantae in KOH plerumque ex rubro rubro-aurantiacae reagentes.

Plants orange-green, red- or yellow-brown, leaves translucent when dry. **Stems** lacking central strand or strand weak, hyalodermis absent; axillary hairs of 4–5 cells, basal 1 brown. **Leaves** 1.2--2.5 mm long, triangular to ovate-lanceolate, intact, strongly reflexed when moist, apex narrowly to broadly acute, keeled and narrowly channeled along the adaxial surface of the costa, margins entire, broadly short-decurrent, revolute in lower 1/2 or to near apex, often apiculate by a conical cell. **Costa** usually tapering to near apex, ending 1--4 cells before apex or percurrent, 4--6 cells across adaxially at mid costa, adaxial superficial cells of the costa quadrate to short-rectangular in the distal half of the leaf, quadrate adaxial costal cells present at the extreme leaf apex lacking a short, boat-shaped window-like groove bottomed by epapillose elongate cells; costal guide cells in one layer, adaxial stereid band usually present, of a few small stereid cells. **Distal laminal cells** unistratose, 13--15 \mu wide, walls commonly yellowish green to reddish in nature at high magnification, very much thickened, papillae usually absent, occasionally simple, 1 over each lumen. **Peristome** erect, short. **KOH color reaction** usually brick-red to red-orange.

Exobryum asperifolium (Mitt.) R. H. Zander, comb. nov.
  Barbula asperifolia Mitt., J. Proc. Linn. Soc., Bot., Suppl. 1: 34. 1869, basionym.

Exobryum asperifolium has leaves adaxially with a narrow medial channel about the width of the costa at least at leaf apex, apex often apiculate by one or more conical cells, costa usually percurrent, margins usually recurved, often to near the apex, laminal color reaction to KOH usually brick-red, occasionally orange. Leaves are strongly reflexed and keeled when moist, papillae when present simple, stem central strand usually absent. Specialized asexual reproduction is absent. Peristome teeth are erect, not long and twisted. This species is widespread northern moist mountainous areas on calcareous or acid rock, moist calcareous soil, peatland, streamside, generally in alpine areas at moderate to high elevations (500–3700 m). It is known for Greenland; Canada in Alta., B.C., Nfld. and Labr. (Labr.), N.W.T., Nunavut, Yukon; U.S.A. in Alaska, Colo.; also northern Eurasia.

—

VINEALOBRYUM

Vinealobryum, a genus for *Didymodon* sect. *Vineales* — *Didymodon vinealis* and related species have always been distinctive by the usually strongly multipapillose leaf cells. The twisted peristome is has been classically uncomfortable in *Didymodon,* which is often considered to have only the short peristome of the generitype, *D. rigidulus.* Many "core" species that probably generative species in *Didymodon* have been placed in *Barbula,* a genus with more commonly elongate and twisted peristomes, but correctly relegated to *Didymodon* s. lat. by the hyaline axillary hairs.

Vinealobryum R. H. Zander, gen. nov. Type: *Barbula vinealis* Brid. Synonym: *Barbula* sect.

Vineales Steere in Grout, Moss Fl. N. Amer. 1: 174. 1938.
> *Barbula* sect. *Rubiginosae* Steere in Grout, Moss Fl. N. Amer. 1: 174. 1938. Type: *Barbula rubiginosa* Mitt.
> *Barbula* sect. *Vineales* Steere in Grout, Moss Fl. N. Amer. 1: 174. 1938.
> *Barbula* subsect. *Vinealiformes* Kindb., Eur. N. Amer. Bryin. 2: 246. 1897. Type: *Barbula vinealis* Brid.
> *Didymodon* sect. *Vineales* (Steere) R. H. Zander, Phytologia 41: 24. 1978. Lectotype: *Didymodon vinealis* (Brid.) R. H. Zander.

Plantae brunneae vel rubro-brunneae. Folia ex breviter ovato longe lanceolata, patentia vel late patentia atque interdum quum madida recurva, apice late acuta vel longe acuminata, folium transversum concava vel carinata atque secus superficiem adaxialem costalem anguste caniculata, marginibus integris vel late crenulatis, leniter in parte proximali recurvis vel recurvis vel revolutis usque ad prope apicem, saepe per cellulam conicam apiculata. Costa plerumque usque prope apicem aeque crassa vel interdum medialiter multo dilatata, percurrens vel in mucronem latum breviter excurrens, cellulis adaxialibus superficialibus costalis in dimidio folii distali quadratis, ad apicem summam folii cellulis adaxialibus costalibus quadratis nullis itaque sulcum brevem, navicularem fenestrelliformem secus fundum sulci cellulis epapillosis, elongatis tectum formans; cellulae ducum costales saepe 2 (-3)-stratosae; stratum stereidarum adaxiale saepe nullum (plerumque pro hoc substereidas substituens). Cellulae distales laminales interdum secus margines folii bistratosae, parietibus vulgo flavido-viridibus vel rubrescentibus, aeque incrassatae, epapillosae vel papillis simplicibus vel irregularibus vel saepius spiculoso-multipicibus. Reproductio asexualis propria ut pote gemmae axillares, interdum apicibus folii fragilibus. Peristomium nullum vel rudimentarium vel bene evolutum atque usque 2.5-plo torquens. Plantae in KOH plerumque ex rubro rubro-aurantiacae reagentes.

Plants brown or red-brown. **Stems** with central strand, hyalodermis absent; axillary hairs of 4–5 cells, basal 1 brown. **Leaves** short-ovate to long-lanceolate, intact, spreading to widely spreading and occasionally recurved when moist, apex broadly acute to long-acuminate, concave across the leaf to keeled and narrowly channeled along the adaxial surface of the costa, margins entire or broadly crenulate, weakly recurved proximally to recurved or revolute to near the apex, often apiculate by a conical cell. **Costa** usually evenly thick to near apex or occasionally much widened medially, percurrent to short-excurrent in a broad mucro, adaxial superficial cells of the costa quadrate in the distal half of the leaf, quadrate adaxial costal cells absent at the extreme leaf apex resulting in a short, boat-shaped window-like groove bottomed by epapillose elongate cells; costal guide cells often in 2(–3) layers, adaxial stereid band often absent (usually replaced by substereid cells). **Distal laminal cells** occasionally bistratose along leaf margins; walls commonly yellowish green to reddish in nature at high magnification, evenly thickened, epapillose to papillae simple or irregular to more often spiculose-multiplex. **Specialized asexual reproduction** as axillary gemmae, very occasional, leaf apices sometimes fragile. **Peristome** absent or rudimentary to well developed and twisted up to 2.5 times. **KOH color reaction** usually red to red-orange.

High magnification might be needed to ascertain the exact hue of the internal distal laminal cell walls. A "marker" character, not always present but unique to *Vinealobryum*, is the absence of the quadrate adaxial costal cells at the extreme leaf apex. This provides an elliptical window (groove or colpos) revealing non-papillose elongate cells.

Vinealobryum brachyphyllum (Sull.) R. H. Zander, comb. nov.
> *Barbula brachyphylla* Sull., Expl. Railroad Mississippi Pacific, Descr. Moss. Liverw. 4: 186. 1856, basionym.

Vinealobryum cordatum (Jur.) R. H. Zander, comb. nov.
> *Didymodon cordatus* Jur., Bot. Zeitung (Berlin) 24: 177. 1866, basionym.

Vinealobryum herzogii (R. H. Zander) R. H. Zander, comb. nov.

Didymodon herzogii R. H. Zander, Bull. Buffalo Soc. Nat. Sci. 32: 162. 1993, basionym.
Vinealobryum eckeliae (R. H. Zander) R. H. Zander, comb. nov.
    Didymodon eckeliae R. H. Zander, Madroño 48: 298. 2002, basionym.
Vinealobryum insulanum (De Not.) R. H. Zander, comb. nov.
    Tortula insulanus De Not., Mem. Reale Accad. Sci. Torino 40: 320. 1838, basionym.
Vinealobryum luehmannii (Broth. & Geh.) R. H. Zander, comb. nov.
    Barbula luehmannii Broth. & Geh., Oefvers. Förh. Finska Vetensk.-Soc. 37: 158. 1895, basionym.
Vinealobryum luridum (Hornsch.) R. H. Zander, comb. nov.
    Didymodon luridus Hornsch., Syst. Veg. 4(1): 173. 1827, basionym.
Vinealobryum murrayae (Otnyukova) R. H. Zander, comb. nov.
    Didymodon murrayae Otnyukova, Arctoa 11: 345. 2002, basionym.
Vinealobryum nicholsonii (Culm.) R. H. Zander, comb. nov.
    Didymodon nicholsonii Culm., Rev. Bryol. 34: 100. 1907, basionym.
Vinealobryum nevadense (R. H. Zander) R. H. Zander, comb. nov.
    Didymodon nevadensis R. H. Zander, Bryologist 98: 590. 1995, basionym.
Vinealobryum tectorum (Müll. Hal.) R. H. Zander, comb. nov.
    Barbula tectorum Müll. Hal., Nuovo Giorn. Bot. Ital., n.s. 3: 101. 1896, basionym.
Vinealobryum vineale (Brid.) R. H. Zander, comb. nov.
    Barbula vinealis Brid., Bryol. Univ. 1: 830. 1827, basionym.
Vinealobryum vineale var. rubiginosum (Mitt.) R. H. Zander, comb. nov.
    Barbula rubiginosa Mitt., J. Linn. Soc., Bot. 8: 27. 1865, basionym.

—

FUSCOBRYUM

Fuscobryum, a genus for *Didymodon nigrescens* and derived species — Although these are largely species of North Temperate or subarctic areas, one of these, *Fuscobryum nigrescens,* is generalist and widespread, while the others are specialized and fairly local, comprising a typical dissilient genus.

Fuscobryum R. H. Zander, gen. nov.
Type: *Barbula nigrescens* Mitt.

Plantae plerumque rubrae vel rubro- vel atro-brunneae. Folia quum sicca appressa, patentia, quum madida excarinata, monomorpha vel dimorpha, ex ovato lanceolata, folium transversum adaxialiter late concava, base quoad formam leniter distincta, marginibus plerumque late recurvis vel usque ad medium folii revolutis vel usque prope apicem, minute crenulatis, ad apicem acuta vel anguste acuminata, saepe leniter cucullata. Costa percurrens vel in 2-4 cellulas sub apice evanida, leniter attenuata, non valde calcarata, pulvillo adaxiali e cellulis composito carens, cellulis adaxialibus costalibus rectangularibus, cellulis distalibus laminalibus in seriebus dispositis, papillis ut videtur nullis sed in sectione transversali ut pote lentibus humilibus, complanatis vel multiplicibus capitulatis, per lumen 1-3, luminibus ovatis, parietibus aeque incrassatis atque leniter convexis perceptibilibus, in utrinsecus laminae leniter convexis, unistratosis. Reproductio asexualis propria interdum praesens ut pote in axilla folii gemmae unicellulares. Dentes peristomii 32, lineares, recti vel 1.5-plo torquentes 100 - 600 um. Plantae in KOH rubrae reagentes.

Plants usually red- to black-brown, occasionally brick-red, or at apex yellow- or orange-brown. **Stems** with central strand absent or present. **Leaves** appressed when dry, spreading and not keeled when moist, monomorphic or dimorphic, ovate to lanceolate, broadly concave adaxially across leaf, base weakly differentiated in shape, margins usually broadly recurved to revolute to mid leaf or to near apex, minutely crenulate, apex acute to narrowly acuminate, often weakly cucullate. **Costa** percurrent or ending 2–4 cells below the apex, little tapering, not strongly spurred, without an adaxial pad of cells, adaxial costal cells rectangular, 2 cells wide at mid leaf grading to 4 below, guide cells in 1 layer; basal laminal cells differentiated medially, walls thick, rectangular, not perforated. **Distal laminal cells** in rows; papillae apparently absent but visible in section as low, flattened to multiplex capitulate lenses, 1–3 per lumen, lumens ovate, walls evenly thickened and weakly convex on both sides of lamina, 1-stratose. **Specialized asexual reproduction** sometimes present, as unicellular gemmae in leaf axils. **Seta** elongate. **Capsule** with peristome teeth 32, linear, straight to twisted 1.5 times, 100–600 μm. **KOH color reaction** red.

Fuscobryum species occur on limestone or limy bluffs, commonly near waterfalls, in northwestern North America and eastern Asia, including the Himalayas, with outliers in mountains of Central America. The crenulate distal laminal margins are characteristic, and with the blackened coloration of the plant, diagnostic.

Fuscobryum nigrescens (Mitt.) R. H. Zander, comb. nov.
    Barbula nigrescens Mitt., J. Proc. Linn. Soc., Bot., Suppl. 1: 36. 1859, basionym.
Fuscobryum norrisii (R. H. Zander) R. H. Zander, comb. nov.
    Didymodon norrisii R. H. Zander, Bryologist 102: 112. 1999, basionym.
Fuscobryum perobtusum (Broth.) R. H. Zander, comb. nov.
    Barbula perobtusa Broth., Hedwigia 80: 194. 1941, basionym.
Fuscobryum subandreaeoides (Kindb.) R. H. Zander, comb. nov.
    Barbula subandreaeoides Kindb., Rev. Bryol. 32: 36. 1905, basionym.

—

Convergence — In Plate 8.6, the morphological cladogram of the genus *Didymodon* was superimposed on a principal components analysis (see Zander 1988 for details), which demonstrated convergence of traits. The species are coded with the first two letters that distinguish the epithets. The species with no autapomorphies (therefore are potential surviving ancestral taxa) are shown with no branch, simply as a sisterless node (e.g., *D. australasiae, D. fallax, D. ferruginascens, D. rigidulus, D. johansenii,* and *D. vinealis.* Only *D. johansenii* is not supported as a surviving ancestral taxon by superoptimization (above). Although longer cladograms may have different dispositions of terminal taxa, the fact that the data set for the cladogram groups are derived from classical study (i.e., match sections of the genus *Didymodon*), lend credence to this particular cladogram.

The groups of taxa in Plate 8.6 show some degree of convergence between them. Using the newly segregate genus names, convergence is apparent between *Didymodon anserinocaptatus* and *Vinealobryum sinuosum* (correctly *V. murrayae*); between *Vinealobryum brachyphyllum* and *Fuscobryum nigrescens*; and between *Vinealobryum nevadense* and *Trichostomopsis revoluta.* Exactly why convergence is apparent between the groups needs investigation. There is a clear association with *Geheebia tophacea* and *Exobryum asperifolium* although each is at the base of different clades.

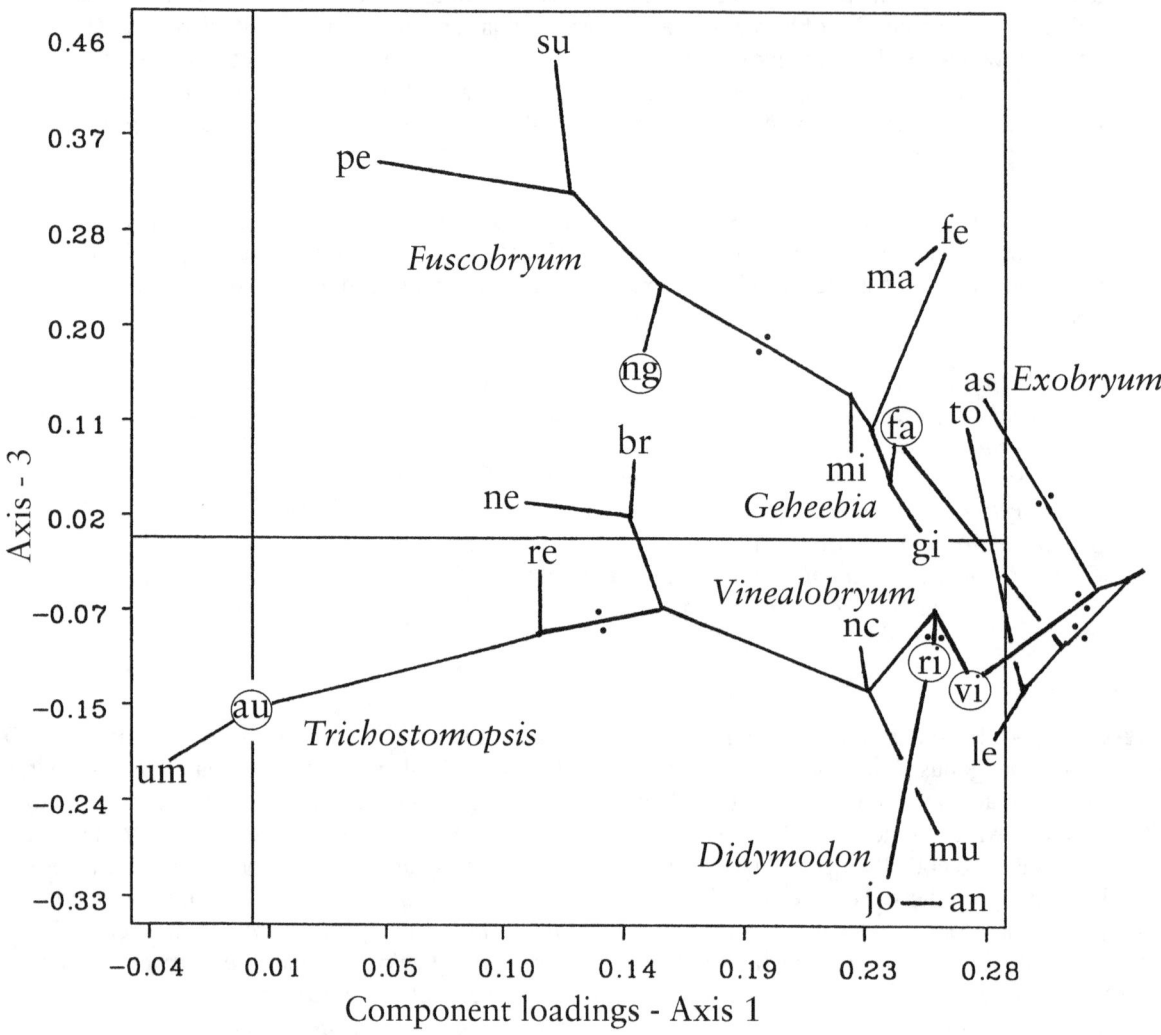

Plate 8.6. — Cladogram of *Didymodon* superimposed on a principle component analysis (Zander 1998). Species epithets are the first two letters that distinguish them. Convergence may be seen between species from all segregate genera recognized here. The five taxa in circles are inferred as intergenerically primitive and generative. *Exobryum asperifolium* may be a remnant of a core-and-radiation group ancestral to *Geheebia*. Four taxa that lack sister groups in this cladogram are those without autapomorphies, and three of these four prove to be also intergenerically primitive (circled). Thus taxa without autapomorphies, at least in this cladogram, well predict inferences of being intergenerically primitive (as deep ancestors) based on nonphylogenetic information. Double dots mark off segregate genera recognized here.

Physical principles and adaptive traits — There is a large, argumentative literature on recognition of adaptational and neutral traits, and here I add discussion of heuristics for distinguishing such. Consider the idea that conservative traits are only conservative (are relatively neutral) in a particular organismal Bauplan such that the trait may appear in multiple selective regimes yet must be tolerable as a burden by the organism. A newly evolved wing cannot conservative in an earthworm as it overburdens the organism. Adaptive traits can be distinguished from conservative traits by a demonstrated or at least correlative functionality limited to a particular selective regime. This is sometimes obvious but often totally opaque. There are many publications (e.g., Bock & von Wahlert 1965; Bonner 1974; Lauder 1981; Nicklas 1992; Thomason 1995) that discuss morphological form and function and attempt to infer function from form in apparently unambiguous cases. There are also modern attempts to use evo-devo (e.g., quantitative trait analysis) to approach some understanding. This includes analysis of epigenetic influences. Also problematic is the possibility of pleiotropy, where a particular morphological structure is linked to a hidden physiological trait, and is not adaptive but is a burden tolerated because of the value of the physiological trait. Likewise, there are cases in which an organism occurs in a particular environment with what seem obvious morphological adaptations to the environment, but occurring the in the same environment is another organism lacking such an adaptation but unfazed. How does the practicing taxonomist avoid speculation when trying to base taxonomy on conservative traits and avoiding adaptive or plastic traits, except to distinguish very close relatives? To determine if a particular structure is adaptive in the manner it immediately suggests, first, one can judge if there are alternative explanations and if so then the sum of those alternative probabilities should not be larger than that of the obvious function, and preferably the obvious function should be massively plausible. A dubious hypothesis may have no rejective alternative, and may prove at least a tentative hypothesis. Second, it would be valuable to demonstrate a causal connection between the form and the function ("look, it uses the wings to fly!"). Often this is also limited to speculation.

Third, a heuristic may be used. For instance, Hugh Iltis' dictum that if a taxon occurs on an oceanic island never connected to the mainland, then that taxon cannot be used in vicariance biogeography analysis because it is demonstrably capable of long-distance dispersal. A similar heuristic is that if one organism has an apparent adaptation to an environment and a related organism with presumably nearly identical physiology lacks such adaption but occurs in the same selective regime, then this negates the form-function inference. An example is the moss *Syntrichia caninervis,* which occurs in harsh, arid environments, and the plants dry out completely during the day. The moss occurs as a small mat of densely crowded stems. In the middle of the mat plants have leaves with long hyaline hair points. But small plants on the circumference of the mat have small leaves and lack the hair points. Thus one cannot simply aver that hair points on leaves of this moss ensure that during the morning and evening dewy times of low insolation, light is guided to leaves, and moisture is wicked to the photosynthetic parts of the

leaves to maximize the photosynthate produced during the short periods of the day of optimum available moisture and light. The adaptation is too simply explained. Apparently crowding is involved in some way, what way is yet to be determined but possibly the mat is a single entity with moisture wicked from the middle of the mat laterally to the juvenile plants around the circumference. That such is important is demonstrated by the desire of some taxonomists to name the juvenile plants as a different species (in my experience), based on only partial collections; mass gatherings demonstrate the physical and taxonomic integrity of the moss mat. The challenge of evo-devo and biophysical experimentation is usually not something the practicing taxonomist wants to do without technical-team aid.

Sometimes recourse to physical principles lends leverage to decisions about adaptation. In a discussion of the geometry of soap bubbles, Peterson (1988: 61) reviewed the area minimizing principle of "a physical system's tendency to seek a minimum surface energy at an interface," for instance, a crystal's "unique equilibrium shape is the one that has the least total surface energy for an enclosed volume." That is, the least energy to enclose a given volume, a kind of non-sphere equivalent of a sphere. Peterson compared the geometry of soap froth (Plate 8.7f) with that of metal crystal grain boundaries. Soap bubbles can meet superficially in three surfaces along the other surface of a bubble and generate angles of 120°, or internally in a bubble mass where six surfaces meet at a vertex of a tetrahedron at about 109°. Stewart (2011: 49, 141, 194) pointed out that the Fibonacci series also results in a minimum energy configuration, as in the double spirals of sunflower seed heads.

Biologically, this translates into the fact that soap bubble cell geometry minimizes photosynthate needed for a given plant structure by minimizing that structure, whether the organs (cells) are squashed spheres or elongate. Plate 8.7 demonstrates a transformation series in the transverse section of a stem from the most morphologically complex and most primitive member of the moss family Pottiaceae, here exemplified by the species *Timmiella anomala*, which has several internal stem features overlain on a soap bubble geometry (figure a), through *Chionoloma latifolium* largely lacking the sclerodermis of supporting tissue (figure b), *Barbula costesii* lacking the hyalodermis of what appears to be superficial water transport tissue (figure c), *Tortula leucostoma* lacking all but the central strand of (apparent) internal water transport tissue (figure d), through to

Aloina bifrons with a stem section of only the basic soap bubble cell geometry. The last is compared with soap froth (figure f). Because the soap bubble cell anatomy is clearly an adaptation for arid environments (yes, habitat and geography correlate with the species distribution), by minimizing photosynthate burden, then any of the primitive features that continue to appear in the same environments are probably also adaptive, being so important that photosynthate is shared with them. Thus, in species not in arid lands, the same features of the stem may be conserva-tive. This is probably part of an heuristic already used by alpha taxonomists, quite unknowingly, to distinguish taxonomically important conservative traits from adaptive, labile traits, but the method should be formalized because of its importance. Relative percent of photosynthate-based tissue directed to an anatomical overlay of the basic soap bubble cell geometry might be a measure of relative adaptive importance of such a feature in a particular arid-land species.

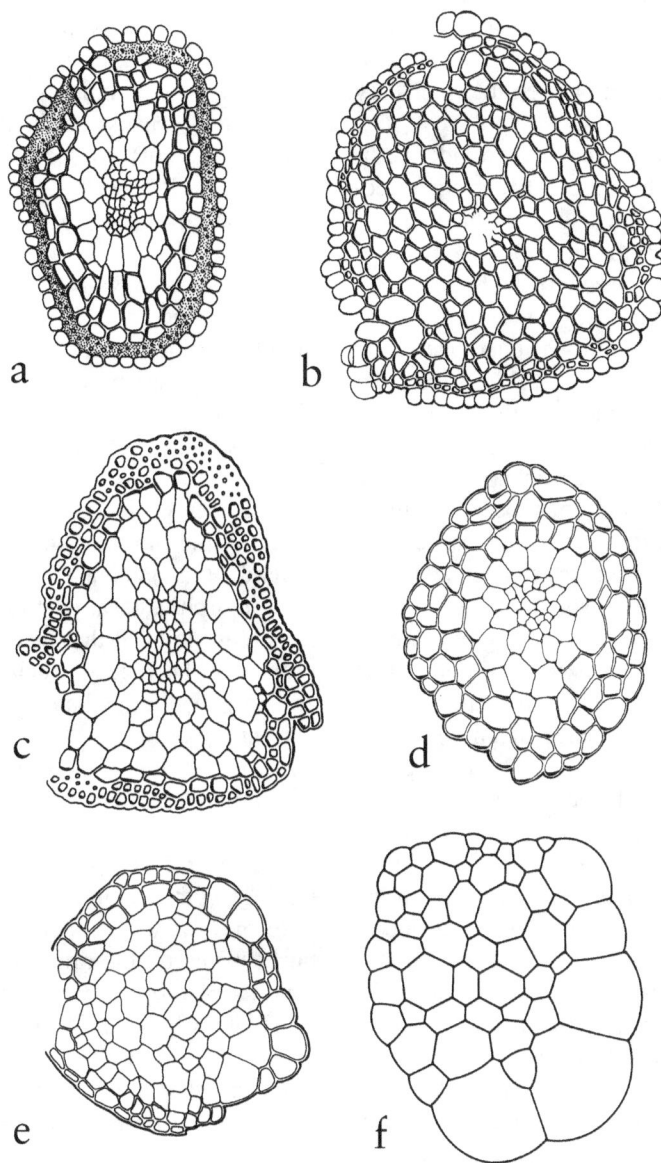

Plate 8.7. — Transformation series in the transverse section of a stem from the most ana-tomically complex and most primitive member of the moss family Pottiaceae, exemplified by *Timmiella anomala* (figure a), with complex features overlain on a soap bubble cell geometry, through *Chionoloma latifolium* (figure b) largely lacking the sclerodermis of supporting tis-sue, *Barbula costesii* (figure c) lacking the hyalodermis of what appears to be superficial wa-ter transport tissue, *Tortula leucostoma* (figure d) lacking all but the central strand of (appar-ent) internal water transport tissue, through to *Aloina bifrons* (figure e) with a stem section of only the basic soap bubble cell geometry. The last (figure f) is soap froth for comparison of tendency to minimum surface. Figures a–e from Zander (1993: 69, 77, 147, 225, drawn by P. M. Eckel), f modified from Peterson (1988: 64).

Other examples of superoptimization — The nam-ing of as many nodes as possible is necessary to maximize parsimony of any postulation of nameless "shared ancestors" that are not assignable to taxa rep-resented by exemplars. This superoptimality is meant in part to reflect the work of several authors (Aldous et al. 2011; Funk & Omland 2003; Gurushidze 2010; Reiseberg & Brouillet 1994) that supports the idea that pseudoextinction (disappearance of a progenitor after generation of two daughter lineages) is rare, while paraphyletic scenarios, including local geographic speciation, is common or even the rule (Frey 1993).

Dollo's Law and biosystematics provide additional data to help judge theoretical diachronic direction of evolution though much ignored in phylogenetic study. The mere recognition of paraphyly as an ancestral taxon is a good example of Dollo's Law in action. The polar bear apparently evolved recently, about 200 Kyr ago, from the brown bear (Demaster and Stirling 1981; Edwards et al. 2011). Molecular evidence indicates it is apophyletic to the brown bear paraphyletic group (Cronin et al. 1991). According to Hailer et al. (2012) nuclear sequences contradict this and indicate, to them, that the polar bear is a more ancient lineage than the brown bear, but no name is given to the basal node assigned as polar bear and brown bear divergence on the cladogram.

Being apophyletic to a brown bear group would be incorrect if there were an extinct (or otherwise unsampled) basal polar bear molecular strain in addition to the terminal one. Evidence against such an extinct isolated polar bear basal lineage is the Dollo evaluation of the clearly advanced and specialized nature of the polar bear's morphological and other expressed traits relative to those of the brown bear (Talbot & Shields 1996). Divergence of brown bears and polar bears likely involved a brown bear ancestor, whether recent or ancient.

Highly specialized traits likewise support the apophyly of the Cactaceae against the paraphyletic Portulacaceae (Applequist & Wallace 2001), because the families might be considered densely sampled taxon-wise by summing the sampling of species in the family. That the apophyletic birds (Aves) are indeed derived from paraphyletic "reptiles" is indicated by fossil evidence of (1) sequence of appearance of fossils in time, and (2) gradual accumulation of bird-like traits in time (Paul 2002). The classification of the groups with strongly divergent adaptational features as separate grades (Mayr 1983) is, however, clear and objective in being process-oriented, as opposed to using strict phylogenetic monophyly (holophyly) as a axiomatic principle.

In the example of Schneider's (2009) fern study discussed above (Plate 7.2), Dollo's Rule involving complex anatomical or developmental traits may be assumed to operate by the authors' note that *Cycas* and *Ginkgo* differ from all other living seed plants in free-swimming sperm and the "nutrient-enhanced ovule" before fertilization. The Marattiopsida (Chris-

tenhusz 2007; Christenhusz et al. 2007) are known from Carboniferous fossils, and have morphologically complex, plesiomorphic traits (thick-walled sporangium with annulus absent, borne on reverse of unmodified leaves, laminal hairs absent or simple, stipules present) conservative for their group. That Cyatheales (Rothwell 1999) is an ancestral taxon for a number of other extant genera is independently supported by fossils from the Jurassic, and several complex, conservative, apparently interdependent plesiomorphic traits (tree-habit, sporangium stalked, annulus oblique, sporangium opening horizontally).

The comparison of morphological and molecular cladograms (plates 5.1 and 7.3) discussed above for the moss family Pottiaceae involves a character-rich (Zander 1993) taxon, *Erythrophyllopsis* (Cano et al. 2010). This genus is accompanied (phyletic propinquity) in the morphological cladogram (Plate 5.1) by several other genera of similar morphology, so the plesiomorphic, basal position is shored up. This is an important point in that clades at the base of a cladogram may be highly specialized through anagenesis and lost of intermediates, but if cladistically nearby taxa are also of the same general morphology, then extreme specialization is no longer of concern and the morphology may be taken to be primitive.

It is not clear now what the function may be, in selection or development, of the plesiomorphic morphological traits, but if they are lasting and conservative, they may be functionless and not particularly burdensome on the organism for the general habitat. That morphologically basal taxa like *Erythrophyllopsis* (Plate 5.1) occur in distal portions of the molecular cladogram (Plate 7.3) among strongly apomorphic, crown genera, indicates strong morphological stasis for the basal taxa and their traits, and Dollo's Rule applies in aggregate.

Superoptimization of *Tortella* and *Trichostomum* — Zander (1993) presented a set of reduction series that apparently obtained for the *Tortella-Trichostomum* group. The series involved reduction in leaf length and shape (long to short, long-lanceolate to ligulate), sexuality (dioicous to monoicous), and capsule complexity (twisted peristomes, shorter untwisted peristomes, peristomes rudimentary or absent). The series is probably only apparent in that several levels of reduction may occur in the same speciational burst from one core generalist species, but the series do signal that an evolutionary concept like that of the speciational burst in genera is a likely model.

Superoptimization in the moss family Pottiaceae

of the *Tortella-Trichostomum* relationships (Plate 8.8) is as follows: *Trichostomum* is a speciose generalist genus, widespread in the world. *Tortella* is similar in morphology and distribution but has the newly evolved conservative trait of echlorophyllose leaf margins of elongate cells—this is also characteristic of *Chionoloma, Pseudosymblepharis* and *Pleurochaete*. A molecular analysis by Werner et al. (2005b) of the Trichostomoideae demonstrates heterophyly in *Chionoloma* implying a deep ancestor generating *Pseudosymblepharis*, even if *Chionoloma* seems morphologically somewhat more modified than the latter. Both *Chionoloma* and *Pseudosymblepharis* are tropical and subtropical taxa, while *Trichostomum* is widespread in both tropic and temperate areas. *Weissia* is separately derived from *Trichostomum*. Although some *Weissia* species are heterophyletic at the genus level, these are unstudied tropical species. The conservative trait (in *Weissia*) of tightly involute leaves and the species' tendency to monoicy further distance this genus from others in the group, thus it may be theorized to be a separate lineage from progenitor *Trichostomum*. *Pleurochaete* is, in the Werner et al. (2005) molecular cladogram, bracketed by heterophyletic species of *Tortella*, and is quite like *Tortella* excepting the advanced traits of pleurocarpous sexual condition, modification of the basal region of the leaf as a second area of distinct areolation, and relatively large size of the habit. There is no Dollo indication, such as chromosome number, that indicates that *Pleurochaete* is more basal than *Trichostomum*. The autapomorphic traits of *Pleurochaete* are definitely unusual elaborations unique in the Trichostomoideae.

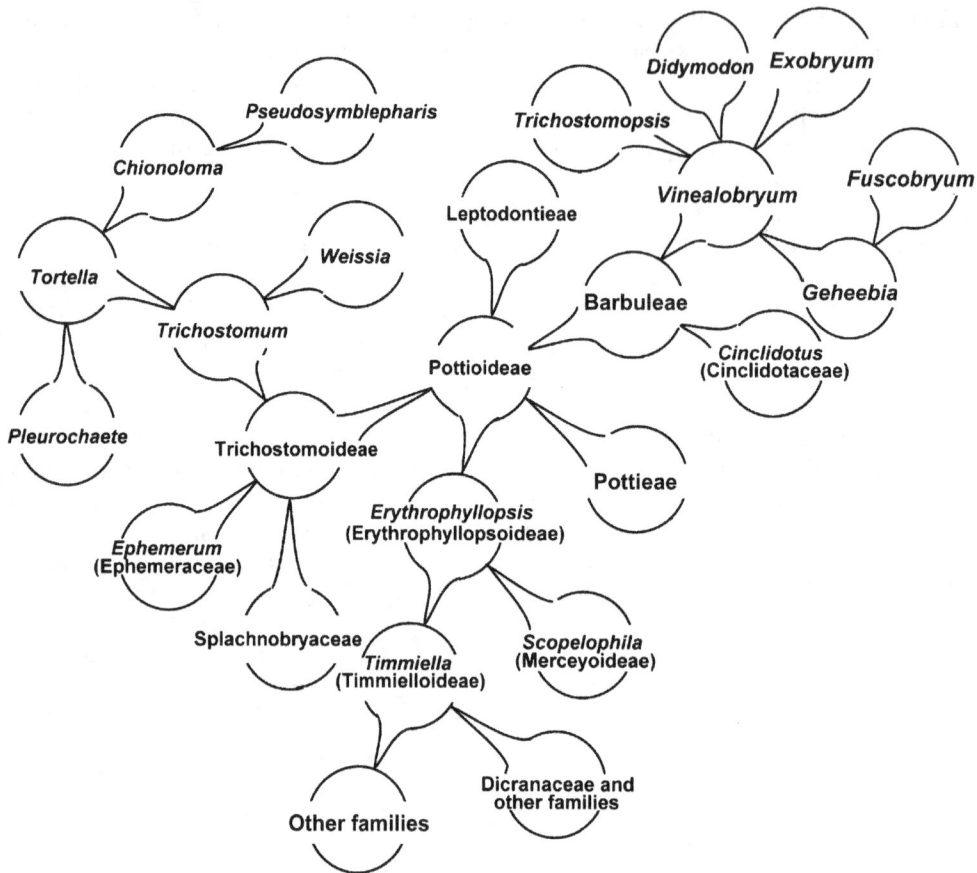

Plate 8.8. — A Besseyan cactus showing salient aspects of macroevolution among some of the ca. 80 genera of the moss family Pottiaceae. New segregate genera of *Didymodon* are derived from the Barbuleae, which is indistinguishable from Pottieae in the Werner et al. (2004) molecular cladograms, but classically a quite distinct tribe. Superoptimization (see text) using non-phylogenetic data provided the macroevolutionary transformational series ((((*Pseudosymblepharis* < *Chionoloma*) *Pleurochaete*) < ***Tortella***) *Weissia*) < ***Trichostomum***. *Erythrophyllopsis* is clearly a primitive genus, but its exact relationship as a supergenerative ancestral taxon with other more advanced taxa is not clear, thus Pottioideae and Barbuleae are inserted as place holders. Note embedded, derived families Ephemeraceae, Splachnobryaceae, and Cinclidotaceae.

Dollo evaluation — In the Dollo evaluation part of superoptimization, all genera are evaluated by considering that an extinct (or otherwise molecularly unsampled) strain more basal than the next lower lineage may have existed. The Dollo evaluation is intended to falsify that possibility. The highly reduced family Ephemeraceae in the Werner et al. (2004, 2005a) cladograms (see plates 7.3, 8.8) as represented by the genus *Ephemerum* was found to be deeply nested in an urgroup named Trichostomoideae. The Ephemeraceae was, prior to the Werner et al. (2004) study, thought to be related to the Dicranaceae. Given that the large patristic distance be-tween where Ephemeraceae appeared in the Werner et al. (2004) cladogram and the base of the Pottiaceae, an extinct lineage of Ephemeraceae basal to the Pottiaceae is not to be probabilistically countenanced. This means we can reasonably discard the possibility of unsampled or extinct paraphyly with an ancestral Ephemeraceae as caulistically basal to the huge family Pottiaceae and ancestral to many of the Pottiaceae lineages. As emphasized throughout this study, macroevolutionary transformational relationships are theories, not "discoveries," and may be changed if additional facts warrant.

CHAPTER 9
Element 6: Linnaean Classification

Précis — The Linnaean classification is simple, and simple is best for classifying organisms in the context of multiple taxon concepts regarding complex, classical relationships and cladistic relationships. It represents differences by analytic lists of taxa, and synthetic groupings by ranks. All other evolutionary or biosystematic information is treated as supplementary, and is not included in the classification.

Both evolutionary and phylogenetic systematics use Linnaean-based classification in most studies. Linnaean classification does not directly represent either sequential macroevolution or cladistic branching because it does not directly represent processes in nature. It is inappropriate for representing phylogenetic monophyly as discussed by Schmidt-Lebuhn (2011). Nesting is not a process in nature, yet serial macroevolutionary transformations are; but neither are directly modeled by Linnaean classification. The limitations of Linnaean classification are its strong points, the analytic (species) and synthetic (higher ranks) results of biodiversity study are clearly distinguished and rendered as hierarchical lists.

Lewis (1965) cogently assigned systematics two quite different goals:

"One objective is to determine relationships among organisms, living and dead, using every means available. The second objective is to formalize the perceived patterns of relationships into a system of classification." "In pursuing the first objective, the taxonomist strives toward perfection in his understanding of the organisms and their relationship, a goal which is inherently unattainable and, consequently, provides assurance to the taxonomist that he need never become obsolete among scientific investigators. On the other hand, when he constructs a classification, his goal is not perfection, but simplification. Classification by its very nature requires that some level of detail be omitted in the designation of classes. The amount and kind of detail to be included in a collective designation depends on the particular classification and its purpose."

According to Wiley et al. (1991: 91), phylogenetic systematists operate under only two basic principles. First, a classification must be consistent with the phylogeny on which it is based. Second, a classification should be fully informative regarding the common ancestry relationships of the groups classified. Note that the first principle fronts up to the inconsistency between various phylogenetic methods and the fact that nesting is not a natural process. The second principle fails because only superoptimization provides fully informative common ancestry relationships.

Linnaean classification provides an optimal, theory-neutral ground for representing a diversity of observed evolutionary relationships or processes, and does not purposefully reject names that cause paraphyly. Its best feature is that it is basically simple although the rules of nomenclature may be tedious and complex.

Hull (1979) has pointed out that genealogy and divergence cannot both be represented in a classification and be separately retrievable. This is acceptable to preserve the neutrality of Linnaean classification, which uses both nesting under higher ranks to signal similarity, and contiguity (that is, separation in lists) to signal differences between similar taxa. Given a series of macroevolutionary changes of one taxon into another, higher ranks are simply evolutionarily regions of the series, as discussed in Chapter Four, Element One.

Linnaean classification is probably the most important heuristic developed in 250 years of alpha taxonomy. The nesting hierarchy of an evolutionary classification reflects the general developmental and environmental constraints on macroevolution implied by descent with modification of well-diagnosed taxa, i.e., a taxon with morphology limited by mutation rates, epigenetic control, and environmental strictures is usually also limited in variation at the level of a new taxon to some similar diagnosable set of traits. Both development and environment determine macroevolution, just as the two scissors blades of formal method and environment determine decision theory (Gigerenzer & Selton 2001). Decision theory is not about inferences but about deciding to take an action (Winkler 1972: 435). A Bayes Solution requires a loss function that helps evaluate risk. Risk in taxonomy affects biodiversity analysis and conservation. Minimizing that risk means that taxonomic decisions are critical and can cannot be passed off as minor with the excuse that taxonomy is easily changed with new information, as has been the practice in the past. Conservation analyses are not the same as white-tower classical taxonomy.

Standard Linnaean classification of groups may be in part modeled or diagrammed by ancestor-descendant "Besseyan cacti" (e.g., Bessey 1915), as exemplified by, e.g., Denk and Grimm (2010), Wagner (1952) and Zander (2008b, 2009). Given pluralist methodology, the standard classical approach to nomenclature is adequate. Such classification serves as a well-hooked framework for a wealth of information from many fields.

Because there are seldom specimen exemplars (OTUs) in morphological parsimony analysis; the morphological cladogram at best reviews and details the relationships of classical taxonomy. While classical taxonomy evaluates "local" evolutionary relationships of individual specimens and traits, weighted morphological parsimony can provide a broad-based summary of all such evaluations, the goal being a detailed natural key or cladogram equivalent. Problematically, morphological parsimony as commonly practiced mixes equally weighted conservative and labile traits, incorrectly posits convergences, and uses an evolutionary model that implies that of every three OTUs, two must be more closely related, yet many processes are involved in evolution (Hörandl 2007). For instance, modeling surviving ancestral taxa yields a tree that is less parsimonious in length (increasing the number of trait changes) but more parsimonious in having fewer numbers of postulated taxa or entities (nodes). It is possible to unite in a robust fashion the scientific, empirical basis of paraphyletic taxa and the evolutionary importance of both expressed traits and molecular sequences, and of both divergence and shared ancestry (e.g., Brummitt 2003, 2006; Hörandl 2006, 2007; Hörandl & Stuessy 2010; Zander 2007c).

Linnaean classification uses, as best possible, hierarchical nesting and concatenate listing of names to represent sequential evolution, is clearly incomplete, and is based on a theory not structuralism. It can, however, preserve macroevolutionarily significant scientific names for use in conservation, biodiversity study, and other fields.

Some taxa that appear basally in a morphological cladogram may be found terminal in a molecular cladogram of the same taxa. These morphologically basal taxa, particularly when grouped with other taxa of similar morphology, are probably primitive. They should not be taxonomically associated with other phylogenetically nearby taxa, except as being possibly ancestral. They are best considered surviving populations of a progenitor with the same morphology (at a particular taxonomic level) as the morphologically basal taxa. Thus, *Erythrophyllopsis*, which molecularly clusters with *Didymodon* in Plate 7.3, cannot be grouped with *Didymodon,* but instead taxonomic grouping must follow morphological inferences (Plate 5.1). These reveal evolutionary relationships more directly because such inferences are based on conservative traits, both adaptive and quasi-neutral, while the appearance of the morphologically basal taxon may be scientifically explained as due to a large self-nesting ladder. *Erythrophyllopsis* is primitive (meaning ancestral and basal on a caulistic tree, e.g., a Besseyan cactus). Conservative expressed traits do not promote false relationships when progenitor taxa are included in a cladogram with descendant taxa.

One of the features that will be noticed in evolutionary classifications based on this Framework (or other pluralistic evolutionary systematics) will be far less nomenclatural and classification changes or modifications of classical taxonomic results. This is because the Framework rejects the classification principle of holophyly (which requires that all included exemplars must be at one classification rank or lower), and offers methods of data analysis and consolidation such as heterophyly (deep ancestors implied by distant exemplars of the same rank), self-nesting ladders (ancestral taxa rise higher in a molecular cladogram by pupping off daughter taxa), pre-selection allowance (if molecular relationships are determined from exemplars preselected from a natural key, then agreement of molecular results with morphology is not support and disagreement is probably bias), and coarse priors (which allow conclusions from classical taxonomy and morphological cladistics to be compared and melded with those from molecular systematics using the Bayes' Formula). This leads to a theoretical understanding of macroevolution at the taxon level that deals with all information available and relevant. Such methodologically pluralistic understanding retains a large fraction of classical descriptive and evolutionary analysis.

CHAPTER 10
Systematics Reviewed and Recast

Précis — Oversimplification and the highly improbable has biased phylogenetics. The value of cladistic study for serial macroevolutionary reconstruction is reduced to—in morphological studies—evaluation of relatively primitive or advanced taxa, and distinction of taxa by autapomorphies; and—in molecular studies—identification of deep ancestral taxa via heterophyly. The Framework is summarized.

Presented in this book is an outline for a new, post-phylogenetic systematics addressing the more salient inadequacies of present methods, but conciliating and consolidating true advances in classical taxonomy and phylogenetics using morphological, molecular, and other evolutionarily important approaches. In botany, for instance, true advances might include all taxonomic novelties and the methods involved in generating them in APG III (Angiosperm Phylogeny Group III 2009) and the recent influential phylogenetic classifications of the mosses by Goffinet et al. (2008) and Frey and Stech (2009), *excepting* the many classification changes based solely on rejection of macroevolution as a process to be recognized in classification, particularly names based on the simplifying but theoretically barren principle of holophyly (strict phylogenetic monophyly).

Classical systematics and phylogenetics — It is clear that classical systematics separates the clustering by similarity of apparently conservative and homologous traits and the intuitive analysis of macroevolutionary trajectories. Classical systematics can profit from the directedness of phylogenetic analyses, with certain provisos.

Morphological cladistics suffers from the idea of Hennigian pseudoextinction as an anchor (Chapman & Johnson 2002) for cladistic theory of evolution and its reflection in classification. Strictly, pseudoextinction is simply anagenetic change in which one species changes into another. Hennigian pseudoextinction requires the ancestral taxon to change into another whenever speciation occurs. This may be the case in many sympatric speciation events in which the Red Queen effect (two evolving taxa compete for resources, Van Valens 1973, 1976; also Sternest & Smith 1984) pressures both ancestral taxon and the daughter taxon. There is, however, no evidence of the *ubiquity* of phylogenetic pseudoextinction in evolution. According to Raup (1981):

"Pseudoextinction is the situation where a single species lineage is transformed by phyletic evolution into a new species. The new species would presumably have been reproductively isolated from the ancestral species had they lived together at the same time but the process is totally different from speciation as studied by the evolutionary biologist. Because pseudoextinction does not represent death without issue, instances of pseudoextinction should be eliminated from the data before extinction is analyzed. This is difficult because it is usually impossible to determine whether a species that is lost from the record actually died out or whether it was simply transformed. In view of the growing consensus in favor of the punctuated equilibrium model of Eldredge and Gould ..., one could argue that pseudoextinction is not a dominant phenomenon, but good numerical estimates of its frequency are not available. Pseudoextinction at supraspecific levels cannot logically occur unless the higher taxon is monotypic and thus the problem is serious only at the species level."

Peripatric speciation is apparently common, and divergence may be only on the part of the daughter species, following the Court Jester hypothesis (Barnosky 2001, 2005) that environmental perturbations largely (or also) drive speciation. With no change, stasis continues.

Universal phylogenetic pseudoextinction is better viewed as an imposition of classification practices on evolutionary theory in that evolution is apparently supposed to occur in the topology of a dichotomous key. The use of irrelevant anchors in decision theory is common (Chapman & Johnson 2002).

Molecular systematics suffers from a number of problems, outlined in detail above, but most importantly in its inability to determine details of branch order of taxa. It can determine details of branch order of exemplars, which when paraphyletic, provide information on deep ancestors, but otherwise molecular systematics introduces major aleatory biases when molecular clustering does not agree with groupings determined by classical systematics. This pervasive problem is explained in the next section.

The highly improbable and its consequences — It has often been stated that any evolution event is improbable (Raup 1981), but associated information may establish phylogenetic relationships anyway. See Taleb (2010) for discussion of attempts to predict (and perhaps also retrodict) highly improbable events that are in fact quite common given the complexity of nature, and are also deeply important. Does overall similarity really imply shared ancestors? I think it does in that there is no other explanation of groups of taxa that seem to be bound by theoretically slowly

changing conservative traits that apparently keep evolution in most cases (pace hopeful monsters) "close to home." In other words, it is the inertia of the phylogenetic constraint of the heavy baggage of conservative traits that militates, theoretically, against the idea of species usually making, not minimal, but short to medium phylogenetic jumps. The phylogenetic drag of the evolutionary ratchet of Levinton (1988: 217) is also relevant here because it may funnel evolution through a narrow developmental restraint. Although improbable speciation events that leave little or no historic data doubtless occur, such events uncorrelated with other data may be treated as minor noise (Taleb 2008: 261) during standard analysis.

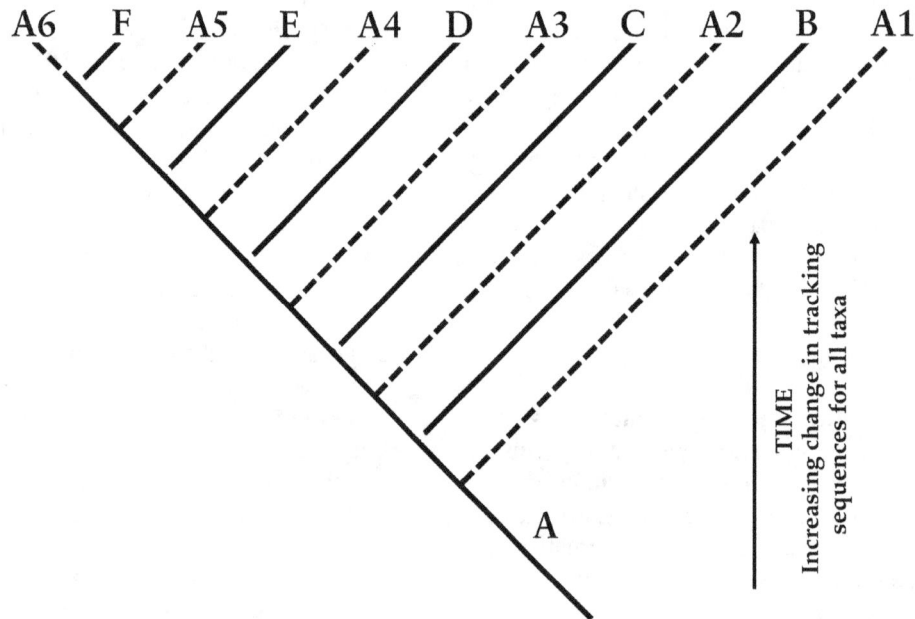

Plate 10.1 — Introduction of uncertainty in branch order analysis of molecular cladograms by improbable but common extinction (or non-sampling) events. **A** is a deep ancestral taxon in morphological stasis which may or may not survive to the present in six lineages or molecular strains—dashed lines. *As time passes morphology of **A** does not change but molecular tracking sequences do.* B through F are descendant taxa, signaled by the break at the base of their lines. Any combination of A1 through A6, and B through F lineages, may be extant, and extinction (or non-sampling) cannot be estimated statistically at this level because of improbability with no relevant data on statistical distribution of extinction of isolated molecular strains of taxon **A**. Molecular estimation of branch order for any group is thus highly dubious.

There are real improbable but common events that genuinely greatly affect systematic analysis. They are not amenable to statistical study because no information is readily available to infer a distribution. An example is the differential extinction of a deep ancestral taxon in morphological stasis as it generates descendant taxa (Plate 10.1). If the deep ancestor (**A**)

does not survive to the present (i.e., lineages A1 through A7 are absent), then the nodes (breaks at base of descendant lineages) may appear to be shared ancestors undergoing pseudoextinction, but are actually the same taxon (**A**) at every node. If one lineage or molecular strain of the deep ancestral taxon survives (say only A1, or only A3, or only A6), then where it branches from the set of descendant taxa determines the sister group, yet all descendants of **A** are sister groups. If two (or more) lineages of the ancestral taxon survive (say A4 and A3) spaced by one or more descendants (e.g., D), then one has paraphyly. If two lineages of the deep ancestral taxon survive but not spaced by a descendant taxon (say, a descendant D is extinct or the ancestral taxon becomes geographically isolated), then one has pseudoconvergence or parallelism (A3 and A4 are extant and molecularly distinct), possibly viewed as molecularly cryptic taxa. If no descendants are now extant for Plate 1.1 (or if there were six isolation events for populations of **A**) and all lineages A1 through A6 are extant, one might decide to recognize six molecularly cryptic species. Although the fundamental transformational macroevolutionary relationships of ancestor and descendants remains the same, extinction or nonsampling results in different classifications. This should be investigated and minimized.

Another improbable event is the generation from a deep ancestor of major taxonomic multiple descendant lineages each further producing descendant lineages of their own (e.g. Plate 7.3). Although the kind of dead-end dissilient evolution like that hypothesized for *Didymodon* s. lat. (Chapter 8) in this book is more likely, with multiple evolutionarily isolated descendant lineages, once an improbable multi-coregenerative speciation event occurs, as is apparently the case with *Erythrophyllopsis* (Chapters 5 and 7)*,* it may form the central feature of a cladogram. This feature may not be recognized as such, however, given lack of extant exemplars or just the inattention given such a possibility associated with current practices.

Overview of the Framework elements — Evolutionary systematics allows all methodological inferences of evolutionary relationships in either nested or sequential orders, using all information, and then bases classification on implied macroevolutionary serial transformations. Cladistic systematics restricts evolutionary inferences to nested orders and allows a principle of classification (holophyly) to influence description of evolutionary relationships. Inference in evolutionary systematics (Fitzhugh 2012) includes abduction (generating hypotheses), deduction and induction, and reasoning by analogy. In cladistics, inference is largely deduction from "discovered" nested relationships. Evolutionary systematics is very cautious about using molecular cladograms in evolutionary reconstruction and classification because so few samples are used for each taxon, excepting certain large families.

The main feature lacking from phylogenetic trees are named shared ancestors that could be obtained through superoptimization (Element 5), or from caulistically mapped deep shared ancestors of all exemplars of one taxon, that is, an ancestor at the same rank as appropriate to include all heterophyletic exemplars (Element 3). The main presupposition of rationality and all matters of fact is "no contradiction" (A is A and not non-A). Phylogenetics fails, then, as discovery process through inconsistency between analyses: (1) Between morphology and molecular analyses. Morphological relationships are considered relevant in phylogenetics but are not influential because morphological traits are mapped to the molecular cladogram or buried in total evidence analyses that simply match directly DNA and morphological data. (2) Between taxa involved in paraphyly and phylogenetic polyphyly on molecular cladograms. Any heterophyly is "explained" by incomplete lineage sorting, convergence, paralogy, hybridization, or recombination, and other molecularly focused explanations. (3) Between phylogenetic (entirely synchronic) and classical (often involving theory of diachronic taxic transformation) classifications. This is because phylogenetics is a method-based analysis that rejects caulistic explanations involving scientific theory conciliating data that is not sister-group informative.

It has been suggested that morphological cladistics has little phylogenetic utility (Scotland et al. 2003). This is because although it does reflect much the same genetic continuity as does molecular trees, molecular trees are of specimens, not taxa, and homoplasy is, of course, a confounding artifact. Although it has been demonstrated that morphological signal can increase phylogenetic signal in combined data sets (Lee & Camens 2009), the increase in support is doubtless in the realm of genetic continuity, but interpretation of the progenitor-descendent relationships of the specimens used in molecular analysis remains problematic.

As a methodology reflecting macroevolutionary theory, the Framework explains and coordinates as a transformative model both synchronic and diachronic aspects of evolution of taxa, and requires a simple but effective classification scheme to reflect this. Linnaean classification, proven effective over 250 years, provides this, see Chapter 9. Phylogenetic theory is not consistent in analysis of morphological and molecular data and in generation of heterophyly in molecular analysis. It applies arbitrary principles in analysis of evolution and reflection of the results in classification, i.e., rejection of macroevolution, focus on synchronic relationships which can be precisely measured, postulation of multiple unnamed, unobservable entities as hidden causes that are unparsimonious in light of high probability of many surviving ancestors, methodological insistence on pseudo-extinction of a progenitor at every node, assumption that morphological data and molecular data directly reflect the same evolutionary processes, and structuralist rejection of theory as mere metaphysical explanatory narratives. The method of evolutionary systematics proposed here is consistent, rejects no relevant data, and matches evolutionary theory. It is the simplest pluralistic method that accounts for both fact and theory.

One might remember, as a rule of thumb in dealing with modern methods of systematics, that one should not "test" facts of sister-group similarity and the trivial assumption that such is due to unnamed shared ancestors, against theories of descent with modification involving serial progenitor-descendant relationships. Facts always trump theories because facts are well-documented observations and theories are more or less well-supported postulates or models of processes in nature, and are seldom proven. There is no qualitative comparison. Thus, one should ask what theories can one make from the facts rather than being satisfied with inconsistent cladograms? Evolutionary taxonomists can advance theories, and good theories always subsume atomized facts.

Oversimplification leads to consistency with other explanations — Although phylogenetics is inherently inconsistent as noted above, because it simplifies to the extent that only integrable results are obtained, it becomes, unfortunately, consistent with "creation science" or phylogenetic baraminology (Gishtick 2006). There are no factors beyond claims that cladogram nodes or nested parentheses indicate unnamed shared ancestors that might be construed as antithetical to creationism (Huse 1983; Kubicek 1993; Poole 1990); .

In systematics, the supreme question of our time is whether to abandon macroevolution as a source of information for classification. Structuralism is attractive, e.g., providing precision, often statistical certainty, and sure, deductive methods of analysis, but it is not good science because it rejects testable theory that conciliates instead of relegates facts. The "shared ancestors" of phylogenetic cladogram nodes imply hidden causes, and unnamable and unobservable entities. Although Einstein (Gilder 2008: 86) stated that every theory includes unobservable quantities, there

is in phylogenetics no effort to name nodes representing shared ancestors beyond that taxon inclusive of all terminal taxa in the clade. The point made by phylogeneticists is that by simplifying to an integrable solution, classification can be put on a firm, "hard science" footing. Without inference of macroevolutionary relationships, nesting ignores total evidence and implies wrong evolutionary relationships. This is because it is trivial, often unjustifiedly exact, or even palpably wrong. In some cases in physics, too many "phenomenological adjustments to make everything fit" (Feynman 1985: 190) can cause a theory to be rejected; in phylogenetics too little attention to expanding theory to fit reality can be problematic because, in fact, everything *can* fit.

There are practices associated with phylogenetics that should be rejected for a new systematics to theoretically consolidate taxonomic methods that separately yield disparate results:

(1) Shared ancestors are not named at the same rank as their derivative exemplars in phylogenetics because this would result in paraphyletic groups, and nodes are simply used as place holders for the next higher inclusive rank. Thus, if followed to an extreme, there are two alterative views possible (a) that species do not disappear at all but their lines are bunched as skeins into the "shared ancestor" of a higher rank, or (b) the "shared ancestor" is a decision made on the Second, Fourth or Fifth Day while the Deity worked to create all life; that is, a Markov chain of Creation would explain why it took all Day. There is no evidence in a phylogenetic classification, or a cladogram, or in the phylogenetic evolutionary analysis that generated the cladogram, against either a species surviving a speciation event (macroevolution with extant ancestral taxa as promoted in the Framework), or of immutability of species (extinction possible but nesting explained otherwise as in creationism). There is no evidence for the second, but plenty for the first.

(2) Mapping of traits on cladograms is commonly referred to as instances of evolutionary change, yet traits do not evolve, species do, so this is extreme atomism and reductionism. Thus, trait changes mapped on cladograms may be used to infer only microevolution (minor genetic changes fixed in a species) and not macroevolution (one taxon evolving from another at the same rank or lower). This yields a microevolutionary rather than macroevolutionary classification. Modeling descent with modification of taxa is avoided in cladistics. Mapping of morphological traits or biogeographic distributions on cladograms is an attempt to transform evolutionary analy-

sis from hypothesis and theory to lemma and theorem, i.e., from both deduction and induction to deduction only.

(3) The phylogenetic rejection of naming tree nodes is said to be due to the fact that if nodes were named at some rank lower than that including all terminal taxa of the clade, then all branches of a tree would need to be collapsed because a taxon cannot directly evolve from another of the same rank, according to the phylogenetic classification principle of holophyly. Classification by holophyly (strict phylogenetic monophyly) is, however, artificial and leads to degenerate (as a return to absolutism) non-evolutionary classifications. Holophyly has no ontological basis as a process in nature, that is, it is not refutable and so is not a scientific hypothesis (Knox 1998; and see Bock 2004). It is ostensibly used for simplifying taxonomy by "naming away" any problems, but in doing so requires one to lump and split taxa that represent macroevolution in classification. Holophyly clearly eliminates representation of ancestor-descendant evolution in classification. Thus, nodes cannot be assigned scientific names (other than a general and trivial attribution as ancestors belonging to the general group of all taxa distal to the node on the tree). Not naming ancestors gives them an ineffable, metaphysical, and recondite substance. It leads to faith-based taxonomy.

These three points (apparent immutability of species, microevolution acceptable but not macroevolution, and ineffable or mystical generation of species) are those of "scientific creationism" or phylogenetic baraminology, which uses phylogenetic software to group taxa by prelapsarian differences, each a single immutable species. This does not instill confidence in phylogenetics.

Using the results of phylogenetic sister-group analysis, one can infer some ancestor-descendant relationships or one can have the principle of holophyly, but not both at once. Since holophyly is not a scientific theory but only a classification principle, it must be cast down. If the early idea that all monophyletic groups must be named and these named hierarchically was abandoned (Wiley et al. 1991: 104, alternatively using the "sequencing convention,") because of a lack of enough named ranks, why not abandon the criterion of holophyly? The prevalence of present-day paraphyly of classical taxa based on expressed traits is good evidence for its prevalence in the past. The needless splitting and lumping associated with the practice of holophyly (Zander 2007b) vitiates the use of phylogenetic classifications by other fields because biases are introduced that may be

"discovered" by evolutionists, biogeographers, ecologists, and others as apparent real facets of evolution even though there is no natural process represented by holophyly.

Multiple test problems, in short — All elements of this Framework are influenced by the multiple test problem. Multiple tests (or multiple comparisons) in statistics refers to mistakenly giving significance to the discovery of unusual results among many observations, when such observations are expected to yield such unusual results probabilistically. For instance, discovering a sequence of five heads in a row when a coin is flipped 1000 times is not a discovery because it is expected as part of a statistical distribution. Bonferroni correction (Hochberg 1988; Holm 1979) or Control of False Discovery Rates (Benjamini & Hochberg 1995) are common ways to deal with clear and measurable instances of multiple tests. In the most simplistic method, one divides the alpha (chance of a false positive) by the number of tests, e.g. if the alpha of accepting a phenomenon is 0.05, then the alpha for accepting two instances is the alpha divided by two, or 0.025.

Multiple test situations are common in modern systematics, and, as problems associated with statistical significance when working with large samples, occur throughout biology (Shrader-Frechette 2008). In classical systematics, traits supporting the best grouping of taxa are the basis of sorting higher ranks, but taxonomists commonly clearly state and discuss the chance that the second best is or is not nearly as good as a basis for a sorting, heuristically avoiding vicious ambiguity (van Deemter 2004). Bootstrapping techniques are helpful in evaluating support for contrary arrangements of cladograms. Problematically, however, when higher taxa are delimited by splits in molecular trees, *expressed* traits involved in selection or as tag-along conservative traits are not necessarily best grouped, nor is there serious evaluation of alternative sets of expressed traits such that the molecular tree may be evaluated in a theoretical context rather than as a discovery of a structuralist, axiomatic pattern. It is now common for phylogeneticists to "discover" a set of expressed traits for a molecular taxon. Combinations of traits considered evolutionarily significant are already at hand in classical descriptions. This variant of the multiple test problem will haunt systematics as long as a new, but very much second-best, morphological description is "discovered" after expressed traits are artlessly mapped to a molecular tree of sparsely sampled specimens, and leads to arbitrary selection of morphological traits in phylogenetically informed descriptions of supraspecific taxa. (See Chapter 15.)

Review of the Framework — Modern phylogenetic systematics provides sometimes quite artificial naming of nested sets in a cladogram, that is, representations of present-day phylogenetic relationships of exemplars. Much of evolutionary theory, particularly concerning macroevolution, is not represented. There is a close horizon of expectations that results in a miniaturization of satisfaction associated with senescent humans (Robinson 1992: 143) and, as is now apparent, deconstructed fields of science. Classical taxonomy, morphological parsimony, and molecular analyses of phylogenetics include much non-overlapping data, but, if these non-nested hypotheses (Cox 1961, 1962) are conciliated, without overemphasis on any one field, by recourse to the unifying caulistic dimension, such additional data can broaden what we infer about evolving taxa and their meaning-rich representation in classification.

The fundamental ground is classical taxonomy, in which conservative expressed traits are identified, and many specimens and taxa are examined for the local evolutionary relationships between similar specimens and their expressed traits resulting in a multidimensional tensor-like data space (Kolecki 2002) that is investigated with heuristic strategies. Numerical analyses (parsimony, phenetic clustering) of morphological data provide a needed broad, integrated view but are variously limited. Molecular data reveal lineage continuity and past isolation events for particular specimens from a limited 2-dimensional vector data set, while molecular heterophyly allows a mapping of inferred caulistic taxa. Differences between morphological and molecular cladograms reveal additional hidden, isomorphic, caulistic taxa as synthetic, emergent properties. Superoptimization minimizes superfluous unnamed and unobservable postulated shared ancestors previously advanced as hidden causes.

Discursive reasoning associated with biosystematics and Dollo's Rule (Gould 1970) along with the evolutionary ratchet of Levinton (1988: 217), help detail direction of evolution. After superoptimization, one may infer a dynamic, process-based view of evolution that translates into a branching caulogram (or Besseyan cactus) rather than the static, hierarchical, classification-like view of a cladogram. Beyond using heterophyly to infer deep ancestral taxa, the three most interesting features of the Framework are: that the heuristics of classical systematics are underlain by physical principles; that self-nesting ladders may impose a backwards or simply arbitrarily molecular tree nesting order; that genera may be recognized by morphological gaps, apparently adaptive autapomorphies, conservative traits, and by clouds of clearly derivative species around intergenerically primitive ancestral species; and that at least in some cases genera are operationally the basic unit of evolution.

The simplest case — Now that we have examined in depth features of the Framework, consider then the simplest case. An aphorism of phylogenetics is that of any three taxa, two are more closely related by shared ancestry. When there is a multifurcation in a cladogram, this gnomic axiom is usually "saved" by suggestions that other data or more data will resolve the problem.

Consider then three taxa, A, B and C, and **A** is inferred as the progenitor (bold-faced by convention) of first B, then C (as determined by, say, superoptimization and molecular heterophyly). The evolutionary formula is $A > ({}^{1}B, {}^{2}C)$. Suppose also that C has reversed one trait of those that supported all three, A, B and C, as a group. Morphological analysis would result then in terminal taxa ((A, B) C), with A and B sharing more traits than each with C. Molecular analysis should result in ((A, C) B) because B was generated first and C last, and thus C shared more of the nearly constantly mutating molecular traits with A than B.

Inveterate morphological cladists might argue in favor of ((A, B) C) as a most parsimonious solution involving traits actually acted upon by evolution. Fervent molecular cladists might throw out the morphological result because there is fewer data and much convergence in morphological analyses. Both are wrong because all taxa are equally related, and no more data can possibly solve the problem when couched as a need to find which two of three taxa are more closely related.

A Possible Paradigm Change — In the Introductory Chapter it was suggested that the cladistically oriented reader test him or her self as to whether reading the Framework protocols might have effected a kind of paradigm change in thinking about the use of information on evolution in systematics.

Questions the reader might now respond to include: Are you alert to your own heuristics? Are you tempted to formalize them, and perhaps judge how generalizable they are to other taxa, and find out why? Are you looking at cladograms now with something akin to superoptimization? Applying coarse priors to morphologically or classically derived evolutionary relationships? Searching for informative heterophyly? Feeling unease that some odd, nonintuitive branch orderings may be due to extinct or otherwise unsampled extended paraphyly? Do you search for self-nesting ladders or serial transformations that explain disparate morphological and molecular cladograms?

If you find that you now at least agree with R. Brummitt that paraphyly is an important source of evolutionary information that should be reflected in classification, or, if by some miracle you find yourself looking favorably on the Framework methods proposed in this book, then you have experienced in the small a paradigm change of some potential importance to systematics. You must, however, also ask yourself if the new paradigm is good science. Not all attractive new scientific solutions to old problems are such. It is here hoped, however, that you will find that this book is a true contribution to scientifically retrodicting evolution and using inferred serial macroevolutionary transformations to inform classification. Given that examples are given, the existence theorem (Reynolds 2007: 444) operates, namely that it is always much easier to find a solution when you can be confident that one exists. You can accept the macroevolution-in-classification meme or reject it, but you as a responsible scientist will never totally forget it.

The Framework is presented as the refinement and consolidation of a set of powerful new and classical methods addressing the complex, environmentally critical, and ever-imposing mysteries of evolution, biodiversity, and classification of nature. Lord Dunsany observed, "A man is a very small thing, and the night is very large and full of wonders."

The future — The treatment of superoptimization and consolidation (Chapter 8) is somewhat incomplete in that the evolutionary tree (Plate 8.2) is derived largely from classical and morphological cladistic studies, the relevant molecular tree being sparsely populated and therefore with little support for contradictory results. The point of the Framework is to provide an overlay or congruence of the serial macroevolutionary transformations inferred from classical taxonomy, morphological cladistics, and molecular phylogenetics. Such congruence is particu-

larly amenable to gauging support with coarse Bayesian priors resulting in coarse Bayesian posterior probabilities for macroevolutionary transformations. Such support would involve coarse but empiric priors from classical taxonomy, cladistic morphology, and molecular phylogenetics. See Chapter 14 for a first attempt to estimate posterior probabilities for molecularly based macroevolutionary transformations.

One might expect, if the method is widely accepted as valid, that software will be developed to help integrate facts (well-documented observations) and good theory (well-reasoned inferences) that hang together (form congruent evolutionary diagrams). This, from a total evidence perspective, can be supported by coarse Bayesian posterior probabilities. Such software must deal with multifurcating natural keys generated by well-informed scientific intuition, superoptimized dichotomous cladograms based on only somewhat parsimonious data, and heterophyly and self-nesting ladders in sparsely sampled molecular trees, and, then, superimpose lines of congruence of macroevolutionary transformations at the taxon level. Then, calculate support for such lineages in the context of a Bayes' Solution (decision theory dealing with risk of being wrong). As noted previously, the maximum parsimony or maximum likelihood solution is for a tree with identified instances of budding evolution, identified previously through superoptimization with morphological traits. Such trees are not simply longer or less likely than present-day analytic cladograms, they are the maximally short or likely possible given instances of budding evolution. All the above is far more complex and imprecise than present phylogenetic practice, yet avoids the pitfalls discussed at length above.

The Framework does not reject phylogenetic methods, instead finding relevant evolutionary information (e.g., primitive vs. advanced taxa, deep ancestral taxa) where previously unappreciated. Thus, modern laboratory methods including DNA sequence analysis and statistical methods are supported as valuable. Jargonistas will revel in new specialized terminology (see Glossary) without giving up their familiar patois. As noted elsewhere, the Framework promotes a win-win positive-sum game (Wright 2001).

CHAPTER 11
Conservation and Biodiversity

Précis — Scientific names of rare and threatened taxa may be fully eliminated, or be buried among a multiplicity of superfluous similar names, because of classification practices associated with molecular taxonomy. Epistemological extinction renders difficult or—for the taxonomically uninitiated impossible—the recovering of sunk names of biodiversity importance or the distinguishing of a truly important taxon in its full range from equally or even more rare minor molecular variants of the same or related taxa. Examples are provided.

As discussed above, modern phylogenetic analysis and classical evolutionary systematics are commonly at odds because phylogenetics is now presented as a kind of "hard science," based on statistics and DNA sequencing, while "traditional classification" is represented by practitioners of the former as intuitive, subjective and arbitrary. Both schools of systematics create modern classifications by evaluating the results of evolution in a group and then basing classifications on that evaluation. Phylogenetics emphasizes highly precise and apparently accurate molecular trees, while evolutionary systematics treats all information equally by applying an overarching evolutionary theory, and, though not a "soft" science, is not as precise as the former. That overarching causally explanatory theory is macroevolution, of taxa changing from one to another through time in a caulistic tree space or diachronic world-line, as opposed to phylogenetic's taxa nested terminally in a cladistic tree space caulistically filled with unnamed, ad hoc "shared ancestors" as analytic placeholders. In cladistics, the nodes are simply dichotomously clustered synapomorphies, in phylogenetics universally treated as pseudoextinction (the analytic software remains the same for both). This practice preserves the precision of statistical analysis but abandons well-supported hyeptheses of taxon transformations based on sets of conservative expressed traits that are available from classical systematics.

There are three patterns that are treated in different ways by phylogenetics and evolutionary systematics. These are:

(1) Classical systematics produces classifications of hierarchical sets of taxa. These are based on taxa distinguished by overall similarity of conservative and apparently homologous traits. These are also clustered by non-sequential distinction of major evolutionary transformations at the taxon level associated with habitat change and other criteria. Conservative traits are those stable at different collecting sites and across different habitats. The difficulty of discerning transformations at the taxon level in all but the most coarse manner may be mitigated in part by phylogenetic methods.

(2) Morphological cladograms present (mostly) maximum parsimony trees of nested sets of trait transformations away from an apparent primitive (plesiomorphic) set of states, the latter usually contributed by selection of an outgroup taxon. Evolution is represented in morphological cladograms by transformations of sets of traits of taxa based on descriptions developed through classical systematics. The transforming traits that are mapped on the cladogram are both labile and conservative characters from classical descriptions but are weighted equally. Conservative traits may be conservative only for certain groups. Additionally, the dichotomous nature of cladograms imposes an artificial structure on evolution.

(3) Molecular trees are variously assembled through maximum parsimony analysis (as in morphology), or maximum likelihood or Markov chain Monte Carlo Bayesian methods. These trees do apparently represent genetic continuity and isolation events associated with the specimens used as exemplars, but not necessarily speciation events associated with the taxa the exemplars represent. This is due to uncertainty contributed by extinction or non-sampling of extended paraphyletic lines or molecular strains. The traits are mostly non-coding sequences usually assumed to be unbiased by differential selection.

These three patterns give views of different aspects of the evolutionary process, and are not equivalent, thus one is not necessarily better at charting evolution through time (as Darwin's descent with modification of taxa) than the others. In modern phylogenetics, morphological traits and traditional classifications are relegated or "mapped" onto a molecular tree, i.e., treated as epiphenomena. The three patterns should, however, be treated as "non-nested hypotheses," that is, models that cannot be obtained from each other by parametric restrictions or as a limit to an approximation (Cox 1961). Actually, one pattern is not a superset of the others, nor are they epiphenomena of any one. The assumption that one and only one pattern is fundamental and other patterns and data may be relegated to that one fundamental pattern is "structuralism," which I have discussed at length as a retrogressive force in modern systematics (Zander 2010). Wishful thinking about advancing science by focusing on the precision of molecular trees and relegating other patterns to one well-supported tree contaminate phylogenetic classifications with unreasonable assumptions and extremely biased representation of the results of evolution in classification. The relationship between phylogenetic analysis and conservation (Mace et al. 2003) must be viewed with concern, as problems in the former must affect the latter.

Conservationists who need general systematic classifications for the plants that are not biased by strict phylogenetic monophyly may found their systematic data sets on the work of Thorne (2007) or of Heywood et al. (2007), following the evaluation of Stevens (2008) who compared modern major classificatory works. See also the fairly recent work of Brummitt (1992) and Stuessy (2010). There are many more directed conservation analyses, like the basal clade study of Moron et al. (1996), that purport to identify taxa important for conservation by their distance on a cladogram. Because many cladogram nodes represent imaginary shared ancestors (and one does not know which represent hypotheses and which placeholders for assumptions of pseudoextinction that ensure complete resolution when there is none), such studies are may be highly biased. There are enough problems in conservation and biodiversity analysis (Ahrends et al. 2011; Costello et al. 2013; Pitman & Jorgensen 2002; Thomas et al. 1994; Tillman et al. 1994, but see Liu 1993) without the phylogenetic impediment affecting dealing with an estimated 713,000 threatened species (Pitman & Jorgensen 2002). Major works are now being published that attempt to advise conservationists on the basis of a mix of informed common sense and dubious phylogenetic analyses (e.g. Purvis et al., eds. 2005).

Costello et al. (2013) have suggested that in spite of increasing numbers of taxonomists and their use of better sampling and analytic techniques, the numbers of new species described per taxonomist are diminishing. They suggest as explanation that about two-thirds of all species have already been discovered. Their study is flawed in that shotgun approaches by pre-1900 taxonomists provided an inadequate set of new species names to compare with the more carefully analytic contributions of taxonomists of the past 100 years, in which new names are commonly evaluated against species recognized in recent monographs. The hoary species of the taxonomists of bygone geographic exploration times had little evolutionary or morphological/anatomical context, and, given the myriad present and potential synonyms, are presently useful largely as preliminary sortings. Thus, both the word "species" and the word "description" are not comparable in the Costello (2013) study in such a way as to estimate a slow-down in species description implying that two-thirds of all species have been discovered (that is, 1.7 million described of 2.5 million total). Why do I mention this study here? Because conservation study is loaded with such tendentious nonsense, and the practitioner must be wary. The same inexorable logic on faulty assumptions leading to false conclusions occurs in any black box solution. A similar problem exists in the case of massive accumulation of atomized phenotypic traits (Burleigh et al. 2013) and molecular sequences that are intended, after computerized sister-group analysis, to "provide a baseline for conservation management plans..." (Stech & Quandt 2010).

A neutral, well-hooked systematic framework — A framework for analytic classification that minimizes evolutionary bias and yields a classification more theory-neutral than that of phylogenetics was suggested by Zander (2010). Evolutionary theory promotes consistency, but the specious precision associated with restriction to synchronic relationships must be abandoned. The complexity of evolution can only be represented by a classification based on both synchronic (clusters of present day taxa) and diachronic (theoretical progression of taxa as descent with modification) relationships.

Taxa of higher rank can be scrambled if macroevolution is rejected in classification through the phylogenetic principle of holophyly. Consider a terminal group in a molecular cladogram: ((A1, B)(A2, C)).... This clade is balanced between two groups, with genus A split with one exemplar closer to the exemplar of genus B and another closer to the exemplar of genus C. Evolutionary systematics would recognize the paraphyly of genus A, and postulate as a progenitor taxon A for all extant exemplars, with classification presenting three genera A, B and C. Phylogenetics would either recognize one genus A with four species, or four genera, with one exemplar of A representing a new cryptic genus. The generation of new genera based solely on splits in a molecular tree does not reflect macroevolutionarily important transformations.

Choice of taxa for conservation concern is difficult but evolutionary evaluation as suggested in this book may provide an additional perspective. Taxa low in a morphological cladogram but high in a molecular cladogram may be inferred as long-surviving ancestral taxa. Given the dissilient species concept (Chapter 8), protection should be extended to rare occurrences of generalist species in a country or regions, and to taxa specialized for presently expanding environments.

Nash equilibrium — The morphological cladograms with the analysis of the moss genus *Didymodon* (Chapter 8) infer probabilistically that the products of dissilience from a generalist ancestral species are usually rather isolated and specialized for marginal or rare habitats. There are many more stirps (descendant species) than descendant generalist species that themselves are ancestral to another collection of stirps. Why is this? Why is there little apparent selection towards generalist, biotype rich species that can invade and equilibrate (e.g. Red Queen effect: Van Valen 1973, 1976) in an optimal habitat (plenty of food, equable climate, maximum photosynthate)?

The Nash equilibrium is commonly invoked in evolutionary game theory to explain features of checks and balances in nature, particularly for selection towards optimal equilibria among and between competing or potentially competing species (Hofbauer & Sigmund 2003; Nowack & Sigmund 2004; Robson 1990; Swanson 1994, but see Mailath 1998).

In the intentional case, if players have a limited number of choices but only one player succeeds if more than one player makes the same choice, then each player maximizes the chance of success by selecting a suboptimal choice. This also assumes that all players are aware of the other players methods of choosing. In this case, if every player vies for the optimal choice, only one gets their choice and all but one fail. In the movie about John Nash, the equilibrium is dramatized by a number of young men and women in a bar. Each man would like to take a woman home with him, but a certain amount of effort at courting is necessary. There is one most-attractive woman in the bar. Nash figures that if each man chooses to court any woman but the prettiest, chances of, well, fornication are maximized for all.

How does this work in a non-teleological case, namely, how is biodiversity maximized through selection or strategies (heritable phenotypes). Apparently three features direct selection away from optimal environments where fitness is highest but intensely limited by competition. (1) Generalism,

where specialized taxa are less likely to generate other taxa of different specialization than are generalized taxa, particularly when special environments are few and far between. (2) Phyletic constraint, where the range of possible viable new taxa is limited by the morphology and physiology of the ancestor. (3) Gouldian "wall" (see Gould's 2002: 893 speciational reformulation of macroevolution) of probabilistic exclusion around already species-rich optimal environments, where successful entry into an already highly competitive environment is far less likely than establishment in suboptimal sites. Biodiversity itself is an adaptation away from optimal environments, in the sense that no curbs are evident to conserve physiological resources that generate variation away from optimality.

The Nash equilibrium supports selection towards non-optimal habitats around Red Queen citadels with Gouldian walls of optimality. Avoiding competition results in a maximization of biological diversity, and that biodiversity has a direct quasi-altruistic positive effect on all organisms on Earth. The dissilient species concept and a world-level evolutionary strategy towards minimizing the struggle for existence may help evaluation, intervention, and maintenance of both biodiversity hot spots and suboptimal but critical reservoirs.

Potential biases in phylogenetic systematics — Although additional information allows better accuracy in statistical evaluations, systematic biases are refractory. Analysis of more biased data does not change the bias (Morrison 2013). Structuralism-related assumptions of phylogenetic systematics may generate biased classifications that can covertly (Law 2011) affect analytic results in other fields that depend on natural classification. An annotated list of some biases and narrow assumptions that are inherent in phylogenetic methods is presented here, at the risk of being accused of building straw men. These are inherent and easily detected in most published phylogenetic papers, and necessarily generate biased evolutionary views and classifications reflecting these assumptions. A source of error of considerable concern to conservationists is that of poor estimation of extinction rates from phylogenetic data, even with complete taxon sampling, due to variation in diversification rates among lineages (Rabosky 2010). Likewise, sampling problems make problematic attempts to use statistical devices in recognizing mass extinctions (Boucot & Gray 1991: 1295).

Bias: *Cladograms are fundamental because evolutionary inferences approximate them just as reality approximates mathematics.* — We all look for patterns, but no single pattern in evolutionary analysis is as fundamental as the axioms of mathematics or certain principles of physics (Bergmann 1949: 195), no matter how limited any axiomatic system is by, say, Gödelian logic or multiple universes (Nagel & Newman 2008; Rucker 1983). A molecular tree is a nesting of present-day specimens (synchronic) inferring the *results* of evolution through time (diachronic) for those specimens, but is not a complete representation of evolution of taxa.

Bias: *All evolutionary patterns are synchronic (in present-time), therefore the possibility of surviving ancestors affecting analysis may be ignored.* — This assumption is method-based, as all evolutionary patterns generated by phylogenetic software are necessarily synchronic because trees are based on a data set of traits. There is no theory presented of one taxon changing into another (macroevolution) but only of trait transformations (microevolution). Thus, if microevolution determines, methodologically, evolution, then evolutionary theory involving macroevolution is methodologically unnecessary in phylogenetic classifications.

Bias: *Cladogram splits are pseudoextinction events, with the ancestor dying off after two descendants are generated.* — Following this assumption, not only may the possibility of surviving ancestors be ignored, but all shared ancestors are unnamable because they are extinct. Actually, paraphyly involving taxa surviving at least one speciation event or even more has been estimated as widespread in extant taxa by Funk and Omland (2003), who indicated that species level paraphyly or polyphyly occurred in about 23% of assayed species. Rieseberg and Brouillet (1994) suggested that at least 50 percent of all plant species and possible much more are products of geographically local speciation, of which half are likely to be not monophyletic, and that in plants "...a species classification based on the criterion of monophyly is unlikely to be an effective tool for describing and ordering biological diversity." Based on simulations, Aldous et al. (2011: 322) asserted, "... for about 63% of extant species, some ancestral species should be itself extant ...." These studies are important, which is why I've repeated their mention three times.

Bias: *All reproductively isolated intraspecific lines and all distinct molecular lineages will differentiate into new species.* — There is now software (Ence & Carstens 2011: 473) that evaluates, on the basis of sampled molecular data within a species, with which intraspecies molecular lineages "can be validated as distinct" in that they have the "potential

to form new species before these lineages acquire secondary characteristics such as reproductive isolation or morphological differentiation that are commonly used to define species." The continued existence of all but the most recent taxa over hundreds of thousand or millions of years indicate this assumption is clearly unwarranted. Such stasis in expressed traits may be due to stabilizing selection, which is probably the ultimate reason taxonomy works at all (Patterson 2005).

Bias: *The principle of holophyly (strict phylogenetic monophyly) is necessary to eliminate hypotheses of macroevolution, which are largely ad hoc intuitions, and therefore only cladogram splits are important in classification.* — According to de Carvalho et al. (2008):

"...only monophyletic units, independent of their rank, must be understood as 'natural entities' (and therefore real, subject to conservation); and (2) that the organismal collectives which are described as species, and that receive formal binomials, do not necessarily correspond to natural, monophyletic units. Species names are intuitive resources indispensable for purposes of communication and organization of information at the species level, but in themselves do not necessarily contain any real scientific value since many species are simply not corroborated as monophyletic...."

This is clearly circular thinking and tendentious in the extreme. The authors further state that "...conservation efforts should be aimed at monophyletic units, not at binomials devoid of real existence." This assumption is based on a phylogenetic principle of classification—holophyly—not a real thing in nature. Here evolutionary analysis and classification practices are conflated to the detriment of conservation efforts.

Bias: *Exemplars may be considered equivalent to taxa, thus there is no need for extensive sampling because sampling either supports the original analysis, or reveals homoplasy and paraphyly or polyphyly, elements of which will then be ascribed to new, cryptic taxa anyway.* — This is a rejection of macroevolution as a theory explaining how exemplars of one taxon can appear separated on a molecular tree due to a progenitor generating, without self-extinction, one or more descendants. This alternative explanation must be seriously addressed. Rejecting macroevolution immediately leads to naming molecular variants distant on a tree as different species.

Bias: *Shared ancestors are anonymous, because naming them as different taxa but at the same rank as that of their derived sister groups would invoke the principle of holophyly and necessarily collapse the cladogram.* — Again, a phylogenetic principle of classification, holophyly, affects the evolutionary analysis needed prior to a sensible, responsible classification, the cart coming before the horse.

Bias: *Postulating multiple unnamed shared ancestors is acceptable because maximum parsimony only minimizes counts of state changes, not taxic transformations.* — Unobservable shared ancestors that may not be named are clearly non-parsimonious ad hoc postulations of superfluous entities. The evolutionarily most parsimonious ancestor-descendant tree may be far less parsimonious than a cladistic tree minimizing trait changes among sister groups. For example, a surviving ancestral taxon may have given rise to several daughter species, and such daughter species may share some new traits; a cladogram constrained to the correct evolutionary tree may be cladistically less than maximally parsimonious. Given that many extant taxa are progenitors of other taxa of the same rank, inability to theorize about taxic transformations is clearly a limitation of phylogenetic methods.

Bias: *Evolutionary theory is unnecessary, because the cladogram is fundamental, like mathematical axioms. Theory is metaphysical in being not a fact but a narrative explanation, while a cladogram is a deduction from a first principle (evolution happens) and is logically as true as the first principle, and as true as the facts from which it is inferred.* — Theory is the basis of science. Although we do have certain axioms of physics, these are subject to change. Even the axioms of mathematics are not immune to criticism (Kline 1980). Theory is necessary to deal with through-time aspects of evolution, while phylogenetics prefers an inappropriate quasi-mathematical lemma-theorem deductive method (Zander 2010). Note that mathematical induction is not the same as scientific induction, it creates a general conclusion from a set of deductions from hypothetical truths. According to Mazur (2006: 160), "We use scientific induction to learn and discover..." but a mathematical induction is a logical consequence of accepted mathematical statements.

Bias: *Genera and higher taxa do not evolve, only species do; therefore, in absence of evolution of higher taxa, such higher taxa may be delimited solely by cladogram clustering.* — There are many more or less well-supported theories of the evolution of genera and higher taxa (Eldredge 1985: 150; Gould 2002; Hubbell 2005; Vrba 1980, 1984). A simple case is of differential extinction and generation of species subject to a single selection pressure across a genus (e.g., increasing aridity or single source predation); clearly the diagnosis of the genus must change over time, and such changes may be different for multi-species groups of the genus in different areas of the world. A more generalized phenomenon is "species selection" (Simpson 2013) in which frequencies of traits among species change over time.

Bias: *Analysis assuming Markov processes needs only present-day data.* — It is true that Markov analysis needs only present-day data to reconstruct the past, but this works only in an ideal, mathematical context. Increasing chaotic uncertainty of prediction in real situations cannot be ignored, and neither Laplace's demon or Markov analysis can predict or retrodict accurately true events in the more distant future or past, respectively. The simulation studies of Cartwright et al. (2011) showed that non-Markovian patterns of ancestral variation contribute to a lack of robustness in molecular phylogenetics. Amorós-Moya et al. (2010) found "experimental evidence for convergent molecular adaptive evolution," highlighting the importance of regulatory mechanisms in evolution. The ahistorical aspect of structuralism is reflected in the now popular Markov chain Monte Carlo Bayesian methods of phylogenetic analysis, yet the Markovian assumption is faulty.

The biases above illustrate the difficulty of thinking rationally or of considering novel ideas when much has been invested in a standard practice, and leads to an inertia of belief. We are all, of course, lumbered with this sunk cost burden to some degree.

How can a conservationist distinguish biased classifications? — Common sense can in many cases signal a disconnect between reality and phylogenetic classification. Take simple scenarios, like an argument that shooting polar bears is not illegal because they have been scientifically classified as brown bears. Birds-of-paradise are now classified as reptiles, so may they be poached? Harvesting natural populations of cacti might be defended as perfectly legal because the cactus family is now submerged in the Portulacaceae family. Similar very strange things have been brought to poorly educated jurists and juries.

There is no direct way to distinguish molecular cryptic taxa (species, genera, families) or absent taxa in new phylogenetic classifications, particularly those that do not give details of the analyses they are based on, and sometimes do not even give synonymy.

When new taxa are proposed, the reasons for doing so are commonly given, but such reasons may not be in the at-hand classification, checklist or biodiversity review. When taxa are eliminated, for whatever reason, the same is the case. Often one must be an expert in a group to identify changes made (1) solely by "total evidence" about sister groups that generate clustering on the authority of classification principles or dogma, and (2) those that are reasonably based on discursive logic and all available information. Worse, changes are often made on the basis of both morphological and molecular evidence, yet the morphological evidence may be that which appears by chance to support the molecular results, not the best combination of expressed traits that supports all evidence (including ecology, biogeography, cytology, Dollo taxon-level evaluation).

An egregious example of epistemological extinction (dogmatically based disappearance of taxa) is the case in the field of the present author's expertise, bryology, particularly the family Pottiaceae, mosses of harsh environments (Zander 1993). In an influential classification of the mosses (Goffinet et al. 2012), the authors eliminate three long-recognized families, Cinclidotaceae, Ephemeraceae and Splachnobryaceae (Arts 2001), lumping their generitypes into the Pottiaceae. There is no synonymy list. This synonymization is done with no discussion, but is apparently based on several recently published phylogenetic studies that show these generitypes to be nested deeply in the Pottiaceae. The thought that these three represent *families* molecularly nested in an ancestral family Pottiaceae is ignored on principle.

Hörandl and Stuessy (2010) indicated that isolated island lineages can quickly become strongly divergent from continental ancestors, yet such lineages may be denied proper taxonomic rank because they are often nested in larger taxa of the same rank. Examples they gave for the flowering plants include the genus *Robinsonia* (Asteraceae), of the Juan Fernandez Islands found to be phylogenetically derivative from the widespread *Senecio*, and *Lactoris fernandeziana,* of the monotypic island endemic family Lactoridaceae, found to be apparently derived from Aristolochiaceae. Hörandl and Stuessy also indicated that conservation of island taxa is threatened by gradual elimination of a proper taxonomic recognition, caused by apparent recent derivation from paraphyletic continental progenitors, because slowly mutating molecular sequences do not match the rate of rapid and major morphological divergence. Padial et al. (2010) review problems in estimation of numbers of species involved in biodiversity studies that are due to restriction of data to single lines of molecular evidence. Stevens (2008), in a review of major changes in flowering plant classification (APG 2009) from past family-level classifications indicated that the "iconic" parasitic plant family Rafflesiaceae are embedded within Euphorbiaceae s. str. and should be reduced to synonymy. The reader needs to be aware of the growing numbers of taxa whose reality is often based entirely or in crucial part on whether the clade nests or not in a larger group of the same rank. These are biases that affect conservation of endangered taxa of major divergent morphology.

Summary of effects on conservation and biodiversity study — Cladistic analysis is axiomatic, and has been long touted as a theory-free, discovery process. The cladogram is considered a discoverable fundamental pattern in nature, following the rationale of structuralism in other fields. All non-phylogenetically informative information is "mapped" or otherwise relegated to the cladogram, following this structuralist procedure. Because the quasi-mathematical method follows the structuralist linguist Saussure in involving only synchronic (present-day) relationships, the model of one taxon being derived from another taxon of the same or lower taxonomic rank is forbidden, being nonaxiomatic scientific theory thus mere explanatory narrative. Microevolution (descent with modification of traits) substitutes for macroevolution (descent with evolution of taxa). This avoids the personal judgment and discursive reasoning characteristic of "soft sciences," for instance using non-phylogenetically informative data in Dollo evaluation of taxon transformation, i.e., macroevolution. Holophyly (strict phylogenetic monophyly) ensures that macroevolution is not modeled in cladograms, which is possible by taxonomic recognition of paraphyletic (the same taxon distant in a cladogram) and apophyletic (taxa nested in other taxa of the same rank) groups. Stucturalism is opposed to theory-based science.

The result is that some taxa that are paraphyletic or apophyletic on molecular trees are threatened with: (a) complete loss of their scientific names (underlexicalization), (b) downgrading of rank to force them into taxonomically and evolutionarily different ancestral or derived groups, or (c) burial among a proliferation of molecularly distinguishable "cryptic" taxa (overlexicalization). Some of these threatened taxa are rare and endangered and all have been considered, by experts familiar with the taxa, evolutionarily distinctive by expressed traits. Loss of scientific names stymies conservation efforts by hiding or masking important taxa, or eliminating them entirely from consideration.

Conservation and biodiversity study, as well as any other field that uses classification of organisms, will be covertly affected by biased, synchronic evolutionary assumptions and artificial principles of classification associated with structuralism. To avoid such biases, a pluralistic systematics is advocated. At a minimum, researchers who expect more from systematics, given new techniques and information sources, are encouraged to found new study on traditional pre-phylogenetic work, and to trust the decisions of alpha taxonomists as complementary to new discoveries. Certainly the major world "hot spots" for discovering new species are the great herbaria and faunal collections, where there remain thousands of types and authentic material sorted and named by classical taxonomists in the past yet never or seldom examined in modern times.

As an example of biased analysis, the work of Jansson et al. (2013) tested the relative importance of tropical conservatism (few tropical clades colonizing non-tropical areas), out of the tropics (many tropical origins for non-tropical clades), and hypotheses of diversification rates in creating latitudinal diversity gradients. They concluded that in the 111 phylogenetic studies they analyzed, most clades originated in the tropics, with diversity highest in zone of origin. Given lineage zone transition analysis, adaptation to new climatic conditions would not be an obstacle to many clades. If it is true, as the present book asserts, that cladistics alone cannot determine monophyly and that a clade is only by arbitrary definition monophyletic, then the above results are highly problematic, and should not be used in conservation research or model-making.

The next 30 years of systematics research, it may confidently be predicted, will include the wearisome task of distinguishing true advances in systematics from wrong and damaging phylogenetic classification decisions that delete or multiply scientific names through holophyly and other structuralist practices. Phylogenetic systematics now cripples our very perception of nature by eliminating or scrambling names of macroevolutionarily important taxa. We must choose between science, where theory explains apparently disparate facts, and structuralism, an extreme and faulty form of fast and frugal heuristics (Gigerenzer & Selten 2002) wherein all facts relevant to one well-supported cladogram are simply mapped to the cladogram. This decision affects how we deal effectively or not with the supreme challenge of our time: the collapse of biological diversity and depend-

ent ecosystems because the earth is passing its carrying capacity for humankind.

A World Flora and heuristics — Heuristics in systematics is central to modern taxonomy. For instance, a World Flora has been proposed by the Global Strategy for Plant Conservation (GSPC) of the Convention on Biological Diversity (CBD) as an identification manual for the species of plants. One might, at first, imagine a standard floristic or monographic approach would be taken: including a list of taxa made, creation of keys to families, genera and species, illustration of at least the most commonly encountered or biodiversity-significant species, dot maps made of distributions world-wide, and so on. This would be an exact, deterministic, integrable (Markov chain-like differentiating keys) well-founded descriptions, complete flora. Yet the date proposed by the GSPC for completion of this very large work is 2020. A simpler method is clearly needed.

There are two ways phylogeneticists create large trees-of-life from multiple research projects. The supermatrix method goes back to the data and reanalyzes all the data at once for all taxa, but the supertree method takes the tree topologies and welds them together (Gatesy et al. 2002). The latter method preserves any special analytic treatment given to each group that addresses unique evolutionary features. When bryologists at the Missouri Botanical Garden convened to discuss a world flora for just the mosses, it became clear that major reliance must be made on floras already published, i.e., a metaflora, a kind of supertree. A checklist of the world's mosses (Crosby et al. 1999) had also been published that identified (with four stars) those taxa that were well-studied in modern treatments. Thus, a world flora of the mosses could be generated that depended on modern floristic treatments of well-known species, which would speed the work immensely and provide a heuristic for approaching plant biodiversity at the highly complex world level, while keeping track of all names of taxa in a grand checklist whether well understood or not. A similar approach might be decided upon for other plant groups.

Apropos of this major undertaking, one of the most pernicious problems for non-taxonomists, particularly informatics specialists compiling databases of accepted species names, is "which name of two or more synonyms is the correct name for the species?" The past rule of thumb for determining the correct name is just use the one in the most recent major monograph or most recent faunistic or floristic compilation. This must be modified because modern phylogenetic systematics allows classification to affect the preliminary analysis needed for evolutionary classification. The new rule of thumb (heuristic) should now be to accept the most recent name in a major work as correct *except* if it is generated following holophyly or based on universal assumption of pseudoextinction (phylogenetics) or sister-group analysis (cladistics). Conservationists can presently deal best with this problem by only using pre-phylogenetics or at least pre-molecular classifications or, better, to team with evolutionary systematists who will identify modern classifications with a minimum of phylogenetic bias.

CHAPTER 12
Scientific Intuition and the Hard Sciences:
The metric dimension heuristic and Gould's macroevolutionary wall

Précis — Heuristics or informal genetic algorithms in alpha taxonomy are founded on identification of patterns in nature using known relationships and principles in mathematics and physics. These heuristics identify values that provide a "tell" to a taxon's identity. They can also signal an "outlier" that negates immediate identification with a known or expected taxon. Here the heuristic involving metric dimensional ranges in descriptions of taxa, in the paradigmatic form (a–)b–c(–d) of three size classes, is analyzed and formalized from data gathered from modern bryophyte treatments. This paradigm aids in taxonomic decisions based on the known and expected distributional dimensional range of examined specimens. The geometric mean and Fibonacci series in powers of the golden ratio are both involved in distributions of informal measurements close to zero (i.e., where the range a–d is a significant portion of the range zero to d) or where ranges comprise much of a magnitude. Inferred ideal proportions for the three size classes involving the golden ratio are $1:(1.6)^2:1.6$, or $1:2.6:1.6$. Scientific intuition establishes dimensional ranges, in the bryophyte literature at least, consistent with physical and mathematical principles. The Fibonacci series (partly reversed) above is apparently also a fundamental rule explaining both S. J. Gould's speciational reformulation of macroevolution and psychologically salient numbers.

This discussion contributes to the literature on intuition in the sciences by examining a particular element of descriptions of taxa, namely the results of estimation of metric dimensions by informal sampling in alpha taxonomy. There are many heuristic methods that in combination are used by alpha taxonomists to identify and classify known taxa, and to distinguish new taxa. Alpha taxonomy is a first pass at biodiversity analysis to provide useful names and distinguishing traits for perceived groups of organisms in nature, while biosystematics, including statistical analysis of variance and investigation of population genetics that well-characterizes taxa, comes later.

Studies of biophysics can model traits and complex organs or behaviors that are important in systematics (e.g., Niklas & Spatz 2012). This chapter addresses a central feature of descriptive taxonomy, metric distributions.

Given the usual, often fierce arguments between characteristically multanimous taxonomists, how has systematics progressed at all over the past 250 years? Classifications produced by alpha taxonomists using classical methods (omnispection, Gestalt, apprehended covariance, and naïve analysis of variance) have been demonstrated time and again to be valuable in predicting the expression of variation among new collections, particularly small samples such as types of new species. They can be used to create a logical, theory-based, pluralist systematic method encompassing results of molecular systematics (apart from strict phylogenetic monophyly), and are thus

effective. Exactly why the heuristics are predictive is not clear, and formalization or explanation of such heuristics, as involved in scientific, informed intuition, is needed.

Formalization is the presentation of a heuristic in the context of physical or mathematical explanation. It is here exemplified by the following contrived very simple heuristic. Consider two propositions A or B that contradict each other. They are similarly supported but there is additional support for A consisting of independent hints. The number of hints it takes to support A over B such that a scientist will act on A instead of B in non-critical situations (e.g., make it a basis for additional theorization) is ... too many to consider (see below). Suppose there are instead a number of somewhat impressive but not decisive independent pieces of evidence in favor of A. The number of such medium well-supported arguments in favor of A to be decisive is ... surely more than two. I think most people would agree with these heuristic guesses in non-life-or-death situations. Formalization involving recourse to relevant mathematics and physics would ensure that one alternative would be "probably" better than another with "probably" meaning more than just anticipated intersubjective agreement (Hempel 1988; Kearney & Rieppel 2006), instead more on the level of logical and empirical probability as discussed by Pap (1962: 195, 213).

In optimization modeling (Martignon 2001), Bayes networks have been found effective in decision theory. Formalization of the metric dimension heuristic can here use Bayes' Formula, with each item of support for A given a probability. For an item of support to be classed as a "hint" it should be distinguishable from 0.50 probability (totally equivocal between two alternatives, like yes and no), let us say 0.60. Using Bayes' Formula with 0.60 as prior and 0.60 as probability yields a posterior probability in favor of A of 0.69. Using 0.69 as prior with 0.60 again as probability yields 0.77. This empiric Bayes' procedure (Gigerenzer et al. 1989: 273) of using the result of Bayes' Formula as the prior for another instance with additional information is continued until the posterior probability exceeds 0.95, a common minimum level of confidence for non-critical conclusions. In fact, it takes eight "hints" at 0.60 probability each to exceed 0.95. The sequence is 0.60, 0.77, 0.83, 0.88, 0.91, 0.94, and 0.96. Requiring eight hints even with no contrary data is practically equivalent to "too many to consider."

In the case of "somewhat impressive but not decisive" support, we will assign each element of the support a 0.75 probability, being half way between entirely equivocal and certain. Two items yield 0.90, three 0.96 posterior probability, thus "surely more than two" is a correct heuristic for good but not decisive support for A over B. It should be clear that, although this heuristic is only a rule of thumb, using many such heuristics involving different kinds of support can "triangulate" a practical solution for one complex problem quite effectively.

This book assumes that nature informs us what our species (or taxon) concept should be for various groups, and will show that abducing hypotheses in alpha taxonomy is built on paradigmatic templates associated with well-recognized principles in physics and mathematics. The simplest species criterion is taken as that of Crum (1985). This is essentially the evolutionary species concept (Simpson 1961; Wiley & Mayden 2000), paraphrased as a biologically unified group with a unique evolutionary trajectory. That evolution is important in modern alpha taxonomy is exemplified by routine searches for geographic or ecological distinctions correlated with morphology and emphasis on homology in selection of traits to study.

Classical heuristics and decision theory — Both morphological and molecular analysis use decision-based heuristics (Gigerenzer, 2001, 2007; Gigerenzer & Selten 2001; Gilovich et al. 2002; Goldstein et al. 2002; Hutchinson & Gigerenzer 2005; Martignon 2001; Stanovich et al. 2008; Tversky & Kahneman 1974) to speed and simplify complex searches to get results that are not guaranteed optimal but are at least close. Examples are heuristic sampling in parsimony analyses and Markov chain Monte Carlo analysis in Bayesian studies.

One of the most basic heuristics in any field is the "covariation principle" (Kelley 1973) in which causal effects are considered potentially involved with two phenomena that covary, but this may by discounted if other equally plausible causes are evident. With multiple observations, a "naïve version of analysis of variance" is initiated with causes treated as independent variables and effects as dependent (Littlejohn 1978: 234). In classical alpha taxonomy, evolutionary relationships are inferred through informal genetic algorithms for rule production (Gigerenzer 2007; Hutchinson & Gigerenzer 2005) as a *heuristically based expert system* (Zander 1982). The quasi-optimal results are used in creating descriptions of taxa when time and funding limit sampling and testing.

We have all heard myriad heuristic guidelines similar to "Once is happenstance, twice is coincidence, three times is enemy action." There are many that are commonly used in everyday life in business strategy but have no familiar associated aphorism. In technical decision theory, Goldstein et al. (2002) offer the "Take the best" heuristic, in which inferences and predictions are based on only a part of the information until a stopping rule ends the search and decisions are made on basis of the cue that ends the search. Such predictions apparently are better than those made by multiple regression, and are based on not allowing less important data to overwhelm highly weighted data, a lesson ignored in phylogenetics when traits are equally weighted. Another heuristic is "Take the first," which contemplates a series of alternatives and ends when an adequate solution is discovered. This depends on memory recall and similar sets of problems.

A familiar example of a heuristic is the practice of "Manual Image Mining" in organism identification (usually by experts). This is more commonly known as "flipping through the pictures in the manual until you find it" method of identification of an unknown whose name does not immediately come to mind. One segregates the section of the identification manual between two hands, and leafs through, marking look-alike illustrations with fingers and thumbs, occasionally an elbow, until one has a solution, usually "That's it! Or maybe this one, or possibly even this one." Then the descriptions and known distributions are read, and lastly technical keys are gone through until one name is found as appropriate. Is this a random search guided by mere intuition? No. If one *formalizes* the method it is clear that the taxonomist is creating a one-time key to the literature at hand and using both technical and "look and feel" traits to identify rapidly an unknown specimen. Digits are the couplets, and sampling is from total evidence as the investigative matrix.

Tells and outliers are here collective names for traits valued as cues (Gigerenzer & Selten 2002: 5) in alpha taxonomy to rapidly identify known taxa or to flag new taxa, or at least signal problematic variation in expressed traits, when one is surveying a sampling of specimens. A "tell" is a trait valuable for identification that is unique to a species or other group, perhaps only found in some specimens of an otherwise difficult to identify taxon. Examples in the Pottiaceae (Bryophyta) are the general absence of a central strand in Trichostomoideae and Leptodontieae; enlarged cells in the medial portion of the stem central cylinder in Barbuloideae; irregular peg-like gemmae in *Gyroweisia tenuis* (Schrad. ex Hedw.) Schimp.; elongated medial laminal cells often present in *Hymenostylium recuvirostrum* (Hedw.) Dix. and *Didymodon tophaceus* (Brid.) Lisa; spherical tubers in *Barbula convoluta* Hedw. (Zander 1993); and, other uncommon traits that strongly aid identification and characterization of evolutionarily unique taxa.

Vavillov's "Law of Homologous Series" (Vavillov 1951; Yablokov 1986: 34) is another heuristic that operates at levels higher than that of species. It states that species or genera commonly have variability with parallel forms in other related species or genera. This is also true at the family level, according to Vavillov.

All "fast and frugal" heuristics exploit regularities in the environment, including those in data, but may not be entirely generalizable (Gigerenzer & Selten 2002) as is the case with the aforesaid examples. "Outliers" are negative tells; see Lim et al. (2012) for statistical methods of identifying outliers. Negative outliers suggest that whatever name immediately has come to the taxonomist's mind as an identification is probably not correct because there is a trait that has not been not recognized previously for the group or there is at hand a dimension out of expected range.

Inasmuch as tells and outliers are not particularly functional in sister-group analysis, being either autapomorphic or commonly lacking precise documentation, modern phylogenetics largely ignores them. They are, however, fundamental to the practice of alpha taxonomy. This chapter formalizes one of these elements, dimensional heuristics; that is, how one recognizes that a measurement is outside what is expected of the range of variation in a known taxon and initiates closer scrutiny and study of additional traits.

There is reason for revisiting alpha taxonomy, even after alpha taxonomy has been deprecated as being intuitive or even instinctive (Hey 2009; Mooi & Gill 2010; Scotland et al. 2003; Yoon 2009). Even if more precise methods have been introduced through numerical taxonomy and molecular systematics, these have proven problematic (Tobias et al. 2010), mainly because of analysis unreflective of macroevolution (Zander 2007a, 2008, 2009, 2010), and multiple equally plausible hypotheses due to possible unsampled or extinct molecular paraphyly and extended paraphyly, which may involve as much as 63 percent of the taxa studied (Aldous et al. 2011). In addition, mathematics does not determine natural relationships (as per Klein 1985) but is simply a hyperprecise "approximation" to real, probabilistic distributions and phenomena for which data are never entirely sufficient for full description (Klein 1980). It is statistics, actually a field of physics, that may provide a better explanatory and predictive picture of reality. For instance, methods of statistical physics have demonstrated a double power law in distribution of extinction sizes from paleobiological data (Sznajd-Weron & Weron 2001).

There is a notion, particularly among mathematicians, that nature follows mathematics (Ekeland 2006; Kline 1985). On the other hand, the perceived real features of nature are fuzzy, somewhat indeterminate, and probabilistic due to the influence of complexity and chaos, plus the fact that no phenomenon is fully described by available data. Mathematics is then hyperprecise for phenomena that are difficult to encompass with a precise answer, and mathematical solutions thus may be inaccurate for all but the simplest, most well-understood phenomena. The bull's-eye is not the target. An exacting bill for services may be wrong. An integrable solution may produce a simple and repeatable classification but the implied evolutionary relationships may be scrambled.

Richard Feynman (Feynman 1985: 70) asserted that he often won arguments by detecting a difference between ideal, mathematical models and real-world examples when puzzles were presented to him. For instance, although in topology an ideal orange may be cut up and rearranged into a sphere the size of the world, manipulations of a real orange is limited in that the thickness of its rind cannot be less than an atom. All integrable (fully solvable) problems in systematics must pass the real-world test, which means that evolutionary models must explain all evidence in a noncontradictory manner.

Systematics is a classic example of the difference between analysis based on optimization and on heuristics in generating accurate decisions among many alternatives (Gigerenzer & Selton 2002). Also, phylogenetics is restricted to integrable problems (Ekeland 2006: 80), in particular Markov chains and parsimony analysis. These, ideally, fully predict the past and future, as a generalized solution, and initial uncertainty is not increased. Speciation and other macroevolutionary events require nonintegrable analysis based on data that fade with time or which lead to chaotic results. Nonintegrable analyses (Ekeland 2006: 97, 103) must involve consideration of individual elements (taxa) with all data available, with regard to what periodicity is available. Metric dimensional range is an example of such periodicity. Given that no one method ensures certainty, a pluralistic approach to taxonomy is needed (Beatty 1994; Giere 2009; Padial et al. 2010; Rieppel & Grande 1994).

Taxonomic heuristics have been explained in general terms (e.g., Zander 1982) in the past. This consists of long-term accumulation of hard-won rules of thumb that are proven repeatedly effective. Gigerenzer (2001, 2007), Goldstein et al. (2002), Hutchinson & Gigerenzer (2005) and Martignon (2001) have well described the genetic algorithm process involved in these apparently idiosyncratic methods, but which are common in many fields. Alpha taxonomic methods are therefore a Gestalt or omnispection process only in the sense that the taxonomist's unconscious collective of useful genetic algorithms have not yet been detailed and formalized as an uncommonly effective set of heuristics. Although intuition can be easily fooled (Kline 1985: 32–34, gives several examples of improperly evaluated mathematical problems), this book takes the position that taxonomic scientific intuition has probably discarded most problematic or wrong predictive intuitive assessments through 250 years of testing and building on the work of others, while formalization of standard practice should enhance the perceived value of intuitive expertise.

Fuzzy logic — In complex systems, precise statements become more and more impossible to make until relevance and precision become mutually exclusive (Jameel 2009; Kosko 1993). Fuzzy logic gives mathematical descriptions of multiple factors affecting membership functions in particular sets such that variables are affected differentially at different times under feedback control. Control systems using fuzzy logic have been built for steam engines, cement kiln operation, water treatment, subway systems, expert medial diagnostics, tunnel excavation, automated aircraft landing, television, electronic eyes, dam gates, automobile cruise-control, and the like. Fuzzy logic is a mathematical expression of heuristics addressing complex systems with an over-whelming number of variables, and may be of value in dealing with historical complexities in retrodiction of speciation events.

Quantum heuristics — Aerts et al. (2010) demonstrated a go-or-no-go theorem involving quantum analysis (three possibilities, computable yes, computable no, and both yes and no but decided only upon actual examination of the results) for dealing with manifest data based partly on hidden variables, and Aerts (2009) discussed the well-structured mechanics of the double layer of human thought that figures in the balance between logic and Gestalt apprehensions of reality. It was shown that heuristics are based on entirely rational processes although involving quantum thinking. In both papers, however, an overarching theory (e.g., Aerts 2009: 22) can reconcile the apparently non-classical disjunctions and conjunctions associated with mesocosmic quantum phenomena in psychological study of cognition. In proper quantum thinking, aspects of description of a phenomenon that cannot be reconciled by common features (Gilder, 2008: 16) are used in concert to deal with or predict outcomes. For instance, the mesocosmic phenomenon of refraction of light in water requires true quantum thinking because no classical theory can deal with the fact that refracted light appears to itself calculate and choose the path shortest in time of travel between the emitter and the eye given different speeds in water and air, while stationarity (Ekeland 2006) is more a clever, though accurate mathematical description for this than a theoretic causal explanation. A quantum explanation (Hanc et al. 2003) involving phase cancellation has been advanced to explain disjunction in perception of direction and distance, but even this does not beat standard heuristics in helping spear a fish. Quantum heuristics is similar to the three-valued logic of Jan Lukasiewica (Jameel 2009), where 1 stands for true, 0 for false, and 1/2 for possible. Heuristics in the present chapter are those of classical, non-quantum theoretic evolutionary systematics that reconciles total evidence under the search for macroevolutionary relationships.

Adequacy of distributions — Santos and Faria (2011) pointed out that alpha taxonomy commonly involves examination of large numbers of individuals, and "small uncontextualized difference in sequences of DNA cannot necessarily define taxa." In many cases, molecular analysis commonly involves sample distributions of one. This refers to taxonomic differences at the species level and higher, not genetic differences between populations in which 20 to 100 individuals is a sufficient sample size (Kalinowski 2005). It is clear that the law of large numbers figures in the desire of alpha taxonomists to examine a large number of specimens to develop a description. This is because estimation of a mean, for instance, keeps getting more accurate the more a distribution is sampled, e.g., the more coins are flipped, the more accurate is the estimate of exact loading on one side or the other (FYI, most coins are heavier on the head side). A fundamental but unrecognized feature of taxonomic heuristics involving the critical central limit theorem (a general rule from physics, not mathematics) is the general statistical rule that about 30 samples are sufficient to ensure a normal distribution of samples from distributions that are not highly skewed or multimodal (Games & Klare 1967: 247–248; Yamane 1967: 146). Smith and Wells (2006) tested the rule, and demonstrated a spectrum of reliability, with 15 samples being sufficient to establish a normal sample distribution in most normal data sets, and 30 for bimodal well-behaved data sets. But not even 300 samples are able to deal with heavily skewed distributions. Using real data sets, however, they found that consistent following of the normal sampling distribution did not begin until 175 samples were made.

Curiously, at least some molecular phylogeneticists are apparently able to infer correct species delimitations with a single sample and 50 DNA loci, or 5 to 10 samples and only 1 or 2 loci (Zhang et al. 2011), including distinguishing cryptic species with no morphological distinguishing features. This is based on degree of reproductive isolation and relative genetic homogeneity of populations following species divergence, according to the authors. The work was based on simulations, and the technique is asserted to be not misled by samples taken from distant areas of a wide-ranging species. Cryptic species recognition and certain narrow assumptions about variance in genetic homogeneity may account for this almost magical taxonomic facility, while assumptions include pseudoextinction (speciation requiring disappearance of ancestor through anagenetic change yielding two descendants), a relaxed biological species concept, and concordance of gene trees across multiple loci to indicate a distinct, stable species.

Alpha taxonomy, particularly in revisions, commonly expects adequate sampling of specimens for each species at this ballpark level, say 30 to 175 specimens. Phylogenetic analysis uses heuristics (e.g., Hastings-Metropolis sampling) to sample multimodal data spaces, yet examples of sampling for unknown modes of DNA sequence data within taxa (i.e. each equivalent to a species description from classical taxonomy) are few. The normalized sampling distribution allows a good estimate of the mean of the sampled, potentially non-normal or even highly skewed distribution. When only one or a few specimens of a species are available, reasoning by analogy is used (Kline 1985: 48) in classical taxonomy such that ranges and modes of variation of morphological traits of similar taxa are assumed to be similar. Such analogy is not unusual, and has been found generally predictive of estimated features when additional specimens become available.

Questions posed — One question asked here is "How does one tell if a specimen's dimensions are so far outside the expected range that an explanation is needed or a new taxon proposed?" Explanations might be found in extreme habitat variation affecting the phenotype, in implied genetic differences, or perhaps macroevolution-based distinction at the taxon level. Informally, an expert in a group can "tell" the negative, or get an uncomfortable (or excited) feeling about unexpected dimensions; but what is the fundamental reason or methodological process in this standard methodology for flagging differences in alpha taxonomy?

If a dimension is found to be greater or less than expected, how was the expected dimension known, particularly when only a few samples may be available? When identification of a specimen involves simply distinguishing it from one or a few other species, the ranges of variation given in published descriptions may be kept in mind, but are such descriptions written in stone? When should the description be modified by new information? What about distinguishing a new species, for example, from a host of other, similar and variable species in a group? A heuristic is involved that allows an expert in the group to recognize when one or a couple morphological dimensions are unexpected, short of extensive scoring and formal analysis.

If the usual range of dimension of a trait is estimated through an informal evaluation of variance based on observation of many specimens, somehow an expert has an idea what extremes might be expected. *What is that heuristic?* If extremes of ranges are known, then the central value can be estimated by using a formula for a central value. Many such "average" formulae are known in morphometrics, including: the *arithmetic mean* (the sum divided by the number of values), used to find that one value which added as many times will also give the total; the *geometric mean* (product of extreme values divided by the square root), used to find that one value which when multiplied as many times as there are values will give the total; and the *harmonic mean* (the reciprocal of the arithmetic mean of the reciprocals), used when values are defined in relation to one unit, such as averaging rates. Given that the range of dimensions are usually given as proportions, is the golden ratio, a particularly common proportion in science and art, involved?

This paper proposes that a taxonomic dimensional heuristic in common use may be formalized by making a general survey of known ranges of metric morphological dimensions, both of usual and extreme ranges, and distinguishing the basis for estimating an expected low or high extreme such that an observed dimension is outside the expected low or high range. Observed is a curve of frequency of measurements against dimensions of measurements. Since frequency is not recorded beyond one size class (the middle) being "most common," in a description the data are rendered as a one-dimensional distribution in three size classes unique to each taxon.

Methods of analysis — Descriptions of taxa involve skewed paradigms of (a–)b–c(–d), where the usual range b–c is larger than the extreme high range c–d, which itself is usually larger than the low range a–b. Investigated are the proportions of this distribution, assuming high and low ranges are tails, and correlation with any fundamental relationships in physics or mathematics that might determine in part these proportions. Descriptions of acrocarpous mosses in volume 27 of the multiauthored Flora of North America Flora (FNA) (Flora of North America Editorial Committee 2007b) and in the first half (pages 1 to 245) of volume 4 of Flora Briofítica Ibérica (FBI) (Guerra et al. 2010) were surveyed by entering into a spreadsheet dimensions of stem length, leaf length, leaf laminal cell size, and spore size. These are, in the author's experience, a good source of tells and outliers for taxa, and are generally considered so important in taxonomy that they are almost always detailed in descriptions. Dimensions recorded were ranges on scales beginning at zero, and were limited to those providing both the usual range and one or both extremes, that is, in the paradigmatic proportional forms (a–)b–c(–d), or b–c(–d), or (a–)b–c. Dimensions giving only the usual range, that is, only b–c, were not recorded because this study is investigating outliers, that is, unexpected extremes as opposed to expected extremes, outside either low or high values. Authors who give only two values (only b–c) on a scale for a range are not clear as to whether the values are extreme or usual values or something in between, but those who give only one extreme value are here taken as meaning for the other far value to be both the usual expected and the extreme observed.

Thus, the range values examined represent what taxonomists will expect and tolerate as extreme variation against the central range of usually encountered dimensions, generally on a scale of zero to 40 (whatever the metric units measured) with precision generally limited to whole numbers or one decimal place. Although it could be pointed out that juvenile features necessarily grade from zero in dimension, the measurements taken from actual descriptions are assumed to bound dimensions of mature parts of the plant.

The dimensions were analyzed in Excel spreadsheets (data available from the author's Web site http://www.mobot.org/plantscience/resbot/misc/Geo Mean.htm). Four columns included the two values for the usual range, b–c, on a scale and a least one extreme, a or d of the above paradigm. When one extreme matched the usual value, i.e., the extreme was not given, the value was entered as identical to the low or high usual value as appropriate. These were averaged by column (Table 12.1).

Data set	Number of samples	Raw data averages (a–)b–c(–d)	Ave. a–d	Ave. b–c	Proportions of total range, raw data, %	Ratio of bold-faced
FNA	305	(5.8–)6.9–10.7(–13.9)	8.1	3.8	13:47:40	
FNA abcd	79	(5.2–)8.2–12.4(–16.8)	11.6	4.2	**25:36:38**	1:1.49
FBI	409	(21.0–)25.0–38.0(–44.5)	23.5	13.0	**17:55:28**	1:1.65
FBI abcd	124	(26.2–)33.5–49.2(–61.4)	35.2	15.7	**21:45:34**	1:1.63
FNA+FBI	714	(14.5–)17.3–26.3(–31.4)	16.9	9.0	**17:53:30**	1:1.77; 1.77:1
FNA+FBI Max. of 10 or less	321	(1.5–)1.9–3.2(–4.1)	2.6	1.3	15:**50:35**	1.42:1
FNA+FBI Min. of 20 or more	177	(44.1–)52.5–79.7(–93.4)	49.3	27.2	**17:55:28**	1:1.65
Poaceae	215	(8.9–)10.8–23.8(–31.7)	20.3	17.3	**11:62:18**	1:1.64
Smith Hep.	166	(38.9–)43.4–82.5(–103.0)	64.1	40.0	07:62:30	

Table 12.1. — Raw scores of metric dimensions given as the range paradigm (a–)b–c(–d) with occasionally a = b or c = d but not both at once, except that data sets labeled "abcd" are truncated to only data with *both* extreme values given as different from the usual values. Note that the range Ave. a–d is a large proportion of the range zero to d. Numbers in boldface are approximately the golden ratio (1:1.618...).

Four additional spreadsheet columns were of those same dimensions but "standardized against the maximum" for unbiased comparison of paradigmatic proportions. The first three columns were of low extreme value or lower bound, a, low usual value, b, and high usual value, c, each divided by the high extreme value or upper bound, d. This is an acceptable standardization method because the minimum possible value is zero. These values were expressed as percentages of d. The fourth column, d, was of course entirely 100.

The four columns of standardized values were then summarized (Table 12.2) by averaging each column to give a single value for a, b, c, and d. The four averages allowed estimating three important dimensional proportions: extreme low range a–b (or b minus a), the usual range b–c (or c minus b), and extreme high range c–d (or d minus c), as applying to all taxa studied.

It was expected that the proportion of these three ranges is the essence of the simple dimensional heuristic developed over time by experts (at least for these studied taxa) as part of a complex set of heuristics for distinguishing known taxa and flagging the presence of new taxa.

Data set	Scores, stand. to the max., %				Proportions, (a–)b–c(–d)		Means						Possible golden mean
	a	b	c	d (max)	a–b:b–c:c–d	total range, %	GM (a–d)	GM (b–c)	Ave. (a–d)	Ave. (b–c)	HM (a–d)	HM (b–c)	Ratios of bold faced numbers
FNA	41	48	76	100	07:28:24	12:**49:42**	64	61	71	62	58	59	1.67:1
FNA abcd	35	50	76	100	15:26:24	**23:40:37**	59	62	68	63	52	60	1:1.61
FBI	49	58	87	100	09:28:13	18:57:25	70	71	75	73	66	70	
FBI abcd	45	58	81	100	13:23:19	**23:42:35**	67	69	73	70	62	68	1:1.53
FNA+ FBI	45	53	82	100	08:29:18	15:**53:33**	67	66	83	68	62	64	1:1.61
FNA+FBI "d" of 10 or less	41	49	81	100	08:31:19	13:**54:33**	67	66	71	65	58	61	1.64:1
FNA+FBI "a" of 20 or more	51	59	87	100	09:28:13	**17:57:26**	71	72	75	73	67	71	1:1.53
Poaceae	33	40	82	100	7:42:18	12:64:24	58	57	64	61	49	54	See discussion
Smith Hep.	44	49	84	100	5:35:16	09:62:29	65	66	69	67	61	62	See discussion
Golden ratio $1:\varphi^2:\varphi$						19:50:31							1.62:1 or 1:1.62

Table 12.2. — Standardized to the maximum scores, being percentages of the high bound d (max), which is therefore always 100. The proportions between scores imply the dimensional heuristic in systematics. The harmonic mean (HM) clearly does not match a–d and b–c midpoints as well as the geometric mean (GM). Numbers in boldface are approximately the golden ratio (1:1.618…). The metric dimension proportion for a Fibonacci series in powers of the golden ratio is given in last row.

The proportion was expected to be skewed to the left in that there is little room between extreme low values and zero (or a developmental minimum size), while high values are free to vary. The discovered heuristic proportion was then compared with the arithmetic mean, the geometric mean, and the harmonic mean with respect to the midpoints of the usual range b–c and the full range a–d.

The data sets were also analyzed for only those data for which all four columns were of different values, i.e., by eliminating any data lacking one or the other of the extreme values. Thus, only those data with the form (a–)b–c(–d), with both extremes given as different from the usual range, were studied (Table 12.2), for FNA abcd and FBI abcd). Additionally the

data sets were divided into measurements near zero, that is metric values of 10 or less and measurements far from zero, that is, metric values of 20 or more (Table 12.2). This is designed to investigate the value of the geometric mean in estimating ranges of a proportional distribution near a magnitude in breadth or close to zero (or to a structure's developmental minimum).

All metric units are here considered equivalent for this study because the method depends on the closeness to zero, or a structure's developmental minimum, of a range on a scale. Close to zero means that the range a–d is a significant portion of the range zero to d. The proportional distribution of measurements can be considered the same whether a range of numbers near zero is measured as in micrometers, centimeters or decameters, as long as the structure measured is properly measured in those units from 1 unit close to zero and small multiples of that unit. The raw data are best evaluated after standardization, and differences in post-standardization proportions are reflections of real differences associated with the commonly exponential distribution.

Results: the proportional heuristic for metric dimensions in mosses — For FNA, a total of 305 dimensional formulae were tabulated, of which 95 were dimensions of the stem; 102 were of the leaf; 70 were of the leaf cells; and, 39 were of spores. The FBI provided 409 data records, of which 52 were dimensions of the stem; 86 were of the leaf; 237 were of the leaf cells; and, 34 were of spores. For FNA, 80 were of the form (a–)b–c(–d), with all data different; for FBI, 124 were of that form. Thus, 0.75 of the FNA and 0.70 of the FBI data sets were of data with only one extreme given, and one end of the usual range assumed as also the extreme. A combined full data set of 714 records was also examined.

The average ranges of the raw scores of metric dimensions (or any unit) are summarized in Table 1 as four columns. After standardizing the four columns against each other by dividing the three lower values by the highest, the average was taken of all the standardized values in each of the four columns. For FNA full data set, the standardized to the maximum values were 0.41 for a, 0.48 for b, and 0.76 for c, against 1.00 for d (Table 1). The differences between these average values were given as proportions. The average proportions between these values are 0.07 for lower bound to low usual (i.e., a–b), 0.28 for low to high usual (i.e., b–c), and 0.24 for high usual to upper bound (i.e., c–d), then converted again to proportions of the usual range, 20:100:81. The rough proportion using integers is 1:5:4. From this clearly skewed proportion, the range of high extreme values can be expected by the taxonomist to be a little less than the range of usual values, while the low extreme values would be about 0.20 of the range of the usual values.

For FBI data set, the standardized to the maximum values were 0.49, 0.58, 0.87, and 1.00 (Table 1). The proportions of these values were 0.09 for low extreme, 0.28 for usual values, 0.13 for high extremes (Table 1). Proportions with usual range at unity are 26:100:47, roughly 1:5:2.5, about the same as for FNA full data set, but with curtailed upper extreme range.

For FNA plus FBI full data sets, standardized scores were 0.45, 0.53, 0.82 and 1.00. The raw pro-

portion was 8:29:18, yielding 25:100:62 with usual range at unity. The rough proportion is 1:5:3, similar to that of FNA.

Discussion — Basic to biodiversity study is recognition, taxonomic analysis, and description of unique taxa, stipulating, however, that some taxa intergrade. What are we describing? Something intuitive or even instinctual as suggested by several modern authors (Scotland et al. 2003; Yoon 2009)? Do taxonomists innately recognize taxonomic patterns (Crum 1985)? If so, how?

Abduction, the devising of a hypothesis, is a central feature of the scientific method. A reason is posited as an explanation for a given observation (Pierce 1903). There may be many abduced explanations, yet, for hypothesis testing, one is singled out as the more worthwhile to test. It can be as simple as educated guesswork or there may be rules for selecting hypotheses for testing. Abduction is usually done through educated guesswork or a system of rules.

A taxonomic description is a set of answers to the application of an established set of heuristics. These also known as rules of thumb or genetic algorithms (Hutchinson & Gigerenzer 2005; Gigerenzer 2007). Commonly, rules of thumb (Parker 1983) are rather trivial because when used alone they may be incorrect by an order of magnitude, though desperate persons value guidance even at that level. On the other hand, when many different heuristics are applied to any one taxon, each heuristic is a sort of triangulation vector helping characterize a real thing "out there" that is (well or poorly) definable because of the nature of evolution. Descriptions are then complex in heuristically guided character dispositions. They are, however, even when based on very small samples, quite accurate in prediction of distinctiveness of a taxon, that is, as measured by continued distinction of newly obtained specimens.

One fundamental heuristic is that taxa with odd combinations of traits are worth further study. I suggest that a formalized basis for this heuristic is Shannon's (Shannon & Weaver 1963) information index, which may be simplified (Brown 2000: 43) as information content = – log probability, meaning that the less expected particular suites of characters are, then the more information they carry, this logarithmically increasing the less probable the trait combinations are. See Pielou (1966) for its use in ecology as a measure of diversity. This heuristic is "common knowledge," yet the information that increases with unusual combinations of traits is usually not clarified.

Another basic heuristic is that descriptions of new

taxa should involve the same characters as are in use for related taxa. Why? Why not a random set of traits? Clearly so that the same traits can be compared and a key produced. This is reasoning by analogy, as unreliable as induction (Kline 1985: 48), but just as basic to science and as much a part of scientific inference as is deduction. Such comparison is, nowadays, commonly based on homology assessment, however diffident or informal, and a natural key based on estimated evolutionary relationships is a desideratum. Some heuristics are obvious, and appear in introductory textbooks on taxonomy, but some, like the Shannon information index, are more difficult to analyze or to explain their application in taxonomy no matter how basic they seem.

Another heuristic that needs formalization is that good taxa have biogeographic ranges similar to other taxa. Many biogeographic analyses apparently support this. A more difficult heuristic is that one unique trait alone is insufficient in most cases to characterize a new taxon. How then this might be formalized or explained is not clear.

Uncovering patterns: relevance of mathematics and physics — Formalizing heuristics involves detailing the structures that underlie heuristic estimation, distinguishing templates impressed on the data by psychology (e.g., "spontaneous numbers") from those impressed by external nature, under the rubric that nature teaches us taxonomic concepts not vice versa. Non-psychological patterns are fundamental, being based on mathematics and physics. For instance, the biology of periodical cicadas involves prime-numbered life-cycles, often associated with populations at the verge of extinction (Yoshimura et al. 2009). In distinguishing patterns, one must avoid structuralism, which is the identification in fields other than mathematics and physics of apparently unassailable and axiomatic patterns in nature and human thought to be so basic and so like Platonic forms that all relevant analysis and theory is then deductive (apodictic). Structuralism, as a "content-free methodology" (Mathews 2001; Overton 1975) was introduced by F. Saussure in linguistics (Balzer et al. 1987; Barry 2002), and spread as a postmodern "rejection of all things past" to architecture, art, anthropology, literary theory, psychology, psychoanalysis, group theory in mathematics, and, as is now evident, to systematics (Rieppel & Grande 1994: 249). The structuralist slogan "theoretical knowledge is knowledge of structure only" can be pitted against the long-accepted completeness criterion, that science must explain how things manifest themselves as con-

tent in reality, namely through an inferred or observed process (Dewey 1950: 12; Giere 2009). Additionally, by reframing systematics as structuralistically dependent first and foremost on molecular cladograms, the phenomenon of statistical certainty for some sister-group relationships is philosophically "saved" as global certainty. As soon as one adopts a pluralist methodology and looks for additional explanation, for instance of a caulistic, macroevolutionary basis of classification, then recourse to needed additional data from morphology, fossils, or biogeography, etc., makes near certainty no longer global and we return from axiomatic structuralism to theoretic empiricism.

Thus, theoretically we take as axiomatic, at least in the mesocosm, many mathematical lemmas (Kline 1980: 263–264, 1985: 224), but as scientists we always assume that one can change classifications and descriptions with additional information and discursive reasoning in the context of a unifying theory (macroevolution) entirely apart from the cladogram. In this spirit, heuristics may be formalized by recourse to underlying structures and patterns but avoiding taking the sister-group patterns of cladograms as fundamental structures in nature.

In the face of the present probabilistic basis for science (Klein 1980), mathematics itself, though exact, is only a hyperexact approximation of real distributions. Although it is logical that if A = B = C = D, then A = D, if all elements were probabilistic distributions then a drunkard's walk from A to D may make A rather different from D. It then depends on an observer to check what logic cannot ensure. It the same spirit, cladograms may in practice be exact but what they represent needs to be clarified at all nodes for any conclusion but the grossest approximation.

Biases and heuristics, fast and frugal — Scientific realism is itself a kind of heuristic. Although we posit "things out there," we expect change in the notions, theories, and scientific laws about nature that we are at present willing to act on, based on new discoveries and new explanations. There are two major schools of psychology dealing with simplified methods of addressing complex problems. There is that of A. Tversky and associates, the "heuristics and biases program," which focuses to a large extent on misinterpretations due to bias, and that of G. Gigerenzer and associates, which investigates "fast and frugal" heuristics that are helpful when time and resources are limited. "Fast" refers to simple methods of information processing, and "frugal" means using little information (Gigerenzer et al. 2002: 561). Fast and frugal is decision-making under uncertainty, the heuristics and bias program is unbiased decision-making under certainty.

Both cladistic morphological analysis and molecular systematics are excellent examples of fast and frugal heuristics. The method is simple and the data are a minor subset of all that could be considered. Morphological cladistics yields good indications of which taxa are primitive (those of similar morphology and found in multiple clades at the base of the cladogram), and which are advanced (those deeply embedded in the cladogram). Superoptimization of cladograms (if necessary weighted to reflect classical groups) helps reveal core generative ancestral taxa and their stirps. Molecular systematics is a fine way to reveal deep ancestral taxa through heterophyly. The problem with the actual use of phylogenetic methods is that the correct and useful point of the cladistic heuristic, revealing macroevolutionary transformations, is ignored, and, instead, nesting of taxa on cladograms is taken as a speedy classification. The considerable random element in both morphological and molecular branch orderings cannot be detected without process-based theoretic insight.

The heuristics and bias program is also of major importance in understanding present-day systematics. Gillovich and Griffin (2002) have pointed out that there are two modes of thought associated with many examples of bias: first, the misapprehension due to some intrinsic psychological skewness, the second a scientific and accurate evaluation. One example is the well-known optical illusion of two equal-length lines each with angle brackets at each end. The line with brackets facing outwards seems longer than the line with brackets facing inwards, even though the viewer is assured that the lines are equal. Another optical example was known by the ancient Greeks, namely the refraction of a stick partly immersed in water—it seems bent though the viewer knows it is not. Examples of bicamerally dissonant perception in phylogenetics include: a dichotomous tree used to model evolution although evolution does not often follow such a model, the principle of two of any taxa necessarily being more closely related than each is to a third even though the reality of paraphyly makes this not at all universal, and the use of gamma-distributed model in molecular systematics even though other distributions are more likely (the gamma distribution is computationally more tractable). Both biased and unbiased empirical data can be dealt with and explained by process-based scientific theory, obviating the present cognitive dissonance of conflicting taxonomic results using different methods.

Some fields of human endeavor clearly condemn cognitive dissonance as highly damaging (e.g., Hughes-Wilson 1999: 252, 262, 302), but there is little discussion of this in systematics. The equally infamous psychological artifact inattentional blindness is relevant. A well-known experiment by Neisser and Becklen (1975; Simons & Chabris 1999) had subjects noting the number of times a ball was passed between actors in a film. Part way through the film, another actor in a full gorilla suit walked through the film. When queried after the test, half the subjects did not remember the gorilla because they were focused on their task. The gorilla in systematics is macroevolution. Why is the psychology of bias-tolerance and single-mindedness relevant here? It is because the monetization of phylogenetic systematics as Big Science coincides exactly with the crazed worship of mammon in economics over the past two decades. Nothing so focuses the scientific mind as the prospect of substantial grant funding. In no way do I accuse phylogeneticists of dissimulation, rather simply not closely examining an apparent bargain that is too good to be true.

Watzlawick (1976: 50) cited psychological experiments of A. Balevas that demonstrated that "once a tentative explanation has taken hold of our minds, information to the contrary may produce not corrections but *elaborations* of the explanation. This means that the explanation becomes 'self-sealing'; it is a conjecture that cannot be refuted." [Italics his.]

In general, because heuristics are not (yet) formalized, they may be compromised by various biases. For instance, confirmation bias is the preference of examples that support one's own view; logic puzzle bias is a preference for simpler explanations even if wrong; motivated reasoning looks more vigorously for flaws in examples we do not agree with; the sunk-cost fallacy encourages continued support for examples (such as particular taxa or methods) in which we have invested much time and effort (Begley 2010; Mercier & Sperber 2011); and future discounting is the excuse that the future can take care of itself, e.g., nomenclature can always be changed, therefore making hasty decisions is corrigible. Additional biases include statistical multiple comparisons (Zander, 2007b), and giving undue emphasis to the unexpected, which is a bias because unusual observations comprise a fairly large portion of a normal distribution. Cognitive dissonance (Festinger 1957) is psychological inconsistency that occurs when a belief does not follow logically from a fact, and resultant psychological tension is then reduced by changing one of the cognitive elements, adding new elements to one side or the other, to imagine one element as now less important, to search assiduously for consonant information, and introduce distortion or misinterpretation of information (Littlejohn 1978: 182). All these biases can affect unformalized heuristics, and must be accounted for in practice. Yet in all, the triumph of taxonomy is its continued facility in predicting the variational integrity of well-studied taxa and, using analogy with variation in similar new, poorly sampled taxa, with continued sampling of nature worldwide. That it matches the results of molecular systematics (short of strict phylogenetic monophyly) to a significant extent is a welcome but unnecessary plus.

Heuristic analysis can give good results quickly but can fail unexpectedly. The results, of course, need to be tested, while any biases associated with particularly heuristics need to be looked for. A particularly fine chart of heuristics and biases was given by Whalen (2012). A rephrased and shortened form is given here in Table 12.3.

Heuristics and Biases		
Name	**Heuristic**	**Bias**
Framing	A view of a problem, nesting represents evolution	Mistaking your view (e.g., cladistic nesting) for the real thing (i.e., serial transformation of taxa)
Anchoring	An irrelevant or insufficient starting point: a dichotomous key, or a phylogenetically informative data set	Domination by starting point such as an evolutionary tree must be dichotomous, or a data set is sufficient
Status quo	Fix nothing that is not broken, phylogenetics is successful	Assuming new is bad
Sunk cost	Resources spent on one solution is an estimate of cost of investigating alternative	Resources spent are a real cost of abandoning an apparently useless solution
Confirmation	Proving a solution you have a hunch about is right	Examination of only supporting evidence can miss a fault
Overconfidence	Decisiveness, assurance	Fooling yourself
Prudence	Conservative estimates of cost	Missed opportunities
Risk aversion	Anything to avoid ruin	Missed opportunities
Selective perception	Knowing what you seek (a cladogram)	Missed opportunities (a Besseyan cactus)
Recallability	That which is not obvious is dubious	Non-obvious features may be important and common (deep ancestral taxa)
Guessing patterns	Distinguishing trends	Seeing patterns that are not real (as may be evidenced in random data or in relationships implied by molecular-strain clustering)
Representativeness	An exemplar is the group	Ignoring a required independence of molecular taxonomy for support or refutation
Most likely	Avoid wasting time on the less probable	Rare or unpredictable events may be very important, e.g. extinction
Optimism	Relentless searches	Nothing is there so opportunities are missed
Pessimism	Duck unpleasantries	Missed opportunities

Table 12.3 — Heuristics and associated possible biases in decision making following Whalen (2012) but slightly modified to fit problems in systematics.

Psychologically salient numbers — Choice of numbers may be psychological. Decision theorists have found that certain numbers have emotional values or are basic to mental processes for comparing data. "Prominent numbers" or "full-step numbers" (Albers 2002) are the series ..., 0.1, 0.2, 0.5, 1, 2, 5, 10, 20, 50, 100, 200, 500, 1000, ... These are integer powers of ten, their doubles and halves. Are these apparently psychologically fundamental numbers selected unknowingly by taxonomists to use for the proportions of the (a–)b–c(–d) paradigm? In Plate 12.1, 5 is not a prominent number. Counting the number of occurrences of the numbers 1 through 20 as given exactly in the raw data of FNA + FBI, the average number of appearances of 1 is 143, of 2 is 137, of 5 is 90, of 10 is 105, and of 20 is 80, averaging 111.4. For all other numbers less than 20, the average is 42.9, with only "3" at 85 instances exceeding the least common of the prominent numbers. The number 50 occurs only 41 times, 100 only 34 times, apparently following a logarithmic decrease. A similar set of psychologically salient numbers are "spontaneous" numbers (Albers 2002; Martignon 2002), basically prominent numbers with inserted "midpoints" based themselves on prominent numbers.

Psychologically salient numbers are summarized to number 20 in Plate 12.1. The intermediate height of numbers 5, 8, 12, 15, and 18, might be explained as instances of interpolated spontaneous numbers. A line drawn through the tops of the columns of prominent numbers and another for the spontaneous numbers shows that 5 must be counted as a spontaneous number in the context of this study. Given the excellent match of the two lines in Plate 12.1 with column tops, it seems—in the present case—more likely that psychologically salient numbers are based on physical and geometrical relationships in nature than that Plate 12.1 demonstrates that the data set is generated from purely psychological notional choices. This is because both the present data and associated salient numbers are close to zero.

The numbers (Plate 12.1) appear to fall off in occurrence slowly as they increase in value, reminding one of Benford's law (Benford 1938) in which lists of numbers from much real data commonly begin with the first digit 1 at about 30 percent, decreasing logarithmically to 9 at about 5 percent, and is valid for ranges of several magnitudes or for distributions of distributions as is the case with the present morphological data. It fails in the present instance in that numbers above 1 beginning with 1 as first digit such as 11 or 100 do not follow this law.

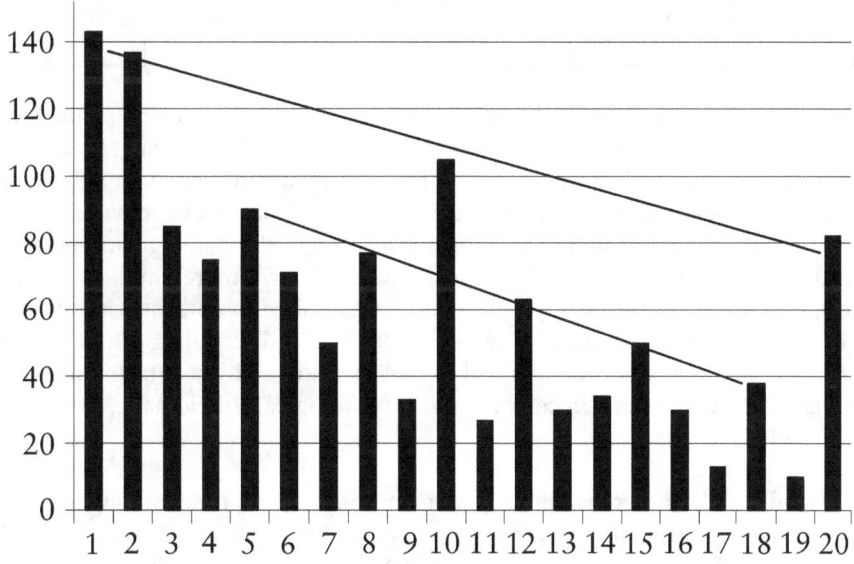

Plate 12.1. — Number of occurrences of numbers 1 through 20 as given exactly in combined data sets FNA + FBI. Psychologically salient "prominent numbers" 1, 2, 10, and 20 are relatively common, while 5, 8, 12, 15 and 18 may be interpolated "spontaneous numbers" of lesser psychological impact. Other numbers trail off approximately.

Thus, the dimensional paradigm is built, in part, with prominent numbers in the raw data set but these reflect what is expected *non-psychologically* from the data, which likewise is close to zero. The raw data averages of (14.5–)17.3–26.3(–31.4) approximate (15–)20–25(–30), which matches no sequential prominent number series. Of course, data consisting of numbers 1, 2 and 3 would be frequently in use, given the measurement ranges of bryophytes, but equally clear is that prominent numbers do figure in the dimensional paradigm, mainly for convenient rounding but also because 1, 2, and 5 are crowded toward zero and 10 and 20 are quite distant, easily of a magnitude difference. Psychologists treat prominent or spontaneous numbers as innate or axiomatic, and unconsciously impressed to a large extent on cognitive tasks involving numbers. This is psychological structuralism and avoids a causal explanation, but a *scientific theory* is possible unifying taxonomy and psychology, namely that the same physical/mathematical phenomenon that forces perceived taxonomic dimensional proportions towards a geometric mean affects cognitive psychology in the same way to establish prominent or spontaneous numbers.

Table 12.4 demonstrates that a series of four psychologically salient numbers do reflect a tendency of these numbers to have geometric mean midpoints of extreme numbers and of middle number more close to each other than arithmetic mean midpoints. This occurs even away from zero. An explanation is that the psychologically salient numbers are spaced, even away from zero, over considerable magnitudes, and these naturally follow the geometric mean midpoint phenomenon as do the data. The psychologically salient numbers thus do not introduce the geometric mean into the data, but respond in the same way as the data to physical reality.

Prominent number series	Range starting at 1, b–c is largest interval	Proportions of total range, %	GM (a–d)	GM (b–c)	Average (a–d)	Average (b–c)
1, 2, 5, 10	(1–)2–7(–10)	11:56:33	3.2	3.2	5.5	3.5
2, 5, 10, 20	(1–)3–13(–18)	6:58:29	6.3	7.1	11	7.5
5, 10, 20, 50	(1–)6–36(–46)	11:67:22	15.8	14.1	27.5	15
10, 20, 50, 100	(1–)11–61(–91)	11:56:33	31.6	31.6	55	35
20, 50, 100, 200	(1–)31–131(–181)	17:55:28	63.2	70.7	110	75
50, 100, 200, 500	(1–)51–351(–451)	11:67:22	158.1	141.4	275	150

Table 12.4. — Psychologically salient "prominent" or "full-step" numbers, namely the series ..., 0.1, 0.2, 0.5, 1, 2, 5, 10, 20, 50, 100, 200, 500, 1000, ... Contrived ranges starting with 1 are given based solely on the intervals between four contiguous prominent numbers, where the largest interval is b–c, the next c–d, and the smallest a–b is here summarized. For instance, for 1, 2, 5, 10, the intervals are 1, 3, 5, translated into the paradigmatic proportions 1:5:3. The geometric mean midpoints of a–d and b–c are clearly more similar than are the arithmetic means (averages). Compare proportions of total range in percent with the proportions of a Fibonacci series in powers of φ, with (a–b):(b–c):(c–d) equaling 19:50:31. Prominent numbers are better considered determined in part by known physical/mathematical phenomena associated with geometric means than as psychologically axiomatic, because they range across magnitudes.

Geometric Mean — The distribution is skewed, i.e., the first portion of the tripartite proportional distribution (a–)b–c(–d), is always smaller than the third part, and is in part explained by the upper range approximating a similar multiple of the lower range. This involves the geometric mean, a measure long used in informal estimation of dimensions (Morrison 1963; Weinstein & Adam 2008: 3). The geometric mean, which is always lower than the arithmetic mean, has been proven effective at estimating central values in dimensional or proportional ranges.

Fermi Problems (Morrison 1963; Weinstein & Adam 2008) are a silly but instructive example of very informal heuristics. "How long is a piece of string?" for instance, would be answered by the physicist Enrico Fermi as follows: The minimum length would be, say, 1 inch (2.5 cm) because less than that is a bit of fluff, and the maximum length would be, say, nine inches (22.8 cm), because more than that is a length of twine, i.e., long enough to be useful. To find the geometric mean, multiply the values and take the square root. So the length of a piece of string is three inches (7.5 cm), which feels is about right as the usual guessimate for this question. Although rules of thumb are often right only within an order of magnitude, this use of the geometric mean can provide a pretty good though facile guess.

So . . . then, "How far is up?" Stop now for a moment and do the heuristic calculation. My own answer, following Fermi's method, is: the minimum height would be a bit above eye level, say, a little more than six feet (about 2 m), while the maximum would be 62 miles (100,000 m), or the official lower limit of space (fide Wikipedia) where there is no "up." The square root of the product of these measures in meters is 447, thus "up" is about 0.28 miles or 1478 feet (447 m) above us.

The possible error in the "up" problem is doubtless greater than in the previous example, but note that the Empire State Building is 1250 feet (381 m) tall (fide Wikipedia) while the Sears Tower is 1451 feet (442 m), and only three buildings worldwide are higher than 1500 feet (457 m). So "how far is up" might be explained as a bit higher than the tallest buildings we know of, which seems sensible in an odd way. Few of us have high mountains nearby with which to challenge and expand our estimate of "up." On the other hand, this is probably an example of the multiple test (or multiple comparisons) problem, in which good explanations are come upon by coincidence or random casting about. What if, for instance, the highest measure was taken to be the height of the highest mountains? the highest clouds? the top of the troposphere? What if I had chosen the distance to the Moon as the maximum height, what would we find at the geometric mean between the Moon and the Earth? Support for an inference must come from another, independently supported inference from different information to avoid the multiple test problem. The correct way to find such support would be an independent psychological investigation of what height "up" is to a sampling of human population.

The multiple test problem in statistics is a recurrent theme in this book because searching for morphological support for inferred molecular evolutionary relationships often finds apparent support but such morphological traits are seldom analyzed to see if they do stand on their own and are significantly better than the best alternative morphological trait combinations. Mutual support requires separate evaluation of two phenomena. Corroboration cannot be had from coincidence. The reader must decide whether or not, as is done in this chapter, the finding

that the geometric mean is a joint explanation or at least intrinsic factor for three physical phenomena (Gould's wall, psychologically salient numbers, skewed dimensional metrics in systematics) is affected by the multiple test problem or is independent of it. (The reader should keep alert for multiple test problems that the present writer may have inadvertently introduced as the rest of this book is read, and the same should be done in the future when reading other systematic papers.)

The geometric mean in morphometrics is used as a standard proxy for overall size (Roseman 2004), such as using the geometric mean of the length and width rather than the maximum dimension to model centroid size (Kosnik et al. 2006). Here, "close to zero" means that the range b–c is a significant portion of the range zero to c. The geometric mean is equivalent to a weighted mean of log-transformed data (Rolf 1990). Dimensions particularly applicable are those that span or almost span a magnitude, such as one to ten, or two to 20, as is common in systematics, and such magnitudes are common on scales near zero, and particularly involving exponential distributions.

Comparison of average percentages of total range for all data sets show data ranges close to zero ("d" equal to or less than 10) have greater difference between a–b and c–d values than ranges away from zero ("a" equal to or greater than 20), indicating that, quite naturally, the geometric mean is less involved in the latter. FNA + FBI with raw data "d" limited to 10 or less had a total range percent ratio of 13:54:33, while the same data set with raw data "a" limited to 20 or more has a ratio of 17:57:26.

Golden Ratio — The golden ratio of ca. 1:1.618... (Livio 2002):

$$\frac{1+\sqrt{5}}{2}$$

is often approximated in the (a–)b–c(–d) paradigm (tables 12.1 and 12.2), where a–b, c–d, and b–c in that order comprise at least in part three elements of its "continued fraction" (Weisstein 1999a). The geometric mean is the only well-known relationship that clearly explains or predicts the dimensional heuristic in taxonomy in that it is commonly used to calculate in many fields central values in proportional ranges. Linked closely to the geometric mean is the golden ratio, also known as φ (phi), as it is associated with a Fibonacci series in powers of the golden ratio: 1, φ,

φ^2, φ^3, φ^4, φ^5, φ^6, φ^7, φ^8, φ^9, …. (Fibonacci means "son of Bonaccio.) For this particular logarithmic series (Gardner 1982: 65), each value is the sum of the previous two, and also the square of each value is the product of two equidistant values. The basic Fibonacci series is:

1, 1, 2, 3, 5, 8, 13, 21, 34, 55, 89, 144, 233, 377, ...

For example, φ^6 squared is the product of φ^7 and φ^5, or of φ^8 and φ^4, i.e., involving the geometric mean. The series may be extended below 1 as reciprocals. Although the match of shared geometric means of a–c and b–c is fairly clear from the data (Plate 12.2), this does not account for the relative ranges of a–d and b–c. For any b–c range, there may be a number of a–d ranges with identical geometric means, and vice versa, e.g. for 4–9 as b–c the geometric mean = 6, while for 1–36 as a–d the geometric

mean also = 6.

The golden ratio stares scholars in the United States in the face every day because common paper size proportions approximate the ratio closely. For instance, 3 by 5 inches = 1:1.667; 5 by 8 inches = 1:1.600; 8.5 by 11 inches = 1:1.668.

It is well known (Stewart 2011) that successive leaf primordial commonly obey the Fibonnaci series. As each primordium is initiated, it pushes the previous two apart to an angle of about 137.5° (the "golden angle"). The Fibonacci series also commonly determine the number of petals in a flower. There another series, the Lucas series, which is responsible for development in four petaled flowers or 4-spiralled cacti:

1, 3, 4, 7, 11, 18, 29, 47, 76, 123 …..

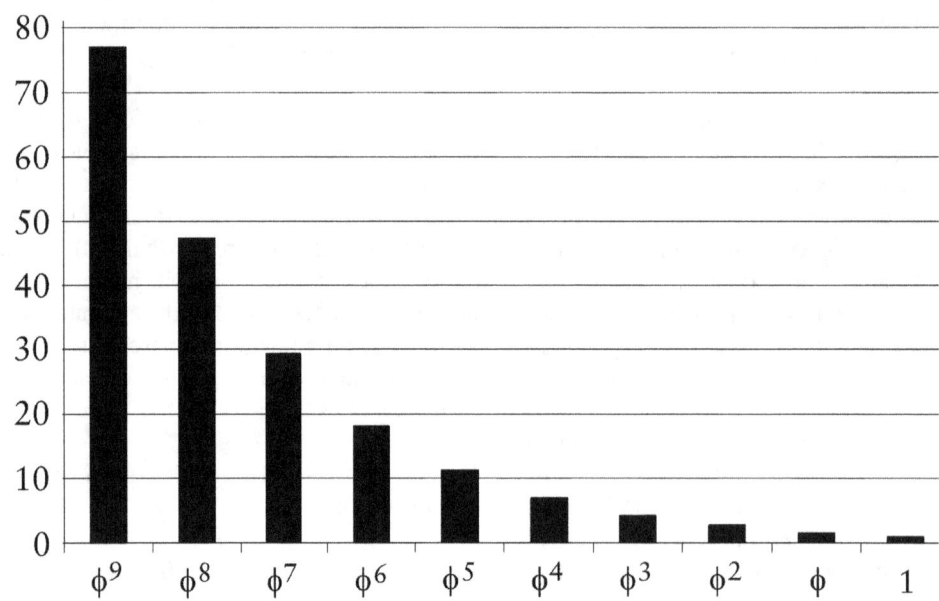

Plate 12.2. — Fibonacci series as powers of the golden ratio (φ, or about 1.618) for …, φ^9, φ^8, φ^7, φ^6, φ^5, φ^4, φ^3, φ^2, φ, 1, reversed from standard order. The curve is logarithmic, each value is the sum of the previous two (reading backwards, to the left, from 1) , and the square of each value is the product of two equidistant values, e.g. φ^6 squared is the product of φ^7 and φ^5, or of φ^8 and φ^4.

The proportions of total range in percent with the proportions of a Fibonacci series in powers of φ, is (a–b):(b–c):(c–d) equaling 19:50:31. The golden ratios as appear in the raw and standardized data (Tables 12.1 and 12.2) are from data that mostly are

never so far apart that they may not be interpreted as contiguous sequences on the Fibonacci φ-power series times a constant. (The constant may be characteristic of an organ or a species.) This could explain the fairly uniform balance of proportions in all dimensional paradigms with a–b smaller than c–d, and b–c largest of all in the present data. Circumstances would be felicitous if the data did show exactly a Fibonacci φ-power sequence in all elements of the paradigm at once, but this is a first attempt at formalization of this heuristic. In Table 12.1, most of the raw data paradigms demonstrate proportions of a–b:c–d approximating the golden ratio, but the ratios of c–d:b–c were ambiguous. Only the combined FNA and FBI data sets showed all three elements to be in the proportion of the golden ratio. Much the same thing is shown in standardized data in Table 12.2, although FNA + FBI only matches c–d:b–c with the golden ratio. When the data are reduced to only those data identified as "abcd" for which all elements a, b, c, d are given and they are different, a–b:c–d show the golden ratio. It may be that when authors do not give a, here assumed to mean that they meant a = b, the authors inadvertently added the data to b–c, making the b–c range larger than it would be if it followed the golden ratio in respect to c–d. This would account for the smaller size of b–c for data from the "abcd" data sets with all four elements given and different.

It is possible to find other standard proportions that approximate, probably by coincidence, those of the above, including the major harmonic proportions in music of 1:2 and 2:3 (octaves and fifths) and the harmonic mean (Table 12.2). Theory should address, however, all physical phenomena that are relevant, where relevant means clearly explainable by that overarching theory, but explanations that involve physical and mathematical relationships also seen in other scientific fields are certainly preferred. It is easy to match numbers with ratios particular to other phenomena because there are only so many large fractions, and numerology must be avoided. Stewart (2011: 54) pointed out that there are exaggerated claims for the golden ratio, any use of it in biological development must point out a "deeper model in which the golden number turns up for solid structural reasons." The correlation with the geometric mean in relative measures near zero, I believe, is just such a basic phenomenon.

Gould's Evolutionary Wall — Gould's (2002: 897) speciational reformulation of macroevolution involving minimum structural constraints on size may be invoked to explain the large extreme range, c–d, compared to the small extreme range, a–b. Gould asserted that apparent directionality of evolution in many cases is simply due to the existence of developmental limits on evolution of smaller size, a kind

of "wall." Mutations occur that equally support speciation towards both smaller and larger size and complexity, yet a basic structural limitation in size favors "drift of a small percentage of species from the constant mode of life's central tendency towards the only open direction for expansion" (Gould 2002: 899).

Data from FNA and FBI together were used to list all standardized a–b and c–d values in two columns in a spreadsheet. The data were sorted in increasing range values first on a–b and secondarily on c–d. A graph (Plate 12.3) of the full standardized data set plotting range as percent of total against numbers of measurements shows a–b at zero on the left leaving zero about mid x-axis and rising towards the right in value. The x axis baseline of zero range represents the Gouldian wall of evolutionary developmental minimum size for the traits mapped. This is because when a–b is near to or essentially as zero (as estimated by the taxonomist) the extreme variation of c–d that is potentially contributory to future macroevolution is mostly larger than the range of a–b. Inasmuch as c–d is also sorted by increasing range, any tendency to match the increasing range of a–b on the right of the plot would have been accentuated, but there is little correlation. There are almost exactly twice (2.02 times) as many diamonds at zero on the left of Plate 12.3 as there are circles at zero on the right of the plate, which reflects a clear tendency of authors to skimp on reporting a–b over c–d.

Biophysical fields of macroevolutionary consistence — Fields in physics are ways to explain or at least describe forces that act at a distance, like electromagnetism and gravity. There are evolutionary "rules" that have been suggested that try to explain or formalize biological tendencies across taxa, such as Glover's rule that highly pigmented animals are more usually found near the equator, that is, in more humid environments; Bergmann's rule that correlates geographic latitude with animal body mass; Allen's rule that animals in colder climates tend to have shorter limbs or at least larger body mass to surface area proportions; and a corollary to Allen's rule, Hessen's rule that animals in colder climates have larger hearts. These rules are often debated and various mechanisms are proposed to account for the observed tendencies.

A corollary then to the idea that there may be many physiomathematical fields affecting taxonomic characters is that certain data may not be randomly generated as once thought, and cannot then be correctly analyzed by assumptions of normal distributions. For example, in some cases, chi-squared analyses must give way to large-sample statistics.

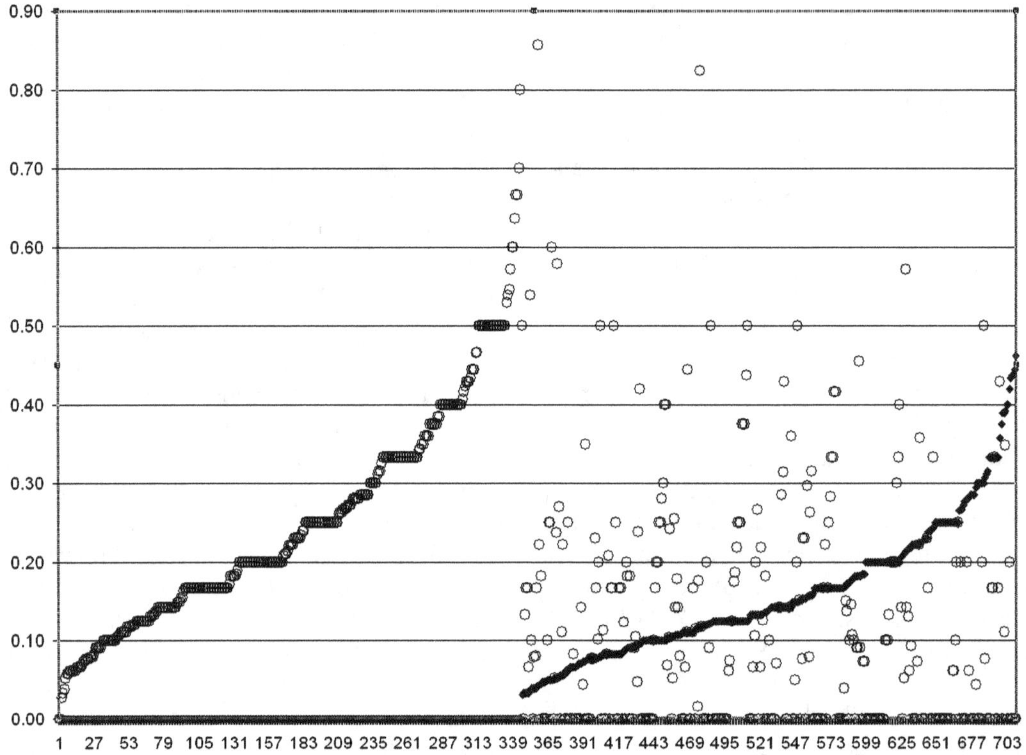

Plate 12.3. — Plot of standardized ranges of a–b (black diamonds) and c–d (open circles) for all data of FNA + FBI, Both ranges are sorted by increasing size, the y axis is percent of total range for that measurement and the x axis is of sequential numbers given to each measurement. On the left, the plotted data points show c–d (circles) as well distributed over a–b (diamonds filling the x-axis) on the baseline of zero range for a–b, and on the right, while a–b (diamonds) rises, c–d (circles) are almost as well distributed, but because the a–b line (diamonds) rises slowly, c–d data points (circles) are fewer between the slowly rising a–b line and zero. The values of zero on the right for c–d seem to be many (actually half those of zero for a–b on the left), but in fact do not affect the calculations of geometric mean. The baseline of zero range is the Gouldian wall of minimal evolutionary developmental size, with c–d free to vary against much restricted values of a–b.

Standard deviation — The actual data appear to approximate the first standard deviation of 66:34 percent for a–c:c–d. The standard deviation is defined as the square root of the variance where variance is the expected value of the squared difference between the variable's realized values and the variable's mean. In an exponential distribution the variance is the square of the mean.

The proportions of the Fibonacci series in powers of the golden ratio (φ) is, for (a–b):(b–c):(c–d), 0.19:0.50:0.31. Thus the middle range of the metric dimension, if it were exactly in powers of φ, is ½ the entire distribution, not ⅔ as with the standard deviation. The ratio of (b–c) to (c–d) is, however, φ, as is the ratio of (c–d) to (a–b).

Discussion of scientific intuition exemplified in the metric dimension heuristic — The discrete geometric distribution (Weisstein 1999b) which affects the proportional distribution of the dimensional paradigm rises rapidly near zero, but becomes flat away from zero. Expected, then, is a tendency to left skewness near zero (i.e., the b–c range is moved left in the distribution). Taxonomists familiar with actual measurements using the paradigm perceive over time that measurements from the actual geometric distribution in aggregate result in a broad range of usual measurements, smaller for the high extreme range, and least for the low extreme range. Both the geometric mean and the golden ratio are fundamental mathematical and physical constructs, thus, theoretically, the dimensional heuristics for the (a–)b–c(–d) taxonomic paradigm are well grounded. All samples of (a–)b–c(–d) are far over 30 in number (and satisfy the normal sample distribution). They are based on trait distribution from scoring, however informal, of multiple specimens of each taxon. Calculation of the means from the large samples minimizes outliers.

The *geometric mean* makes the high range of any series, particularly ranges of large spans, a similar multiple of the low range, establishing a midpoint usually lower in value than that of the average of the lowest and highest values. For instance, if the midpoint is 10 and the lower bound is 5, then 10 is 2 times the lower bound and the upper bound should be 2 times 10, or 20. Mathematically, the geometric mean is the square root of the product of the extreme values. Checking the first calculation, five times twenty is 100, the square root of which is 10.

If a measurement is inside the a–d range, it is a "tell", and if outside, it is an "outlier" and indicates that study of other taxa is necessary; if inside the b–c range then confidence is warranted for this trait, if outside but within a–d, additional traits might be checked. Because an experienced taxonomist can give an estimate of the usual range of dimensional variation of a structure from the variation within a collection or between a few collections, the dimensional heuristic may be expected to be accurate.

Given that doubtless many heuristics are used for taxonomically significant morphology, and that these "triangulate" to well-characterize a taxon, an experienced alpha taxonomist has at hand a set of tools that should not be deprecated.

The main point of this discussion of formalization of taxonomic heuristics is that if one knows what proportions or ranges of (a–)b–c(–d) have been established by experts (or oneself), then for a new specimen examined, one can predict from even a few traits whether such experts might consider the specimen to be a new taxon or not, and perhaps encourage location and investigation of more specimens, or a description as new.

The only really clear statistical explanation for these ratios is the geometric mean, which is known to operate in distributions near zero where the distribution is a large proportion of the extreme and zero, or large portions of a magnitude in range, and which is commonly used in calculations involving proportions. The data strongly suggests that the dimensional heuristic is of two parts. (1) A skewed logarithmic distribution with two tails is detected reflecting distributions of samples. (2) The informally observed skewed distribution is then represented in descriptions by (a) sliding b–c to the left within a–d. This approximates a left-skewed distribution with matching geometric mean midpoints of b–c and of a–d, which is entirely expected for ranges near zero or comprising much of a magnitude, or both. (b) The relative proportional sizes of a–b, c–d, and b–c approximate a constant times each of three contiguous values of powers of the golden ratio (φ or 1.618...) as given in the Fibonacci series ..., $1/\varphi^3$, $1/\varphi^2$, $1/\varphi$, 1, φ, φ^2, φ^3, φ^4, φ^5, φ^6, ... These intuitive judgments, although newly presented here as important theoretic possibilities, are supported by the evidence, and may be entirely expected because they are associated with physical and mathematical relationships found in other scientifically investigated natural phenomena.

The bryological data, at least, match the heuristic well. The hyperprecise ideal or bull's-eye measurement for dimensions near zero has the proportions 1:2.6:1.6 reflecting the two golden ratios (a–b):(c–d) and (c–d):(b–c). This is about (1–)2–4.6(–6.2) in metric dimensions along a gradient starting at 1. Developmental differences unique to a taxon may cause a change in the ratios particular to the organ measured, e.g. for mosses, the ideal ratio for leaves would be (0.5–)1–2.6(–3.4) mm along a gradient. Also any significant differences in expected ratios may signal developmental, evolutionary, or researcher bias effects that could or should be examined. Casual preliminary analysis of several multiauthored treatments of Poaceae (Flora of North America 2007a: 1–284) yield the raw data ratio (Table 1) of 11:62:28, with 11:28 near the golden ratio, and standardized proportions of 12:64:24, which match the golden ratio only if 12 and 24 are added, which yields 1:1.78 as a ratio for a–b + c–d:b–c. A treatment of hepatics (Smith 1990: 1–164) was also casually investigated, and although the raw data yielded no illumination, the standardized data ratio of 9:62:29, when a–b and c–d are added, gave 1:1.63 as ratio of the added ranges and b–c. This may indicate that all the authors did follow the golden ratio as well as the geometric mean shared midpoints (Table 12.2) but (unconsciously perhaps) used only one pair of values on the Fibonacci φ-power series, in any case clearly signaling a rather steep observed curve with small tails.

This is a first pass at formalization of the dimensional heuristic in taxonomy. Further work, if experimental, might involve a set of naïve taxonomists each duplicating the study of sets of the same taxa so biases of taxonomist and of dimensional variation from the normal unique to a taxon or trait might be analyzed. Failing this, quasi-experimental (Cook & Campbell 1979) analysis like the present study might be extended to published papers on other taxonomic groups, individual traits, and a variety of taxonomists. Given the paucity of taxonomists, and even the possible bias introduced by reading the present paper towards a "correct" proportion, dimensional heuristics may be not investigated successfully in the future, but actual scoring and plotting of sampled observations in a variety of taxa and traits may provide a guide to what may be expected in reporting perceived dimensional ranges in standard descriptions.

Empirical phenomena, not mathematics, informs science. Mathematics, in my opinion, only "approximates" with extreme precision the probabilistic and fuzzy data of empirical phenomena. The precise central framework of math, however, is a guide to physical reality. It is our innate human ability to perceive and appreciate fundamental physical relationships, like the geometric mean and the golden ratio, that allows us to do alpha taxonomy. Although psychologically salient numbers are involved in the dimensional heuristic, it is clear that the decisive template in the dimensional heuristic of taxonomy involves perceived physical, not innate psychological, rela-tionships. It may be incredible that taxonomists follow to such an accurate extent the geometric mean and golden ratio in the dimensional heuristic but, because these physical guidelines are essentially least effort troughs in a curved data space like the world lines of relativistic space, any other result should result in a search for bias.

An additional observation is that if alpha taxonomy, as an abductive science, is formalized to any great extent, criteria are then available for review and judgment of taxonomic papers, including description of new taxa and new combinations, as opposed to the present practice of "the expert knows best," relying on a proven authorial track record, or overall plausibility of the hypothesis. This also addresses the "qualia problem," which asks do different persons perceive the same thing the same way? Is there a fundamental difference in perception that might affect acceptance of scientific realism as a general philosophy of scientists? If physical fundamentals are involved in data collection, then certainly different people view the same things except for psychological biases, which can be documented and allowances made.

CHAPTER 13
The Macroevolutionary Taxon Concept

Précis — The macroevolutionary taxon concept is a sleeve or sabot that gives any taxon theoretic justification along the lines of descent with modification. It extends the macroevolutionary species concept to all taxa of any rank capable of being mapped on an evolutionary tree as progenitors of extant species or genera of a different taxon of the same rank or higher. Arguments are marshaled in support of Darwinian evolution of taxa at the rank of genus or higher, though this concept is not necessary for the present Framework to provide macroevolutionary insights. This is a generalized treatment extending the supergenerative principle of Chapter 8 beyond the genus *Didymodon* (Pottiaceae, Bryophyta).

The macroevolutionary *species* concept (MSC) — A macroevolutionary species is any species concept capable of allowing a classical, morphological (or other expressed traits fairly easily viewed) description, and also demonstrating macroevolutionary transformations on a caulogram. A caulogram is a cladogram in which all nodes are named at the lowest rank possible using both phylogenetic and nonphylogenetic information. The highest rank possible is the lowest rank inclusive of all exemplar taxa in the clade distal to the node; the lowest rank is the name for an inferred ancestral species represented by one or more exemplar specimens. The macroevolutionary species concept (MSC) differs from the evolutionary species concept methodologically in being restricted to only those species actually demonstrable as involved in macroevolutionary transformations, that is, one species inferred as giving rise to another species or a taxon of higher rank. There are multiple reasons for speciation (budding, sympatric, isolation in various ways, strong selection, founder effect, polyploidy and neo- or subfunctionalization, etc.) and multiple species concepts that may apply to some taxa and not to others. There are ways of describing speciation that may show valuable research directions, such as "symmetry breaking" (Stewart 2011: 204; Stewart et al. 2000).

The inference of macroevolutionary species may start with any species conceived according to many standard concepts associated with alpha taxonomy for which a formal taxonomic description based largely on morphology can be generated. The data for these species can be presented in a transformational context using cladistics and some optimization method, e.g., maximum parsimony or Bayesian Markov chain Monte Carlo. Morphological cladistics provides natural keys to taxa when appropriately weighted and/or divided into subsets with locally conservative traits. Molecular phylogenetics is presently limited by being restricted to analysis of one or few specimens per taxon (of the lower ranks) and provides inference of genetic and isolation events that result in nesting of those specimens, but not necessarily revelation of speciation events. Inference of speciation events requires both dense sampling at the taxon level, and evaluation of non-phylogenetically informative information, i.e., inference of progenitor-descendant relationships.

The standard cladistic tree is reconceived as a caulogram where nodes are named at the lowest possible rank given phylogenetic and nonphylogenetic information on direction of evolution at the taxon (not character) level. Non-phylogenetic information includes biosystematic data and taxon-level Dollo evaluation together termed "superoptimization." Macroevolutionary species are the fundamental taxonomic units distinguishable as species that are recognizable as taking part in theoretical taxic transformations on the caulogram. This requires careful distinguishing of pseudoextinction and budding evolution at each node. The most revealing information about macroevolution comes from superoptimization of cladograms and from morphological and molecular strain paraphyly, presently suppressed in phylogenetics as antithetical to classification by clades. Support for macroevolutionary species may be calculated in part from relevant clade support.

The macroevolutionary *taxon* concept (MTC) — If that which holds species together is not clearly demonstrable as cohesive or balancing forces consistent with the biological species concept (BSC), then such forces are unknown, probably some combination of developmental constraint signaled by conservative morphology, and ecological constraints. It is evolutionary stasis, enforced by some combination of long-lived habitat and stabilizing selection, that is just as important an evolutionary force as is adaptation (Patterson 2005). Therefore, since such forces are possibly plural and now not well understood, the MSC is applicable to any genus or higher rank that is amenable to analysis of macroevolutionary transformation, e.g., on a cladogram. Thus, we may consider a macroevolutionary taxon concept (MTC).

Evolutionary stasis may not be a function of lack of selection, say in microenvironments, but may be a kind of reverse Red Queen hypothesis. Instead of constant change to keep up with the changes on competing organisms, an organism in evolutionary stasis may be following a kind of Nash equilibrium (Maynard Smith & Price 1973; Stewart 2011: 219) in which no mutant can successfully invade a parent population.

The simplest MTC is simply Darwinian selection of species in a genus or genera in a family as environments change; e.g., with dryer climate, those species sensitive to aridity die out and others evolve that are resistant, but having the same conservative traits (which are not affected much by environment within limits) and same developmental constraints that promote stasis at the genus level. The differential extinction of species may also occur in groups of species of the same genus occurring in different geographic localities with different environmental conditions. The environmental influences that keep taxa stable are adaptive in trimming new traits that are less effective in competition or fatal in the stable environment. This concept, involving selection towards and away from stasis functioning in unique environments for particular taxa, may be termed an "envirosome" (i.e., the paragenetic regulator of Bock 2003) in analogy with chromosomal control of stasis and change in panmictic species suitable for the BSC. The effect, given Darwinian selection of separate taxa, is the same. This is a catch-all concept, and does not replace specific explanations for particular processes unique to certain species and species groups.

Genus and higher taxonomic ranks are operationally the same as the macroevolutionary species when they, too, can be detected in a caulogram. Thus the macroevolutionary species concept can be extended to a macroevolutionary taxon concept, whether or not arguments that higher taxa are only human constructs are valid. The MTC is a theoretic interpretive and guiding jacket, sleeve or sabot that works with any species concept (e.g., biological, ecological, evolutionary) that can generate a standard taxonomic description and also is amenable to analysis of macroevolutionary transformation involving higher taxa. Most species concepts are definitions of species-in-themselves. The MTC is a definition invoking a process relationship involving two or more taxa. Macroevolution is the guts of evolutionary systematics. A hypothesis of a macroevolutionary transformation is equivalent to finding "hidden variables" in physics. This is not a trivial comparison because Darwin's explanation of evolution is as fundamental a scientific theory as any in physics.

The macroevolutionary *genus* concept (MGC) — Cladistic analysis involving morphological trait changes is not much better than phenetic cluster analysis in that parsimony fails as a discovery process. This is because, although morphological traits may be generally independent and uniquely distributed as homologues, preadaptation of traits to new environments links them through selection. Thus, if three traits are needed for survival in a new biorole, three must be forthcoming, and they are then equivalent, parsimoniously, to one trait. There is evidence that rates of speciation are greater than rates of ecological change and adaptation, with internal parasites (Brooks 1985) and insects (Ross 1972). Although this indicates that much speciation (with attendant diagnosable traits) is associated with isolation mechanisms that are not associated with environmental or geographic isolation (e.g., allo- or autoploidy) and then are neutral or tolerably counter-adaptive for that niche, it also indicates that considerable genetic diversity merely awaits the isolation mechanism and predates niche openings.

Taxa at higher levels than species have long been considered merely convenient, subjective groupings. According to Lindley (1853, fide Coggon 2002: 18), "But as the Classes, Sub-classes, Alliances, Natural Orders, and Genera of Botanists have no real existence in nature, it follows that they have no fixed limits, and consequently that it is impossible to define them." This is not necessarily so.

Nature teaches us our taxon concepts. The superoptimization of the moss genus *Didymodon* above found that the genus was the basic element of (dissilient) evolution for this group. Thus the macroevolutionary genus concept was paramount in its taxonomy and classification. Other groups may have different concepts as basic to their evolution and classification, but any concept limiting pseudoextinction militates against the fundamental phylogenetic method requiring that two of any three taxa at the same taxonomic level must be more closely related. This last is immediately falsified when an ancestral taxon may be inferred for two or more daughter taxa.

Both mutations in *cis*-regulatory sequences and in gene-associated tandem repeats (Frondon & Gardner 2004) have been associated with rapid evolution of phenotypic traits. The conservation of such gene-associated orthologous tandem repeats across mammalian orders despite high mutation rates have been shown to be indicative of strong stabilizing (non-neutral) selection. Thus, we have the theoretical potential of an abundance of pre-adapted, pre-speciation phenotypic traits that confound any exact probabilistic expectations of parsimony analysis.

Plate 13.1. Stylized evolutionary tree of two genera with the dissilient genus as the operational basic evolutionary unit. The dendrogram exemplifies genus-level speciational bursts. A supergenerative core species with a radiative set of descendant species each specialized in habitat, range or morphology compromises one genus; another genus of similar complexity is evolved from the first core supergenerative species. The core maximally preserves generalized traits, which leads adaptively to enhanced evolvability. The order of speciation is given by numbering the ancestral and descendant taxa. The dissilient genus is not a fully generalizable concept because it is nature that teaches us both species and higher taxonomic concepts, but it is useful for some groups.

According to Arendt and Resnick (2007), because genomic analysis has demonstrated that the same genes may be involved in the same phenotypic adaptation in quite distant groups of animals, while different genes are apparently the source of the same phenotypic adaptation in related groups, the usual distinction between parallelism and convergence (parallelism expected to be based on the same genomic pathways, and convergence on different) breaks down. The authors recommend that "convergence" should be the general term. Given the findings of Arendt and Resnick, evolution of the phenotype based on static expressed traits may be quite disconnected from evolution of the genotype though remaining based on it, and this is a rather different perspective than changes in mostly non-coding traits used in tracking phylogenetic relationships.

Just as a species may be operationally defined as the basic unit of taxonomy, a genus may be defined, however annoyingly vague or unexpectedly exact, as the basic unit of groups of species. In those cases in which evolution must be analyzed by examining groups of species for a central evolutionarily active core species, then operationally, in these cases, *the genus is the unit of evolution.* The reason the genus may be considered the unit of evolution in some cases is that the whole radiative complex needs to be examined to determine the ideally generalist, Dollo-primitive, ancient-habitat dwelling, widely dispersed progenitor, or to hypothesize one. Examination of the evolutionary tree of *Didymodon* given in Plate 8.1 provides evidence that some genera can be defined as a cloud of derived species around one super-generative progenitor core species, in addition to morphological distinctiveness. That is, there is clearly a clearly definable second level of organiza-tion in some groups above the species level—a genus is not an ad hoc grouping of convenience in those cases.

Given the idea of descent with modification of taxa, genera evolve from *species* of other genera when a new species of one genus is sufficiently evo-lutionarily distinct to be flagged as an evolutionary novelty (by adaptive or neutral traits that imply a dif-ferent evolutionary trajectory at the same level of novelty as the genus that generates it). In the case of *Didymodon,* each radiative complex acts as an evolu-tionary unit, with transformations between species restricted, in most cases, to one progenitor core spe-cies. Genera, if complex one-way (Dollo) trait trans-formations are conserved at the taxon level, are gen-erated from the core character-rich species.

But does this mean that genera can be seen as evolving from other *genera*? Caulistic evolutionary trees imply that taxa evolve from taxa at any rank. Is this an artifact? Gould (Gould 2002; Hubbell 2005) and many others have decided that supraspecific taxa do evolve though not perhaps from or out of each other. Eldredge (1985: 150) wrote that there is a "dis-tinct possibility of some higher-level sorting principle in nature" that affects higher taxa, and this may be due to differential species survival (1985: 172); he (1989: 183) also considered that "higher taxa are co-herent pools of genetic information." Vrba (1980, 1984) attributed macroevolutionary change not to properties of species but to attributes of organisms, particularly a cascade of specialization and in short

(geologic) time-span in certain lineages, called the "effect hypothesis." The phenomenon of "macroevo-lution lag," in which the origination of a major group is followed by a quiet phase that itself precedes an increase in diversity (Jablonski & Bottjer 1990), im-plies selection-like pressures on the supraspecific taxon. Barraclough (2010) argued that both species and higher level phylogenetic patterns can be ex-plained or at least described by equilibrium explana-tions.

When species of a genus or of a geographically isolated part of a genus are affected by changing cli-mates and habitats, sensitive species die out and newly evolved species of that genus are adapted in various ways. Clearly the genus or part of a genus evolves "anagenetically" in this case, and one genus or part of that genus changes into another. This fol-lows the Court Jester Hypothesis that changes in the physical environment instead of biotic interactions can be initiators of major changes in organisms (Barnosky 2001).

Chase et al. (2000) pointed out that there is a "pronounced tendency for close relatives at the fam-ily level to develop traits in parallel"; their word is "develop" implying necessarily de novo speciation events, yet a better and less bold hypothesis is simply shared expressed traits of a joint ancestral taxon.

There are problems with premises in evaluating classification based on morphology alone, largely evinced in relationships between closely related spe-cies (e.g., within a genus). Chesterton (1956: 156) has discussed Thomas Aquinas' point that choice of correct first principles ensures true deductions. In phylogenetics, if one's premise that morphological traits are largely not adaptive (i.e., are spandrels, sensu Gould & Lewontin 1979), then fixation in a new species is governed by the Central Limit Theo-rem and indeed when evaluating parallelism or con-vergence, parsimony correctly gives the best, most probabilistic choice of ((AB)C) when A and B share more advanced traits than do B and C, or A and C. On the other hand, such a premise flies in the face of observations that clear cases of parallelism and con-vergence apparently involve adaptation, such as se-lection of similar traits associated with arid or hygric environments, competition, r and K selection, repro-ductive modes, and the like. For instance, A and B may share three advanced traits while B and C share only one, but the three advanced traits may clearly be associated with one particular adaptation, and are fixed as a unit; thus three synapomorphies yielding (AB)C may not be a more parsimonious solution than one synapomorphy giving (BC)A. Certainly convergence resulting in cryptic taxa at the genus or family level is improbable.

A better premise, null hypothesis, or "state of nature," then, is that salient advanced morphological traits apparently requiring parsimony analysis because there are multiple interpretations of relationship due to possible parallelism and convergence are at least in part subject to selective pressures. Diverging infraspecies commonly are diagnosed by two or more traits, and parsimoniously settling questions of possible parallelism by choosing those sharing the most traits is incorrect because selection for a particular environment or biorole is not simply through gradual accumulation of evolutionarily neutral traits (see discussion by Bachmann 2001). Mutations of expressed traits may well arise at different rates in a population, but fixation in a new species is dependent on isolation through a number of mechanisms, such as polyploidy or the availability of a new ecological niche or adaptive zone (Hutchinson 1957; Whitaker 1972). Stebbins (1959) reviewed fundamental discoveries showing that hidden genetic complexity in multiple somewhat isolated intraspecific lines contributes to maximum evolutionary flexibility, in particular citing Clausen et al. (1940, 1958), who quite long ago demonstrated that many microspecies of overlapping constant "races" contribute to a species' gene pool of potential complex and immediate adaptive or preadaptive response to the opening of a niche (or other selective challenge). If a niche becomes available that requires four mutations, then four must be available for success, i.e., isolation samples any number of traits as portions of the genetic complexity of an ancestral species. Only if a trait is uncommon or apparently evolving slower than appearance of isolation events that successfully establish a new species can the Central Limit Theorem be applied to judge expectations of trait combinations. Thus, for common expressed traits a synapomorphy of one step is much the same as of several since the traits may evolve as an adaptive unit.

Preadaptation of traits in a genetically and phylogenetically complex ancestral species might easily provide the exact number of traits needed when a niche opens, including adaptive traits appropriate for cases of sympatric allo- or autoploidy or gradual splitting because hybrids among genotypes in a populations are less fit, while selection is ultimately the shaper of species if morphological traits change anagenetically, as merely linked to a physiological trait, or through drift. It is the minimal requirements of the new environment for a newly isolated species that determines the morphological diagnosis of a new

species having any particular number of traits, not chance gradual accumulation of such traits. Such preadaptation or exaptions are may include (Bachmann, 2001; Caporale 1999, 2003; Zander 2006) silenced clusters of traits that have greater immediate adaptive value than sequential mutations of all the traits.

Cladistic analysis of morphological data has an additional problem. Assumed in such terminology as "sister groups" and "shared ancestry" is the idea that products of evolution must be new, and there are no or very few surviving ancestors. Yet surviving ancestors (as species pairs with one derived from the other) have been inferentially demonstrated in the past through biosystematic and cytogenetic analysis, for instance, by Lewis (1962, 1966), Lewis and Roberts (1956) and Vasek (1968) in *Clarkia* (Onagraceae). Evolution may be inferred in cladistics as morphological trait changes over a tree, or in molecular systematics as nucleotide base changes, yet evolution as descent with modification involves organisms, not disconnected characters. The proper evolutionary tree on which to base classifications is that of lineages of living things, or ancestral taxa, as best inferred from extant taxa and fossils, and necessarily involves surviving species, genera and families. This is in agreement with Farjon (2007), who emphasized that "...taxa must evolve from other taxa...." A taxon tree, contrary to phylogenetic sensibility that no extant taxon is to be represented as derived from another extant taxon, is probably similar to a "Besseyan cactus" (e.g., Bessey 1915), exemplified by Wagner (1952) as pointed out by Stevens (2000). Each Besseyan cactus "pad" represents an evolutionary cluster with no detailed derivative structure.

Published cladograms of multiple exemplars in individual species, e.g., that of the domestic cat (Driscoll et al. 2007), demonstrate considerable internal phylogenetic complexity and infraspecific morphological complexity preadapted for species-level fixation but no clear evidence of the particular morphotype of the ancestor of each or all subspecies or indication of the morphotype of any future species that might evolve from such subspecies through gap formation, for instance following extinction of all but one or two of the subspecies. This in spite of the general agreement of phylogeneticists that a species is only a lineage, or better a segment of a lineage, not a clade of multiple lineages (Wiens 2007; de Queiroz 2007).

The same lack of clear connection of ancestral morphology and internal molecular complexity is true of subgenera and other taxonomic levels that are through time more strongly distinguished by gaps caused by selection of largely quantitative changes. Evolutionary divergence through gradual divergence of populations, geographical races, semispecies, sympatric species, and genera is discussed at length by Grant (1971), this updated by Levin (2001), and it is clear that convergence of taxa on the basis of differential traits of portions of species or genera that are promoted to species and genera *by selection* are not, at least for small numbers of traits, more or less probable as judged by relative numbers of traits.

Clustering properties of morphological and molecular data sets are alike such that somewhat similar groups of exemplar morphology or DNA loci commonly are obtained by any method, but the particular inferred tree or other evolutionary structure through time may at times be quite different (Lyons-Weiler & Milinkovitch 1997). Although neutral evolution (Kimura 1968, 1983; Nei 2005; Ohta 1992) is assumed, this may not obtain because there is abundant evidence of selection (and possible convergence) at the molecular level (Gillespie 1991). Hillis et al. (1996: 11) strongly advise that phylogeneticists should state that neutrality is an assumption in their studies. Selection at the gene locus level is apparently locus-specific, but (1) although locus specific, many genes each convergent across taxa towards a different developmental adaptive norm must contribute an element of chaotic uncertainty, and (2) swamping of small correct data sets by large ones reflecting convergence, plus support of the wrong tree of the large data set by data in a contrary small data set through Simpson's Paradox (Barrett et al. 1991; Getesy et al. 1999), are still problems.

Statistical phylogenetic analysis rests on many assumptions and emphasizes the speculative (Zander, 2007a). Simplicity arguments or point estimations are now abandoned in favor of credible intervals and parsimony with bootstrap support, but conclusions through optimality alone remain with us in sequence alignments (Redelings & Suchard 2005), model selection (Alfaro & Huelsenbeck 2006), and other choices. Multiple-test problems abound; for instance, if two lineages are monophyletic each at a probability of 0.95, then the chance of the two being "reciprocally monophyletic," i.e., both true at the same time, is the product, or 0.90 (Zander 2007a). A confidence level of 0.95 (or 0.99 in problematic cases) is standard in fields (psychology, ecology, evolution, population biology) using statistics, but present-day emphasis (seldom admitted) is on statistical discriminatory power and avoidance of Type 1 error (false positives) at the expense of reliability, with confirmation

of speculative lineages left to others (Zander 2007a). Bayesian analyses in systematics seldom follow the Bayesian philosophy (Winkler 1972: 393) of assigning a likelihood to every factor that may add uncertainty, and after analysis making a bet (an action or Bayesian solution, such as recommending a particular lineage to biogeographers and other scientists as reliable for their own work) only in the light of the risk if wrong. Bayesian phylogenetics is instead underlain with a host of problematic assumptions (Zander 2005, 2007a) even though Bayesian phylogeneticists (e.g., Huelsenbeck et al. 2002) commonly define the credible interval derived from the Bayesian formula embodied their software as the actual chance of the lineage being correct.

This disconnect leaves systematics with a vast literature in traditional taxonomy evaluating morphological evolution (e.g., as reviewed by Mayr & Provine 1980) as best as can be reconstructed given few fossils. This situation, however is not resolved by a phylogenetic paradigm change involving traditional taxonomic categories that are now based on, lumped, or split among molecular lineages simply because molecular phylogenies are far more detailed than what we can know of morphological phylogenies. What molecular phylogenies detail may have little to do with speciation involving selection and drift of expressed traits. It is perfectly acceptable that an evolutionary classification reflecting what can be inferred about evolution of expressed traits be far less resolved than molecular trees. Molecular trees, however, in spite of the generally speculative nature of published phylogenetic analyses may contribute significantly to diagramming evolution of taxa diagnosed by expressed traits without resorting to the present practice of simply mapping morphological traits on a molecular tree as though such atomized traits (Burleigh et al. 2013) were the stuff of evolution, not descent of taxa with modification.

The genus as a unit in biological reality — Stevens (1994) has reviewed natural classification of higher taxa in botany beginning with Jussieu's work expecting continuity between taxa, and the following monographers like Cuvier, Charles-François Brisseau de Mirbel, and Augustin-Pyramus de Candolle, who found discreet groups in nature, largely after 1812. He also reviewed the conservative nature of systematics through distrust of theory, emphasis on instinct and observation, an apprenticeship system, and constraints on change by lay users of taxonomy, problems that continue in traditional taxonomy to this day. All this, he averred, is in the absence of a "well articulated theory" of relationships. This is tendentious in light of Stevens' well-known long support of the phylogenetic perspective (e.g., Stevens 1985), because Darwinian theory is well-articulated, and is presently unfortunately challenged as a basis of classification by advances in understanding of a quite different thing, that is, patterns of molecular lineages. Stevens reviewed reticulate (web, net) and tree arrangements of natural or evolutionary relationships of higher taxa offered by botanists of the late 1800's.

Papers concerning the concept of genus from the mid-1900's (Anderson 1940; Cain 1956; Bartlett 1940; Camp 1940; Greenman 1940; Sherff 1940) generally concurred with the dictionary and nomenclatural code definitions that a genus is a taxonomic category ranking below a family and above a species, being a group of species with similar characteristics. As a taxon, it was whatever a specialist in a group determined it to be, as far as included species. Clayton (1972) found that numbers of species in genera of 19 families of vascular plants varied well with an expected logarithmic distribution and such distributions were similar in different families, arguing for genera being real entities (there was, however, considerable excess of monospecific genera). Sokal and Sneath (1963) promoted an entirely phenetic approach to classification, as overall similarity using equally weighted traits, seeing the genus as a primary cluster of species, but without detailing a particular definition. Legendre (1972) defined a genus as "a group of species which cluster after a chain is formed on pairs of species between which there is a calculated possibility of occasional hybridization," following Löve's (1963) conclusion that hybridization is possible between species of a genus but not between genera, and combining phenetic cluster analysis and estimations of out-crossing; see Grant (1971) for a summary of fertility relationships among species in a genus. Mayr (1969) used the operational definition of a genus as "a category for a taxon including one species or a group of species, presumably of common phylogenetic origin, which is separated from related similar units (genera) by a decided gap, the size of the gap being in inverse ratio to the size of the unit (genus)." A general review of the concept of genus is provided by Sivarajan (1991), with the traditional view basically being a group of species with some level of natural affinity and separated by a gap from closely related other groups.

With the paradigm change from general affinities to phylogenetics and tree-thinking (Baum & Smith 2012), Wiley (1981) defined genus as "a mandatory classification category to which every species must belong and which contains one species or a monophyletic group of species." Gill et al. (2005) characterized the phylogenetic genus as monophyletic, reasonably compact, and distinct as to evolutionarily relevant criteria: ecology, morphology, or biogeography. Cantino and de Querioz (2000), responding to phylogeneticists struggling with a range of problems in finding an acceptable definition of higher taxa in the context of molecular systematics and tree-thinking, proposed an entirely rankless classification. Contrariwise, Crawford (2000) rightly pointed out that phylogenies are interesting only when viewed in the context of other data. There is, in all, a general acceptance that *higher taxa have some kind of biological reality* (Chase et al. 2000).

Clearly there is a lack of methodological sympathy between practitioners of traditional and of phylogenetic systematics. Linnaean systematics remains valid as an observational and quasi-experimental (Cook & Campbell 1979) science, being a 250-year study by generations of researchers.

The genus concept suffers the same problems reviewed by Lam (1959) as to what a species is in nature, beyond being "the basic unit of taxonomy." The object of the brave, new (for taxonomy) field of biosystematics (Camp 1951; Camp & Gilly 1943), to define natural biotic units and develop a nomenclature reflecting limits, relationships, variability and dynamic structure, has been replaced by a different dimension, as it were, focusing on inferred lineages of changing morphological and molecular traits. The paradigm change in systematics is, however, the substitution of a tractable problem (determining molecular lineages of sister groups) for the wearisome, difficult problem of finding an acceptable evolution-based classification in the general absence of fossils and the prevalence of parallelism and convergence in expressed traits. But substitution of a different basis for classification, namely phylogenetic lineages of traits, for the evolutionary classification we have been pursuing since Darwin, solves no problems.

Molecular lineages do not directly reveal changes in expressed traits that follow selection and drift, nor do they substitute for a taxon tree showing descent of one taxon from another with modification, i.e., lineages of taxa. The latter may not be recoverable in anything like the detail that lineages can, but we must find some way to preserve what has been inferred about evolution of taxa with hints from fossils and heterophyletic trees of molecular strains, and not fragment the results of past morphological analyses in conformance to the focused enthusiasm of the moment. A paraphyletic morphologically based surviving ancestor is an evolutionary hypothesis that is not falsified by even well-supported reciprocal monophyly of the molecular lineages involved, and may be the most parsimonious solution for morphological homoplasy on molecular trees.

Genera have predictive value. Taxonomists rely on the integrity of genera in collecting, sorting and identification of species. If genera were truly random clusters of traits, this would soon prove useless (see Clayton 1972). Given that in the 250 years of Linnaean omnispection there have been few complaints and much positive activity in describing genera, one might assume that taxonomists are seeing biologically real genera. In describing new species, the new species either fit an old genus or a new genus must be described. Over the past 250 years, new genera based on expressed traits are commonly perfectly acceptable and useful once well examined, and are apparently not artifacts of randomly sorting intermediate species. One way to look at descriptions of genera is that these are suites or "libraries" of traits expected for species in the genus, with various combinations of particular traits coupled with one or more unique or at least phyletically isolated traits considered conserved or slow to change over time. Like species (Pianka 2000: 366), genera have their own "hand of cards." Given that a genus' library of expressed traits, particularly of morphology, is limited in observable numbers of traits compared to the potential library of DNA base changes, then saturation (crowding leading to overwriting and reversals) is probable. Because unique traits may be reversed in evolution, a polythetic genus concept is commonly necessary. The sum of all expressed traits may be governed developmentally, and may reflect some limitations of particular combinations of expressed traits (as phyletic constraints). It is clear, however, that genera, large and small, may be described to include most but sometimes not all intermediate species. Libraries, or suites, of expressed traits appear to be real at the genus level.

Beyond the fact that the Evolutionary Taxon Concept includes the EGC as one of the taxonomic ancestral sleeves that are implied by macroevolutionary analysis, the EGC is similar to the Evolutionary Species Concept (Simpson 1961; Wiley & Mayden 2000) in its generalist nature, and to the Ecological Species Concept in the idea of an associated niche (Anderson 1990; Grant 1992; Pianka 2000) that may limit the genus trait library. Simpson (1961) defined the Evolutionary Species Concept as an ancestor-descendant sequence of populations evolving separately from other species, with its own biorole (or ecological

niche), and having particular evolutionary tendencies. Wiley and Mayden (2000), more recently, asserted that the Evolutionary Species Concept reflects both pattern and process. "An evolutionary species is an entity composed of organisms that maintain its identity from other such organisms through space and time, and that has its own independent fate and historical tendencies." They advanced the somewhat rigorous guidelines that a genus may be only partially tokogenetic, particularly when populations may be isolated in time then remixed; species must be lineages; ancestral populations may be present in a cladistic analysis and may have no autapomorphies; and, infraspecific groups with diagnosable morphological traits that are genetically isolated should be recognized at the species level. Wiley and Mayden insisted on strict phylogenetic monophyly and asserted that lack of gene flow requires assumption of present or imminent speciation (albeit allowing that diagnosable traits are a plus), but the EGC described in the present paper focuses on inference of consensus or actual ancestral morphospecies to the degree resolvable given present data.

Phylogenetically complex species — In molecular analysis of DNA loci, an inferred split in gene lineages can be well supported as reflecting population isolation (or with organellar DNA, individual isolation plus purifying selection). As time goes on, an ancestor if defined molecularly must be different from molecular lineages continuing after the split because of continued DNA mutation of the sequence of interest. Thus, the idea that *molecular* ancestors disappear after generation of sister lineages is quite acceptable in *molecular* analysis, because DNA is necessarily different given some clocklike or quasi-clocklike change in non-coding DNA, an anagenetic process. It is not necessary, however, that expressed traits (morphology and biorole) change in parallel and tempo with changes in molecular non-coding traits, even when these signal probable isolation of ancestral populations. It is quite possible (as a null hypothesis) and perhaps even probable that stabilizing selection keeps an ancestral population stable in expressed traits whether it becomes phylogenetically internally complex or even after a daughter species of different expressed trait combination is generated. One or more isolated populations may develop into new species or may retain the expressed traits of the ancestral population, even though molecular phylogenetics reports having tracked (inferentially) many phylogenetic splits.

Past phylogenetic splits without species-level differentiation in expressed traits, evinced as inferred phylogenetic complexity, may well characterize many extant species. Mitochondrial haplotype analysis of many samples of subspecies of the domestic cat (Driscoll et al. 2007) showed complex infraspecific phylogenetic relationships. Thus, any one extant taxon may be phylogenetically complex, and inferences via molecular or morphological analysis cannot determine this on the basis of a few exemplars.

At this point, one may ask whether the complex cladistic relationships within a gene tree belong to a medley of ancestors, or to a single ancestor with static or only slowly changing expressed traits. In fact, a cladistic tree based on morphological traits is liable to the same question. At what point does one reject the idea that all ancestors associated with nested branching in a tree are not just one phylogenetically complex species? Persistence of evolutionarily static populations or species (Guillaumet et al. 2008; Leschen et al. 2008; Shen et al. 2008) are implied by the increasingly supported punctuated equilibrium theory of Gould and Eldredge (1993), which may be valid for many or most taxa, though apparently rare stepwise transitions have been demonstrated (e.g., Deméré et al. 2008). The generalist consensus morphotype of a genus is the best resolution obtainable for an ancestor of two or more extant closely related species, each of which could speciate from the other.

Inferential demonstration of the exact morphotypes of (at least) one ancestral species (in absence of fossil information) involves identification of surviving ancestors. This can be done (1) by biosystematic and cytogenetic studies, particularly in the case of "quantum" or local evolution (Grant 1971; Levin 2001; Lewis 1962), the budding of a daughter species from a peripheral ancestral population, and including the more recent method of Theriot (1992) inferring a surviving ancestor in a group of diatoms by evaluating a morphologically based cladogram and biogeographical information; or (2) the somewhat more simplistic and problematic selection of a surviving ancestor as one lacking autapomorphies on a cladogram; or (3) the method of heterophyly introduced here. The heterophyly method uses demonstration of two exemplars of one morphospecies separated on a molecular tree by an exemplar of a different morphospecies, implying a shared ancestor of all three that is identical or nearly identical to the two isomorphs. An example is the increasingly common discovery of what are generally viewed as phylogenetically isolated fully cryptic species (e.g., the 14 cryptic or nearly cryptic species of bryophytes recently

found in mosses, Shaw 2001) but which are more probably surviving disjunctive populations of one ancestral species or (alternatively but less probably) the results of two different speciation events from one ancestral species into isolated but identical niches with fixation of identical expressed traits. In both cases the two populations accumulate different DNA mutations over time but are morphologically static through time through some process like stabilizing evolution. In fact, demonstration of homoplasy of traditional groups, sometimes ingenuously characterized as "massive," in any molecular cladogram signals a possible surviving species, genus, family or other taxon.

Inferential demonstration of the morphotypes of *two* ancestral species (Fig. 13.2) is the presence on a molecular tree of two pairs of such individually phylogenetically isolated isomorphs, implying two different ancestors. With enough data, this can be extended to identification of many ancestral taxa in the genus. Such demonstrated ancestral morphotypes may then be nested or otherwise ordered by reference to molecular lineages. The Evolutionary Genus Concept combines morphological and molecular data to reveal as best possible both morphological and molecular evolutionary relationships.

Thus, two kinds of ancestral taxa can be inferred for the taxon tree. (1) A poorly resolved consensus taxon, which is basically the diagnosis of the highest taxonomic category including all terminal taxa to the level of genus. Higher taxa may be addressed in the same way, for instance, a consensus taxon with the diagnosis of the subfamily. A consensus ancestral taxon may also be implied when there is a gap in a molecular tree and a genus or family is split but the taxa at each side of the split are different species or genera; in this case one can infer an ancestor only with consensus traits of the next highest rank in the cladogram. (2) Surviving ancestors inferring an identical or nearly identical paraphyletic ancestor of the same rank. This would be, for example, one or more ancestral species in a genus, or ancestral genera in a subfamily or family. All such morphotype ancestral taxa may be arranged as progressive evolutionary diagrams with the help of reliable molecular trees. This concept reflects Cronquist's (1975) contention that parallelism among closely related taxa is of no import or consequence in taxonomy, and can be ignored, but may also deliver evolutionary information.

When morphological and molecular data agree, they may simply be subject to the same bias (selection and convergence) or lurking variables (LeBlanc 2004: 303), but when they disagree (and not simply due to error) then valuable new information may sometimes be available through cross-tree heterophyly. Paraphyletic taxa provide evolutionary data just as parsimoniously informative traits provide phylogenetic data. When taxa are renamed to enforce monophyly, evolutionary data are hidden, thus reinterpreting cladograms with somewhat older nomenclature that reflects expressed traits alone will best preserve informative molecular paraphyly.

Examples of macroevolutionary genera — Evolutionary diagrams or taxon trees combine consensus ancestral taxa with those inferred from heterophyly or other means, which are better resolved. The ancestral taxa are linked using phylogenetic information from molecular analyses. Examination of published molecular trees and mapped traditional taxa demonstrates cladistic splitting of what are probably ancestral taxa at various ranks surviving to the present (evolutionary paraphyly). With increasing numbers of exemplars of particular taxa in recent molecular analyses, this should become more evident. The authors of such analyses commonly have chosen to rearrange or otherwise modify to some extent taxonomic categories in the studied groups rather than investigate evolutionary and systematic implications of inferable descent with modification of taxa. It is important not to try to explain away evidence of oddities in descent with modification of taxa as merely cryptic or otherwise problematic genus identities or by attempting to enforce monophyly by finding a particular analytic method or gene sequence with a data set that by chance does so.

The superoptimized morphological evolutionary tree of the moss genus *Didymodon* (Plate 8.1) shows clustering at the caulistic level of several ancestral taxa, each with a cloud of derived species, and such clusters can be (and are above) described as genera. This is an exact definition of one kind of macroevolutionary genus (as a category just higher than species), and works for at least some groups. Given survival of any of the species, the *genus* then may be in longer morphological stasis (defined as prolonged existence of conservative morphological traits) than most of the species.

A test of the method of heterophyly for inference of surviving ancestral taxa is the degree to which paraphyletic morphotypes are nested on a molecular tree as not expected by chance alone, allowing for occasional true convergence or reticulate evolution. Sufficient data for such a test are not yet available but papers published to date seem to fill the requirement of general overall nesting of paraphyletic taxa, these

often of broadly distributed generalist species.

Some examples may be given: Zander (2008) found after analysis of a study by LaFarge et al. (2002) that the moss family Dicranaceae was the paraphyletic ancestor of two phylogenetically disjunctive and well-supported clades, the Dicranaceae s.str. and a lineage of generally small-sized taxa commonly referred to in phylogenetic literature as the "Rhabdoweisiaceae" but with no agreed morphological diagnosis to distinguish it from the Dicranaceae.

An ITS analysis of the Trichostomoideae (Pottiaceae, Musci) by Werner et al. (2005) demonstrated several ancestral taxa. *Chenia leptophylla* is sister to *Tortula inermis*, with *T. muralis* lower in the tree, implying a joint ancestor diagnosable as *Tortula*. In the same manner, *Hymenostylium hildebrandtii* is sister to *Tuerckheimia svihlae* and both are subtended on the clade by *T. valeriana*, indicating all three share an immediate ancestor diagnosable as *Tuerckheimia*. *Pseudosymblepharis schimperiana* is embedded within *Chionoloma* species and thus shares *Chionoloma* as an ancestral taxon. Both *Pseudosymblepharis* and *Chionoloma* are further embedded in *Trichostomum tenuirostris*, which implies that this last, widely distributed species is ancestral to all. *Pleurochaete squarrosa* is embedded among several species of *Tortella*, indicated it shares *Tortella* as an ancestor. Although *Tortella tortuosa* and *T. fragilis* are clearly close ancestors of *T. densa*, *T. rigens* and *T. inclinata*, there is an overlap on the molecular lineage that implies a double ancestor (silencing of a gene complex, perhaps of the odd propaguloid apex of *T. fragilis*, or hybridization and subsequent divergence). The same is true among *Weissia* species with many species bracketed by *W. controversa* on the molecular tree, but with sufficiently short branches (and overlap with *W. condensa* as an implied ancestor) that additional analysis with more data would be worthwhile. The evolutionary taxon tree of Plate 7.3 combines data from Werner et al. (2004) and Werner et al. (2005).

In a study of the moss family Hypopterygiaceae, Shaw et al. (2008) analyzed six nuclear, plastid and mitochondrial nucleotide sequences of 32 exemplars and found of four species of *Cyathophorum*, two were paired near the base of the strongly supported cladogram and two were buried deeply among 10 exemplars of the genus *Hypopterygium*. Although Shaw et al. chose to simply transfer the two species bracketed in *Hypopterygium* to that genus to preserve monophyly, it is clear that the cladogram is also evidence of *Cyathophorum* being a paraphyletic genus ancestral to several genera bracketed by it in the cladogram (*Arbusculohypopterygium, Canalohypopterygium, Catharomnion, Dendrohypopterygium, Hypopterygium, Lopidium*) while itself has one of these (*Hypopterygium*) as ancestor of two of its species. This scenario was described by myself (Zander 2009) as a "double ancestor" occurring along part of the cladogram but its particular interest lies in examination of various mechanisms that explain such contradiction (such as silenced genes and hybridization). There are two reasons not to accept the transfer of the two species of *Cyathophorum* to *Hypopterygium*: (1) preserving phylogenetic monophyly is insufficient reason not to recognize differences at the genus level (when the same are recognized elsewhere in the cladogram), and (2) reversion from silenced traits is an explanation far more plausible than total re-evolution of several major morphological traits to converge at the genus level, or multiple convergence from a *Cyathophorum* ancestor to lineages of *Hypopterygium*. An explanation is possible using extended paraphyly, but that requires multiple generation of

several different genera by a *Cyathophorum* ancestor, or by a *Hypopterygium* ancestor. The alert reader will note that my discussion above is of the "arm-waving" variety, characteristic of someone who really wants to come up with an explanation but none is terribly convincing. Lines of research, at least, are evident. If the isolated *Cyathophorum* species are genuinely of that genus, the problem becomes very interesting for future study.

In another study of mosses (Hernández-Maqueda et al. 2008), genera of the Grimmiaceae and related families allowed inference of a rather unambiguous evolutionary taxon tree (Plate 13.2) of several paraphyletic taxa. This tree is a joint structure of classical study, and morphological and molecular cladistic analysis. The Grimmiaceae cladistically brackets Campylosteliaceae and Ptychomitriaceae (by *Jaffueliobryum* and *Indusiella*) while *Grimmia* brackets *Schistidium* and *Coscinodon*. The latter brackets *Hydrogrimmia*.

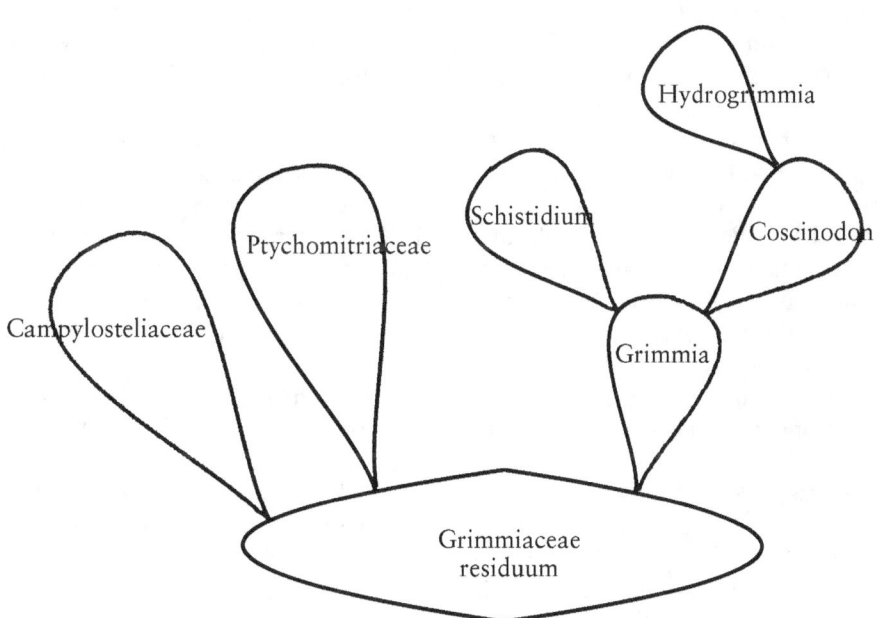

Plate 13.2. —Grimmiaceae taxon tree based on Hernández-Maqueda et al. (2008). The Grimmiaceae brackets Campylosteliaceae and Ptychomitriaceae (by *Jaffueliobryum* and *Indusiella*) while *Grimmia* brackets *Schistidium* and *Coscinodon*. The latter brackets *Hydrogrimmia*.

In other literature, a casual survey will commonly reveal well-supported paraphyly that implies ancestral taxa. In plants, for instance, the study of the

Knoxieae (Rubiaceae) by Kårched and Bremer (2007) detailed a well-supported phylogenetic disjunction of exemplars of species of *Otomeria* (separated by species of *Batopedina* and *Parapentas*), of *Pentanisia* (separated by species of *Neopentansia* and *Calanda*), and of *Pentas* (separated by species of *Carphalea*, *Knoxia*, and *Placopoda*). The molecular tree indicated that the phylogenetic relationship of these three ancestral genera is (*Pentansia*, *Otomeria*) *Pentas*, while the actual evolutionary relationship (one may be the ancestor of another) indicated by the molecular tree is better conceived as a Besseyan cactus with *Pentas* budding off the two genera *Otomeria* and *Pentansia*. Exemplar species of *Parapentas* are widely disjunctive on the molecular tree, separated from *Otomeria* and *Pentanisia* by several genera, and if re-examination supports this as true homoplasy (and not better interpretable as two different genera) then *Parapentas* would be an intermediate on the evolutionary cactus between the rather basal *Pentas* and the other two surviving ancestral genera.

Four well-supported ancestral genera can be perceived at the crown of a fern phylogeny (Schuettpetz & Pryer, 2007: Eupolypods 1, part 2) with exemplars of surviving ancestors *Ctenopteris* and *Lelingeria* arising from *Terpsichore*, which arises from *Grammitis*. In the Senecioneae (Asteraceae) relationships (Pelser et al. 2007) within the subtribe Othonniae clearly demonstrate *Othonna* as a surviving ancestor, and *Senecio* the same in the subtribe Senecioninae. The appearance of exemplars of certain genera (e.g., *Curio*, *Dendrophorbium*) with exemplars themselves phylogenetically disjunctive within a range of *Senecio* exemplar branches indicates that the molecular lineage may document either parallel development of the same genus from a consensus *Senecio* ancestor, or a switching back and forth among ancestral genus morphologies (as ancestral species of consensus morphotypes), perhaps via silenced gene complexes. These lineage splits are, in some cases, well supported.

A study of the Coreopsideae, Asteraceae (Mort et al. 2008) showed paraphyly of *Coreopsis*, exemplars of the genus appearing in scattered groups throughout the molecular cladogram, implying that this genus is the surviving ancestral morphotype for most genera of the tribe, including *Bidens*.

In an investigation of lava lizard phylogeny, Benavides et al. (2007) demonstrated phylogenetic disjunction on a nuclear DNA tree of *Microlophus quadrivittatus*, which bracketed *M. atacamensis*, and *M. tigris*, which bracketed *M. peruvianus*, although this is contradicted by the mtDNA clade. The consensus tree supported the nDNA tree, which, according to the authors, also makes better geographical sense.

All the above estimates of ancestral taxa must be viewed as speculative and just a first attempt at inferring a taxon tree, in spite of commonly high support values, because of the general paucity of same-taxon exemplars in all studies.

Discussion of evolution at higher ranks — Phylogenetically disjunctive major clusters of phenotypic traits ("massive homoplasy") on published molecular cladograms apparently contravene Dollo's Law (Gould 1970; Hall 2003) that sets of complex traits cannot be expected to re-evolve. These are evidence of deeply buried shared ancestral taxa, though this phenomenon is usually mis-interpreted as calling for massive rearrangements in classification. At the genus and species level, phylogenetic nesting on the basis of non-coding molecular traits may be accurate, but any one ancestral species may be phylogenetically complex. There is a kind of Library of traits characteristic of a genus. These traits commonly evolve faster than the opening of major niches. Evidence that this is so consists of (1) sympatric but closely related species, and (2) any subspecies or varieties, or diagnosable microspecies. Infraspecific genotypes make up the Queue waiting for an appropriate niche and selection pressures rewarding change by enhancing fitness. Only a complete fossil record will allow inference of the actual ancestral patterns of species within a genus. Nesting at the species level depends on statistical analysis of niche openings, and this is not now amenable to sampling or even acceptable characterization. Niches may be easily defined post hoc, but are problematic in predictive theory.

Apologists for molecular phylogenies commonly invoke such explanations as "massive homoplasy" or "cryptic species" for what may be interpreted as shared morphotaxon ancestry involving surviving ancestors (species, genera, etc.). Although it is good theory that evolution acts on fitness first at the individual then at the population level, that it may involve a few or many genetic traits, and it may or may not affect a whole species (Funk & Omland, 2003; Reiseberg & Burke, 2001), a genus conceived as a real thing in nature may well prove valuable as in diagramming evolution on a taxon tree. In many cases cryptic species discovered with molecular analysis such as barcoding (Hebert & Gregory 2005; Kress et al. 2005; Newmaster et al. 2006) may be associated later with diagnosable morphological or life history traits (Hebert et al. 2004; Hillis et al. 1996: 519). One must be wary, however, of multiple-test problems, in that there may be two alternative morphological sets of traits associated with recognition of two different taxonomic groups but molecular support that is randomly generated (or, better, indistinguishable from random generation at some level of confidence) is not support for one of them simply because morphology and DNA by chance agree (Zander 2007a, c). But in many cases the species remain fully cryptic, that is, phylogenetically disjunctive isomorphs (e.g., Elmer et al. 2007).

Regarding DNA barcoding, classical alpha taxonomy has a much wider range of sample space, and one can identify taxa that are not known for the flora. What about species that are new to science? How are these identified in a flora? By a percentage sequence difference, say 3.2%? These questions continue to be problems in all barcoding studies. This is because to really assure scientists that DNA barcoding has serious practical advantages, a separate alpha taxonomy of molecular data must be compared with the alpha taxonomy of morphological data. If they are the same, based on similarly large sampling, then indeed one is interchangeable with the other. Even using barcoding as an aid to morphological identification requires a molecular study with dense sampling of the problem for barcoding to help solve a morphological problem. Although it is not necessary to reexamine the fundamentals of a research program for every study, in the case of barcoding the statistical basis for the research fundamentals is not well settled.

Presently, the emphasis in phylogenetics is on using parsimony or coalescent theory Bayesian analysis on data sets of randomly mutating mostly non-coding DNA sequence traits that (1) theoretically track lineages of morphological speciation, or (2) are considered sufficiently and effectively evolution itself. Although molecular lineages apparently are statistically demonstrable, when conflicts occur with the results of morphological analyses or when resolution is needed in morphological analyses, the relationship between a molecular split and a speciation event is commonly based solely on the biological species concept requiring speciation after an event of isolation (e.g., as criticized by Rieseberg & Burke 2001). This does not apply to many if not most groups of organisms.

Systematics is gradually becoming based primarily on this molecular foundation, isolating itself from fields, such as population biology, ecology, evolution, and biogeography, that investigate or use theories of evolution based largely on expressed traits. Theories in science, particularly quasi-experimental or historical fields whose assumptions and results are not directly verifiable, are often easily generated and may be sustained by pure reason in the absence of empiric data. This is particularly true in systematics where the theoretic scaffolding for progression of evolutionary change in species is poorly resolved or understood because of a lack of facts and thus fact-based theory. Pieces of the puzzle are easily filled in by appeal to simplicity, a fancied similarity to the Principle of Least Action in physics. The latter, however, is quite solidly based in observation, while parsimony of tree length, or probabilities of branch coalescence, are at a remove from corroborative observations of details of descent with modification of taxa, e.g., budding evolution, as to their expressed traits.

It is here hoped that the difficult task of retrodicting evolution of taxa diagnosed by expressed traits is not abandoned by a paradigm change substituting molecular phylogenies for taxon trees. Declaring a difficult problem solved by changing the matter under scientific investigation because a different problem is easy to solve is not valid science. Certainly molecular phylogenetics can help to some extent, but only as revealing of relationships of ancestral taxa based on expressed traits important in survival and fitness or at least neutral within the bounds of phyletic constraint. In the future, exemplar isomorphs in a molecular tree will be most probably evolutionarily informative if selected from geographically isolated populations that are isomorphic at the species level, or as that plus unusual morphology at higher taxonomic levels.

Stevens (1994: 263) indicated that there is a struggle between evaluations of continuity and discontinuity of higher taxa, between process and pattern. In my view, speciation events involving taxa diagnosed by expressed traits represent the process, while lineages based on apparently non-coding DNA traits are the pattern. Superimposing inferable ancestral morphospecies (or higher taxa) on gene trees allows information about evolution of real entities without simply mapping expressed traits on a gene tree or assuming every molecular split affects evolution of expressed traits. Stevens (p. 265) asserted that many now believe that the "genealogical integrity of groups is of paramount importance." Genealogy, however, is far more than lineages of change in noncoding DNA bases of molecular strains but involves the ancestral-descendent relationships of actual species. In the absence of evidence that multiple ancestors in a supraspecific taxon actually exist, phylogenetic nesting is here relegated to a form of cluster analysis (as per P. Legendre, pers. comm.) and could be advantageously replaced by, for instance, cluster analysis that emphasizes rare morphological traits and genus trait libraries. Although knowledge of molecular lineages is valuable, I submit that evolutionary classifications based on paraphyletic groups that reveal ancestral taxa are more practical and more meaningful scientifically. I have called in the past (Zander, 2007c) for systematics to adopt (actually re-adopt) a basic unit that is the same as that of biodiversity studies. The MTC I believe is a step towards that goal.

The Macroevolutionary Taxon Concept postulates:

(1) that taxa, not traits, are the topics of interest in descent with modification, and lineages should ultimately be expressed in terms of taxa;

(2) that expressed traits involved as sets in selection are the focus of evolution;

(3) that because assumption of neutrality in expressed traits allows derivation of information through statistical analysis (selection confounding requirement of independent and random generation of traits), such assumption must be treated as an alternative hypothesis to the null of selection and preadaptation of any number of traits at least at the species level (which confounds morphological parsimony analysis);

(4) that two phylogenetically disjunctive (patristically distant) exemplars on a reliable molecular tree imply that they are surviving populations of a shared, paraphyletic ancestor with the consensus traits of whatever the exemplars represent, and that otherwise the best resolution of an ancestor are the consensus traits of the taxon (species, genus, family, etc.) or some other means of determining surviving ancestral taxa;

(5) that the genus is a real thing in nature not the least because it has utility in macroevolutionary analy-

sis, and is not just a requirement of Linnaean nomenclature or an artifact of random clustering;

(6) that the best evolutionary diagram is something similar to a Besseyan cactus;

(7) that ancestral taxa must have diagnoses or morphological descriptions to be placed on a taxon tree;

(8) that genera may be recognized by the combination of morphological gaps, apparently adaptive autapomorphies, conservative traits, and clouds of derivative species around supergenerative ancestral species in morphological cladograms, i.e., dissilient genera;

(9) and, that molecular trees if reliable help arrange the evolutionary diagram through their heterophyly.

Many of the concepts presented here are not particularly original but scattered in a large literature of criticism of cladistics and statistical phylogenetics. What I hope is new is the offering of an alternative to basing classification on lineages of trait changes. The MTC and the taxon tree incorporate advances in molecular systematics but go beyond phylogenetics to begin to chart descent with modification of integral, ecologically coherent living taxa. Papers by other authors contributing ideas essential the MTC and the taxon tree concepts include those of Alexander (2006), Brummitt (2003, 2006), Caporale (1999, 2003), Farjon (2007), Hörandl (2006, 2007), Lee (2005), Nordal and Stedje (2005), and Sosef (1997), among others.

This chapter is somewhat lengthy but is important, in my opinion, to developing scientific insight into evolution-based classification

CHAPTER 14
Support Measures for Macroevolutionary Transformations

Précis —Separate support measures are necessary for clades and for the taxa they represent. A taxon may be heterophyletic on a molecular tree, either with exemplars distant on the molecular tree or between morphological and molecular trees, and such heterophyly implies a deep ancestral taxon of the same name as the heterophyletic OTUs. Support for inferred macroevolutionary transformations involving such deep ancestors may be estimated from either the amount of present-day paraphyly in densely sampled, related groups, or from clade support and nearest neighbor interchange.

Uncertainty associated with exemplar branch order — Dayrat (2005) demonstrated that Darwin's Tree of Life was based on taxon-based progenitor-descendant transformations, not sister-group relationships. Although sister-group relationships can be often identified with some certainty in the hypothetico-deductive, cladistic context, uncertainty is contributed in evolutionary systematics by the use of induction in generating scientific theories of caulistic macroevolutionary transformation. Induction also increases the chance of false conclusions from true premises (Sober 1991: 20). Increased uncertainty associated with this "total evidence" analysis, on the other hand, because it uses deduction, induction, abduction, and reasoning by analogy, must be tolerated for a complete scientific theory.

Lurking variables (LeBlanc 2004: 303) are non-obvious influences on statistical analysis that may simultaneously affect two variables and simulate a correlation, or obscure the effect of one variable and mask a true correlation. Take the example of tossing a coin, a Markov chain of one step. Ideally a "fair coin" is defined in statistics as one with equal probability of landing heads or tails. In reality, most coins are heavier on the head side, and so come up tails slightly more often. Less equally well known is the fact that the side that is uppermost when the coin is flipped has to turn 360° to be uppermost when it lands but the underside of the coin only has to turn 180° to land uppermost, giving the lowermost side a slight advantage. If a coin is flipped without randomly arranging one or the other sides uppermost, the advantage of the tails side is then lessened. If the coin is set so that exactly 50 percent of the time heads is uppermost before tossing, then heads will come up more often in a fair coin. One can determine the exact percentage of times tails comes up in a heads-heavy coin only by recursively arranging the proportion of head and tails being uppermost before flipping. The proportion is changed until one side coming up more often than the other does not increase or decrease

depending on initial conditions. These are called "lurking variables." In molecular systematics, there are no error bars on nonparametric bootstrap support values or Bayesian posterior probabilities even though many problems and assumptions are well known.

Zander (2007) suggested a penalty of 1 percent in credibility support for each internal branch of a molecular cladogram to allow for unaccounted assumptions. It was pointed out that the final result of a Bayesian analysis is properly not the posterior probability, which reflects only the data set, but the Bayes' Solution (Kendall & Buckland 1971), which minimizes risk by taking uncertainty into account, in this case contributed by assumptions and data not dealt with in the phylogenetic method, model, or data. Sources of uncertainty, all familiar to phylogeneticists, include alignment, wrong gap costs, differential lineage sorting, hybridization, polyploidy, recombination, non-clocklike behavior, rates other than gamma distributed, differences between the results of "total evidence" and evaluations based on separate gene studies, possible strong selection pressure on non-coding promoter sequences, persistent pseudogenes, too few exemplars, endogenous retroviruses, gene conversion, self-correction of flawed DNA, paralogy, codon or nucleotide composition bias, chloroplast capture and other horizontal gene flow, novel clades, saturation, third codon bias, wrong identifications, long-branch attraction (see Kolaczkowski & Thornoton 2009 for Bayesian long-branch attraction in particular), model insufficiency, and other problems affecting the Bayes' Solution. Most phylogenetic analyses ignore the Bayes' Solution philosophy and it is asserted that the often highly probable results are "conditional" on some assumptions (occasionally some few assumptions are listed). Because the assumptions are actually many, this is like saying "I win the bet conditional on not having lost it."

The argument for a 0.01 penalty on the credible

support (i.e., multiply the support by a correction factor of 0.99) is based on the following rationale: if only 10 of these assumptions affected one branch support one out of 1000 times, then the jointly contributed uncertainty is 0.01, or 0.20 of the 0.05 window of reliability. This may seem a burden on statistical power, but it can be dealt with by empiric Bayesian analysis of multiple sequence studies (Zander 2007). The joint probability of more than one branch is the product of their individual credibility support values. In the case of phylogenetic trees with all branches supported at 0.99 credibility, only five contiguous internodes (chained clades) anywhere in the tree are acceptable as having branch order being acceptably correct at a joint probability of 95% (that is, 0.99 multiplied by itself five times). The 0.01 uncertainty is compounded because each node is a separate solution to the integrable analysis.

The Bayes' Solution — The Framework attempts address inconsistencies between classical systematics, morphological cladistics, and molecular analyses to offer a Bayes' Solution (Kendall & Buckland 1971), which formally incorporates all sources of uncertainty involved in the various methods used, and of certainty from neglected information. In decision theory, a Bayes estimator is broader than just using Bayes' Theorem, and is essentially a decision rule that minimizes the posterior expected loss; in other words, it maximizes the posterior expectation, given tolerable risk. In evolutionary analysis, a Bayes Solution is accomplished less formally by placing all data and inferences in a macroevolutionary context, and weighing risk (of, say, inadvertent extinction) as in decision theory by minimizing it as best possible (by, say, requiring high Bayesian posteriors for classification changes). Risk is discussed in this book for the biodiversity problem in Chapter 11, and for the case of multiple test problems in Chapter 15.

Classical systematics may include observations and inferences that are very well-supported, or "uncontested." For instance, an inference like taxon A is almost certainly the progenitor of taxon B because of biogeography, cytology, specialization to habitats, etc., may be the basis for a particular classical classification. This statement cannot be directly refuted by cladograms because cladograms do not offer serial models, only hierarchical, nested diagrams of evolution. Only if a cladogram separates taxa much farther than seems reasonable (e.g., moving it far from its traditional family) is there evidence against a particular serial macroevolutionary inference using heterophyly (see Chapter 6). Classical systematics, then,

has the advantage of a homogeneous reference class (Salmon 1971).

Although cladograms are often precise, various analyses of the same taxa often produce contradictory relationships, and fine resolution may be viewed suspiciously in some cases as merely precise, not accurate. Molecular data may conflict among sources, or between molecular data and morphological data, or be simply incorrect in certain situations (Avise 1994: 314; Philippe et al. 1996; Seberg et al. 1997; Sites et al. 1996). Nei et al. (1998) wrote: "We suggest that more attention should be given to testing the statistical reliability of an estimated tree rather than to finding the optimal tree with excessive efforts."

Cladograms may be published with some portions poorly supported. A possible explanation of why poorly supported arrangements are tolerated or accepted is "statistical relevance" (Salmon 1971: 11). Statistical relevance is the philosophy-of-science version of the Bayes Factor, recently much promoted by Bayesian statisticians (e.g., Aris-Brosou & Yang 2002; Suchard et al. 2002). The prior understanding, in this case, is that there is no or equal support for a particular hypothesis, and this is replaced after analysis by some statistical support which demonstrates what appears to be a relatively great and perhaps significant *increase* in support. This, however, is only apparent, and based in part on neglecting contrary evidence from classical systematics or morphological cladistics; a particular absolute level of support, based on all relevant data, is required for the arrangement to be accepted as due to shared ancestry. As Huelsenbeck et al. (2002) have pointed out, although the Bayes Factor has value, such as in model selection, it is the posterior probability that genuinely reflects the chance of an arrangement being correct. Also, a similar attitude known as "clinical relevance" (Hopkins 2001, 2003) is valuable in practice when an effect is demonstrated as not entirely reliable (e.g., p-values of 0.80 or 0.90) but the chance that it is helpful far outweighs the risk, e.g,. of using a harmless drug to treat a dread illness. In betting our science, however, the loss on failure can far outweigh any benefit on success.

Another easy way to wrongly transform Bayesian credibility intervals into certainties is to take the stance that a result that is far more probable than any one of a large number of contrary results must be correct because of that large difference in probability. For example, consider an icosahedron, with a contiguous ten of its 20 identical sides painted red, the others painted each a different color. Rolling the die will come "up" red 0.50 of the time. This can be tested in the frequentist fashion and a prediction made that future rolls will average about 0.50 red. But the frequentist viewpoint may be lost in Bayesian analyses. For instance, the icosahedron may be sliced into a hemisphere with one large side painted red. and the remainder with 10 much smaller sides. Thus, the die has one side similar to that of a coin (0.50 chance). Rolling the die will actually result in the red coming "up" (actually down) far more often than half the time because the somewhat rounded side "rolls." This is an instance of a heterogeneous reference set (Salmon 1971). The large red flat side does not represent 10 red flat sides, it is one side of a coin, and is not comparable.

Bayesian and frequentist statistics are both based on empiric observations of frequency, no matter how disguised. Consider two statisticians, a frequentist and a Bayesian. Both enter a casino that has a large number of (unloaded) games of chance. The frequentist plays one game, analyzes it by a number of tests, then predicts that the loss on continued play will be 12.5 percent. The Bayesian analyzes all the games by their physical appearance and predicts as a prior that each game provides a 12.5 percent loss. Given that the games are fair, they will each lose an average of 12.5 percent if they play one game a long time (the frequentist) or many games a long time (the Bayesian). In addition, the empiric Bayes analysis calls for using the posterior probability of one analysis as prior for the next, which is clearly empiric sampling and akin to frequentism.

Since the introduction of mathematical methods of reconstructing phylogenies more than 30 years ago, taxonomists have puzzled over just how results of maximum parsimony and, more recently, maximum likelihood and Bayesian Markov chain Monte Carlo analyses, particularly based on molecular data, should be incorporated into their classifications (e.g., Abbott et al. 1985; Adoutte et al. 2000; Jenner 2004; Lipscomb et al. 2003; Mallet & Willmott 2003; Seberg et al. 2003; Zander 1998a, 1998b, 2001a, 2003), short of just accepting them. It is the promise of molecular phylogenetic estimation and associated statistical methods that well-supported phylogenetic theories will be presented as alternatives to previously uncontested (Russo et al. 1996) morphologically based hypotheses, and either shore up or present new, reliable theories for past puzzling relationships. This work has appeared impressive because of the wealth of molecular data, sufficiently ample to allow statistical methods, and published reports detailing fully or very well-resolved trees with branches commonly having high bootstrap or Bayesian posterior probabilities. The opening of the phylogenetic black box as detailed in this book, however, may be illuminating.

Inferred macroevolutionary transformations — Frey (1993) found that paraphyletic scenarios, including local geographic speciation, are common or even the rule, while Gurushidze (2010) considered pseudoextinction (disappearance of a progenitor taxon after generation of two daughter taxa) to be rare. Thus, there should be many taxa that are paraphyletic but have one heterophyletic branch unsampled, for instance because of lack of research time, funding, or computer limitations. The lineage may be extinct or represented only by very old specimens or those rare in nature. The sampled lineage only appears to be phylogenetically monophyletic.

The percentage of paraphyletic taxa at each split in a cladogram (as measures of the past) may be judged by the percentage of paraphyletic taxa in the present. If 10% of the taxa represented by exemplars in a cladogram are paraphyletic, then the chance of any presently apparently monophyletic internode in the cladogram representing a paraphyletic taxon is 10%, one branch of the paraphyly assumed unsampled. That percentage, however, is not the same as the percentage of taxa being initially paraphyletic, which may be nearly all of them that generate new species by geographic isolation, or the percentage of taxa with static macroevolutionarily generative paraphyletic lineages that are unsampled today. Given little or no information on the percentage of true pseudoextinction events, an attempt to infer an ancestor of a particular taxon name at each and every node in a cladogram seems justified (see Chapter 8 on superoptimization). Albert Einstein said to W. Heisenberg (Gilder 2008: 87): "It is theory which first determines what can be observed." Phylogenetic theory is blind to caulistic macroevolutionary transformations at the taxon level because such cannot be modeled cladistically.

At times it is correct to postulate unsampled paraphyly in a molecular tree to explain incongruent relationships among morphological taxa. Morphological taxa are often well supported. The amount of data, as traits multiplied by the number of specimens examined, is commonly comparable between morphological and molecular studies, or even much larger in morphological studies, contrary to assumptions in the literature. It is not appropriate to simply ignore the morphological cladogram or to map traits atomistically on the molecular cladogram. Scientifically, no violence is done to logic and no information is lost if one postulates an unsampled paraphyletic branch of A occurring below B on the molecular cladogram, supporting the inference from superoptimization of the morphological cladogram that A is the ancestor of B, and also C. Such theorization yields testable hypotheses of major import.

Support measures for macroevolutionary transformations — The present-day molecular branch order of exemplar specimens is valuable to the extent it helps infer macroevolutionary transformation of taxa with paraphyly is evident. Given that good sampling of taxa (e.g., many specimens from many localities and habitats that sample a possible multiplicity of intrataxon populations) is rare, we must use rules of thumb for the prevalence of paraphyly. Paraphyly may be simple, with one included lineage of a different taxon (Plate 14.1a), or extended, with two or more included lineages (Plate 14.1b) by a heterophyletic pair of lineages or molecular strains of the single taxon that establishes the deep ancestor of the same name. In general, at least simple paraphyly is here considered extremely common in the past and mostly is unsampled. Extended paraphyly (or phylogenetic polyphyly) is probably less common but seems not uncommon as judged from published molecular trees in the literature, and superoptimization of *Didymodon* in Chapter 8.

In the case of *present-day* paraphyly, support measures for inferred macroevolutionary events at the taxon level can be derived from standard support measures for exemplar specimen branch order on a molecular tree. In the contrived rooted cladograms provided here (Plate 14.1a,b), all nested clades are supported at 0.95 posterior probability, commonly taken as a measure of good support. In simple paraphyly, as in Plate 14.1a, support for specimen B (representing taxon B) being derived from caulistic deep taxon **A**, which is implied by paraphyletic specimens A1 and A2, can be inferred from nearest neighbor interchange (NNI). NNI between clade (B A1) and exemplar A2 would not affect **A** > B (i.e., ancestor A macroevolutionarily giving rise to exemplar B) if we interchanged A1 with A2, but would if we interchanged B with A2. Because the support for (B A1) is 0.95, we can estimate half the uncertainty, or a support value of 0.975 for **A** > B. This is a simplification done in lieu of evaluating all the most probable branching patterns. The chance of B interchanging lower in the cladogram (given more clades towards the base of the cladogram) is small, at most one half of 0.05 times 0.05, or 0.0013. This, though finite, does not much lower the support for **A** > B.

One may note here that support for an optimal cladogram branch arrangement by comparing the support for the optimal arrangement with only the two NNI branch arrangements may not be appropriate for *morphological* cladograms in which convergence is common and the next shortest tree may result in a major branch rearrangement, not an NNI. The only exception is the use of NNI within morphological local groups that have elements that are well clustered and match classical analyses, and major rearrangements that break the cluster unreasonably are not expected. In that case, a probability can be calculated as follows:

The appropriate null hypothesis is that the support for the optimal branch arrangement and support for each of the two possible alternatives (through analysis after constraint and nearest neighbor interchange) is equal and randomly generated as neutral unlinked conservative traits. That is, if lineages attendant on an internode are termed A and B (for the sister groups terminating the internode), C for the basal branch, and D for the outgroup, then in a rooted tree support in numbers of steps for taxa A and B (here designated AB) is equal to support for A and C (= AC) and for B and C (= BC), or AB = AC = BC, and all variation is random. Thus, if AB is significantly larger than 1/3 the sum AB + AC + BC, then that support may be assumed due to shared ancestry, while all support for AC or BC is due to convergence between these two pairs of taxa. This calculation may be done with a variety of statistical methods, here using VassarStat's (2012) Exact Binomial Probabilities calculation, which treats the data as a Bernoulli trial. The minimum level of acceptable probabilistic support, here at least 0.95, for a reliable reconstruction of a single internode is a ratio of 3:3 of AB:(AB + AC + BC). This ratio would occur only 0.04 of the time if generated randomly. Other levels of support would be a minimum of 4:5, 5:7, 6:9, 7:11, all of which would provide at least 0.95 probability. But a level of 0.50 probabililty is attained by much less support, such as the following minimum ratios of steps: 2:3, 2:4, 3:5, 3:6, 3:7, 4:8.

That is, for example, if the number of steps supporting (AB) of three terminal clades A, B, and C is 4, and those supporting BC and AC total 4 (random if AB is true), then the chance of AB having 4 steps or more out of 8 is 0.26, thus the probability of AB being unexpected by chance alone is 1 minus 0.26, or 0.74. Using the traits supporting BC or AC in support of a molecular optimal for either BC or AC must be less than 0.50, and therefore must reduce the chance of the molecular relationship being true. The empiric

Bayes' analysis, with BC as, say maximally 0.26 probability as a prior, and BC from molecular analysis 0.99 posterior probability is, is 0.97, which is still acceptable, but with molecular analysis at 0.95, the solution is 0.86, not acceptable. See the Silk Purse Spreadsheet at:

http://www.mobot.org/plantscience/ResBot/
phyl/silkpursespreadsheet.htm

for easy Bayes' Formula calculations.

In the case of the highly supported *molecular* trees used in determining heterophyly, using nearest neighbor interchange (NNI) analysis is a good measure, particularly when the local group is clearly well clustered in the classical sense, and omnispection determines that there is no particular reasonable alternative clade on the cladogram that may be involved by moving to the NNI position, i.e., no expected major rearrangements in reasonably longer cladograms. In any case, it is *this* particular cladogram that is analyzed by NNI, not another, even though different longer cladograms are compared to it.

Alternatively, in the case of extended paraphyly (Plate 14.1b), both exemplar specimen B and clade (C D) are bracketed by A1 and A2, these last being two exemplar specimens of inferred deep ancestral taxon **A**. There are then two non-**A** lineages that apparently descend from the implied deep ancestral taxon A. Firstly, the support measure for the macroevolutionary transformation **A** > B depends on NNI as inferred from support measures between A1 and B, and A2 and B. NNI between B and A2 eliminated support for **A** > B, but interchanging A2 with the other lineage ((C D) A1) does not. Thus, 0.975 is support from this one NNI analysis. Switching of A1 with B also eliminates support for **A** > B, but switching (C D) with B does not, thus 0.975 is the support from the second half of the analysis. Both must be true at the same time for **A** > B, so the joint probability of **A** > B is 0.975 times 0.975, or 0.95.

A second apparent macroevolutionary transformation (Plate 14.1b) is then **A** > (C D). The NNI analysis is that if (C D) is interchanged with B, there is support, but if A1 is interchanged with B, there is no support, thus support for **A** > (C D) is 0.975 (half the 0.05 uncertainty of the support value). Taxonomically, the name of the implied caulistic deep ancestral taxon as the immediate shared ancestor of C and D would be the lowest ranking taxon that includes both C and D, unless C or D is identifiable as the progenitor of the other, then the ancestral name is more exactly identifiable.

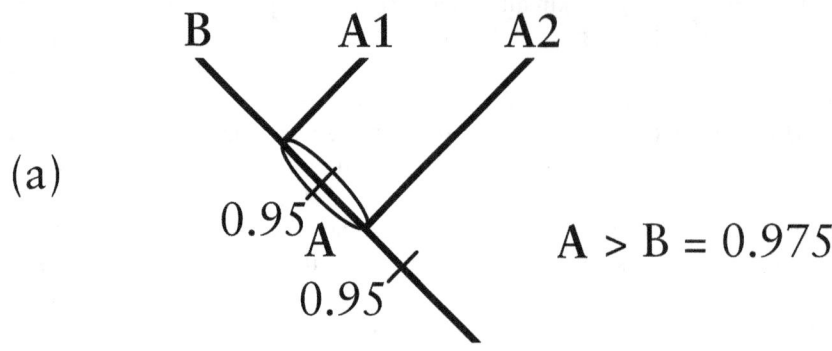

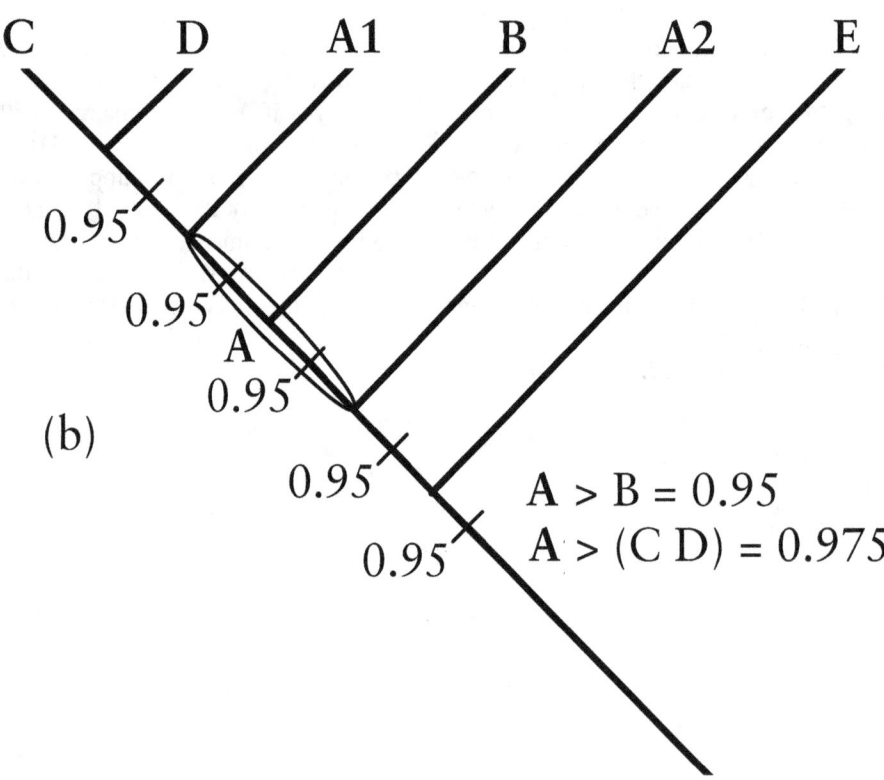

Plate 14.1 — **(a)** *Simple paraphyly:* Support for macroevolutionary transformation **A > B** through nearest neighbor interchange analysis has one-half the uncertainty of support for sister-groups (B A1). **(b)** *Extended paraphyly:* Both B and (C D) are "bracketed" by A1 and A2, and thus are both descended from implied deep taxon A. Support for transformation **A > B** is the product of one minus the uncertainty contributed by two calculations, the nearest neighbor interchange of A1 and B, and of B and A2 (see discussion in text). The support for **A > (C D)** is calculated as per **A > B** in (a).

A deep ancestral taxon may be inferred by cross-tree heterophyly (Element 4) of a distal exemplar on a molecular cladogram and that same taxon more basal in a morphological cladogram of the same taxonomic group. This is done by postulating an un-sampled (extinct or missed in sampling) lineage inserted into the molecular cladogram in the same more basal position as in the morphological cladogram. An example is the position of the bryophyte genus *Erythrophyllopsis* (Pottiaceae) very low in a morphological cladogram (Zander 1993: 47) but rather high in a molecular cladogram (Werner et al. 2004). The level of clade support for the position of an inserted theoretical lineage must be high (say, 0.95 posterior probability using coarse priors) if the morphological cladogram otherwise roughly approximates the molecular cladogram and there are basal taxa morphologically similar to the taxon in question (implying a similar generalist ancestor, as is true with *Erythrophyllopsis*) in the morphological cladogram. Propinquity of similar basal taxa on a morphological cladogram supports a primitive interpretation of those taxa because one might expect a set of shared ancestral taxa that explains such similarity though not in the sense of pseudoextinction. This avoids Crisp and Cook's (2005) otherwise good arguments against simply accepting an extant taxon that is a basal branch on a cladogram as an ancestral taxon. In addition, if there are no alternatives that contradict insertion of a paraphyletic branch, then Cohen's (1994) arguments against unremitting calculation of superfluous support values applies. Calculation of support for evolutionary transformations from an inferred deep ancestor can be based on the two heterophyletic branches. Remember that the molecular cladogram is an incomplete theory because it details only inferred genetic continuity and isolation events, not descent with modification of taxa. The addition of inductive inferences may limit degree of certainty but is scientifically sound and theoretically complete.

Consolidation with coarse priors — The support measure for heterophyletic inference of a deep ancestral taxon, e.g., **A** > **B**, or **A** > (**B**, **C**) in Plate 14.1, in the context of Bayesian analysis of total evidence, cannot stand alone. Coarse priors from classical taxonomy (natural keys) and from morphological cladistics (either nonparametric bootstrapping converted to posterior probabilities, see Zander 2004, or expert intuitive estimates of reliability superoptimized cladograms) must be addressed with the Bayes' Formula.

For instance, with a measure of **A** > **B** of 0.95 (as in Plate 14.1b) for a molecular clade, suppose a morphological cladogram found that an alternative branching pattern had a support of 0.95. The morphological results also imply that there is support for **A** > **B** but is low, 1 – 0.95, or 0.05. Bayesian analysis of these two support values for **A** > **B**, one high and the other low, yielded a posterior of 0.50. This is true for both alternative branching patterns when they are calculated separately.

Suppose we have from *different data* two clades as results, **A** > **B** and an alternative. The support for **A** > **B** is 0.99 and that for the alternative is 0.80; the alternative therefore supports **A** > **B** (there being no other reasonable results) at 0.20. The posterior for **A** > **B** by Bayes' Formula is 0.82, a lowered value.

CHAPTER 15
Multiple Tests and "Discovering" Morphological Support

Précis — A consideration of various solutions to the multiple tests (or multiple comparisons) problem demonstrates that there is a tendency to choose increase in power of discrimination as opposed to reliability. This may be problematic in considering cladograms as a whole or even in part. The probability of all of a set being true at once requires the multiplication of the probabilities of each member of the set. Finding apparent support from morphological analyses for molecular analyses that are contrary to accepted classifications by searching for alternative shared traits may be incorrect if such support is less than 0.50 probability. The low probability of the alternative morphological shared traits being ancestral will *reduce* the posterior probability of the molecularly inferred evolutionary relationships. After a molecular analysis, newly discovered supporting morphological traits doubtless seldom really support a molecular analysis given a Bayesian context. All contrary morphological relationships lower the probability of a molecularly based cladistic relationship whether there is alternative support for the molecular relationship or not.

Consider looking for remnants of Noah's Ark by examining photographs of the Mt. Ararat region (Noorbergen 1977). Will you find, say a big rock that looks like a boat? Sure you will, among thousands of big rocks. How about evidence for Atlantis and Lost Lemuria among the ancient literature (Scott-Elliot 1968)? You can find it if you delve persistently enough. How about morphological support for odd molecular results? Will the Noah's Ark Effect operate in science?

Felsenstein (2004: 299) asked whether selecting for the most probable branch arrangement might not involve multiple tests. Multiple test (or multiple comparison) problems should be of major concern to phylogeneticists. A simple example is flipping many coins several times each to test for a loaded coin. Eventually, a fair coin will come up 10 heads in a row by chance alone, so multiple tests alone fail to probabilistically identify a loaded coin. An example of multiple tests in cross-correlation of genes and characters is given by Stewart (2011: 120). Technically, the multiple test problem is the potential increase in Type I error occurring when statistical tests are used repeatedly.

Note: In frequentist literature, α is the level of significance (Type I error), the chance of being wrong that you can live with, usually 0.05. The *p* value is the actually attained α in your study.

The chance of two branch arrangements in a cladogram each at 0.95 CI (confidence interval) or (1 minus p) being both true at the same time is the product of the CIs or 0.9025. (The CI is used here instead of Bayesian posterior probability because the multiple test correction is usually made in the context of frequentist statistics.) The simple Bonferroni correction, commonly used in phylogenetics, is that a pretest α of 0.05 requires that the CIs of the two branch arrangements be at least α/k (k being the number of tests), or a CI each of 0.975 to indicate that each arrangement is actually 0.95 CI at the same time.

Experiment-wide (or cladogram-wide) reliability is not a particularly ardent goal in phylogenetics as nested sets may have poor internal or external reliability yet the one branch arrangement of interest may be both relatively independent of the rest of the cladogram, and well supported (e.g., a cladogram of bats and birds). Also, evaluation of a molecularly based cladogram may be somewhat reassured by the fact that many or most branch arrangements match those of morphological analysis (representing, for example, a high Bayesian prior). It is a fact, however, that the chance of all branch arrangements being correct (cladogram-wide or locally) is the product of the CIs, and that problematic branch arrangements or those contrary to accepted hypotheses require more support than being embedded in an otherwise well-supported tree.

In psychology and ecology, where an entire set of models must hang together for it to represent a decent theory, Bonferroni correction is important. It has been pointed out (Moran 2003; see also Nakagawa 2004) that the chance of obtaining many tests all resulting in very high CIs (and low *p* values) is very low, and most if not all results should be correct, yet, their assertion goes, both simple and sequential Bonferroni correction mask that and are overly conservative. On the other hand, a cladogram of 20 internodes (subtending 20 branch arrangements) each at 0.95 CI will have one branch wrong and which branch it may be is unknown. A paper with a paragraph about each arrangement will have one paragraph out of 20 discussing a wrong result. García (2004) pointed out that if Bonferroni correction is frustrating researchers with results involving many multiple tests, there are other methods of evaluation, including multivariate ANOVA, sharpened Bonferroni, false discovery rate tests, re-evaluation of independence, and repeating the experiment.

The problem, in my opinion, is between researchers pushing the bounds of knowledge and willing to let future study confirm new hypotheses, and other researchers requiring highly reliable hypotheses. Increasing power of the test is important to phylogenetic theoreticians to distinguish between hypotheses of moderate probabilities, but taxonomists, evolutionists, and biogeographers must base new studies on previous very low Type I error (in phylogenetics, this means the possibility of accepting branch arrangements generated by chance alone). There is good reason to not require correction for multiple tests or to use a weak correction procedure if the purpose is to examine extreme theoretical possibilities, but correction is necessary in all other cases. Theoreticians may be quite happy with 4/5 of the branch arrangements correct, and shrug off confirmation to the attention of future study. Pragmatists, on the other hand, will be disappointed with 1/19 of the arrangements wrong, as with CI = 0.95, preferring to focus on their own biogeographical, evolutionary, or ecological studies, which may themselves not be very well supported. Such discrepancy in research context is understandable and acceptable if acknowledged in print.

With simple Bonferroni correction, in the case of a cladogram of 20 branch arrangements each needing to attain 0.95 CI at the same time, the required corrected α is 0.05/k or the requirement of all 20 CIs to attain or exceed 1 minus 0.0025, or 0.9975, a tough requirement (α is $1 - CI$; k is number of support values). With a subclade of five branch arrangements, using simple Bonferroni correction, α/k is 0.01 requiring a corrected α yielding 0.99; the product of CIs of, say, 0.995, 0.99, 0.98, 0.97 and 0.96 is only 0.899, and no branch arrangement is significant at an 0.05 α in that subclade.

Although the requirement that a cladogram be entirely correct is justifiably relaxed in exploratory phylogenetics, the evaluation of monophyly is nevertheless impacted. In particular, for a cladogram with, for example, all branch arrangements at 0.95 confidence interval, only *one* arrangement is statistically allowed

to be considered at the same time monophyletic and reliable at 0.95, and that ideally must be a hypothesis selected to be tested beforehand to avoid multiple comparisons. If you test two hypotheses, then for two monophyletic lineages to be reliable at 0.95, they must each be 0.975 CI.

For a cladogram with many 0.99 CI branch arrangements, only 5 at most can be viewed as monophyletic and reliable at 0.95 CI, since the product of 0.99^6 is 0.94. And these 5 should have been selected beforehand as hypotheses to be tested. Most phylogeneticists ignore this basic multiple-test problem when presenting cladograms with BPP, decay or bootstrap values.

The sequential Bonferroni correction (Holm 1979; Hochberg 1988) is intended to relax the α such that the stricture of attaining extremely high CIs in multiple test situations of many tests does not invalidate the entire study, even when most results are alone significant. Basically, one calculates the α/k, then compares the results to this required significance level beginning with the highest CI. If that CI has a p value larger than α/k, it fails, also no other CI is acceptable. If the α is smaller, the resultant CI is taken as attaining, say, 0.95. Then the α/k is increased by lowering k by one (the required CI is lowered), and the calculation is repeated with the next highest CI.

Note that the Bonferroni correction does not require independence of data (though the similar Bernoulli test does). The sequential Bonferroni was devised to increase power of the test, but in fact this is empty resolution. The key idea is that once a particular test passes, the null hypothesis is rejected and the test is no longer part of the "family" of multiple tests requiring correction (because the hypothesis is true). The hypothesis is, however, not true (and therefore not part of continued testing) but is accepted as a working theory with such low chance of Type I error (the long-run probability of rejecting the null hypothesis that is in fact true) that can be acted upon given the risk of being wrong. The uncertainty, though small, remains. Eliminating a hypothesis to be tested is perfectly allowable when some descriptive or absolute feature isolates the hypothesis from other hypotheses being tested, but this cannot be done with probabilistic data; parts of a cladogram may be phylogenetically isolated by uncontested morphological or chemical descriptors that agree with probabilistic molecular results, but not by molecular data alone.

Strict Bonferroni: Select α as maximum tolerable chance of results being wrong (due to chance alone). Divide α by number of tests (k). All *p* values (1 minus nominal CI) that are smaller than α/κ are acceptable at the level of α (the corrected CI). Alternatively, the product of all CIs accepted in the tests must reach or exceed the required CI (1 minus α), such as 0.95.

Sequential Bonferroni: Select α. Do as for strict Bonferroni. After identification and segregation of *p* values that are lower than and therefore acceptable at α, recalculate k as equal to the number of tests left. Accept any *p* values that are now lower than α/κ , then recalculate, continuing until no *p* values pass.

Control of False Discovery Rate: Select α. Rank *p* values of tests in order of smallest to largest, and do strict Bonferroni. Of *p* values remaining, test next *p* value in order against α times the number of that test (n) in order divided by full original number of tests (k), e.g. 0.05 × (4/15) (i.e. the α times fourth in order of 15 original number of tests divided by original number of tests). If the *p* value is smaller, then accept and go to next test in order. Stop when correction fails, i.e., when *p* value is no longer less than α (α/κ) (Benjamini & Hochberg 1995).

Table 15.1 — Simplified descriptions of the three major correction methods for multiple tests. CI is confidence interval.

A now popular work-around for the strict Bonferroni correction method, likewise attractive to theoreticians in search of discriminitive power, is the Control of False Discovery Rate method (Benjamini & Hochberg 1995). Like the sequential Bonferroni, this requires the acceptance that the researcher will tolerate a higher proportion of Type 1 errors in a trade-off for accepting a larger proportion of true alternative hypotheses (correct branch arrangements). The FDR (false discovery rate) correction increases power of the test by allowing a higher proportion of "false discoveries" than sequential Bonferroni correction, and this proportion can be selected. Thus, for FDR, a correction at the 0.05 level would result in 0.05 of the results judged significant after correction being Type 1 errors (false positives for phylogenetic signal). For more exact control, a more complex "sharpened" version is available that depends on the fact that p values of true null instances follow a uniform distribution, which true alternative cases do not (Verhoeven et al. 2005).

The bottom line is that with strict Bonferroni correction all acceptable hypotheses are in almost all cases correct and Type 1 errors are almost entirely avoided, but, with a number of low CIs in a cladogram, no useful hypotheses can be countenanced. The lack of correction for multiple tests in published cladograms can falsely represent reliability when many conclusions are made from multiple branch arrangements with high CIs. With control of FDR, when the α is set for 0.05, then of every 20 significant results, one is incorrect as a Type 1 error. That theoreticians eschew correction entirely and pragmatists use strict Bonferroni is probably an extreme and deplorable situation. In fact, a control of FDR correction for theoreticians and use of sequential Bonferroni correction for pragmatists might be a better across-the-board solution. The problem is mitigated by the following practices: (1) partitioning the cladogram into phylogenetically isolated groups based on a combination of uncontested absolute descriptors (morphology, chemistry, etc.) and probabilistic molecular traits, which greatly lessens the size of the "family" of tests requiring correction, (2) preselecting hypotheses to test (one at 0.95 CI, up to 5 at 0.99, or sometimes more with sequential Bonferroni) which reduces the size of the multiple test family to a minimum (i.e., the hypotheses tested), and (3) using sequential Bonferroni correction or even increasing α to 0.10 (doubles Type 1 error to 1 in 10 instances but much increases detection of phylogenetic signal).

Although preselection of hypotheses can lead to bias, preselection is necessary to avoid choice of one hypothesis with a high CI from a number of tests. Of course, such a choice can be tested with later studies with different data and any choice of a randomly generated high score will be revealed as anomalous, but unless such later studies are available, preselection is the only alternative to rejection of all CIs by Bonferroni evaluation of many tests (nodes), including those without high CIs. Once preselection is done, then those high CIs that match the preselected tests are valid after Bonferroni correction at the much lower number of tests (the number of high CI arrangements rather than the number of all arrangements).

The steps that may be suggested for dealing with support values then are:

1. Divide molecular cladogram into as many phylogenetically isolated partitions as possible.
2. For each partition make a prioritized list of hypotheses to test.
3. Convert BPs (nonparametric bootstrap probabilities) to CIs, see Zander (2004).
4. Penalize each support value 0.01 for unaccounted assumptions.
5. Combine internodes less than 0.95 CI into one composite internode (see Zander 2007a) at 0.95.
6. Test prioritized list against 0.95 CIs, one per partition (or two per partition if α is increased to 0.10).
7. If many are available or if 0.99 are plentiful, redo composite internode correction to 0.99 CI, and test for up to five 0.99 CI hypotheses (or up to 10 0.99 CI hypotheses if your α allows 0.90 CI).

The ability to correct for particular subclades (when each subclade is considered fairly independent of other parts of the cladogram) determines the limit of accurate resolution of reliability determining optimal sister group placement in cladograms. Simply making sure that the products of the CIs of branch arrangements in a subclade exceed the preselected CI determined by α is a quick Bonferroni correction.

Support from morphological study — In many cases in the literature, an optimal molecular cladogram that contradicts morphological cladograms of maximum parsimony or "accepted" classical classifications is asserted to be supported by morphological shared traits discovered after they are searched for. For instance, according to Smith and Clark (2013), "Phylogenetic analyses derived from molecular data that are independent of the difficulties of morphological convergence have been especially useful to resolve monophyletic groups that can then be examined for morphological characters that unite the species in each particular clade...." (Note that such "con-

vergence" is often identified by difference from molecular results.) Although there is "support" for the molecular relationships from such discovered, non-classical traits, the contrary traits used in classical analyses may be overwhelming or so cogent that the probability of correctness of the classical arrangement of taxa is clearly of probability 0.50 or greater. Thus, *in the context of Bayesian analysis* the morphological support for the molecular analysis is less than 0.50, and when the two probabilities are combined in an empiric Bayesian analysis, the probability that the molecular result is true is necessarily reduced. Not evaluating the probability (no matter how imprecise) of the discovered new support from morphology for molecular results is a common and egregious error in multiple tests or multiple comparisons in phylogenetic analysis. The use of coarse priors advanced in this book allows evaluating morphological evidence against molecular patterns in the context of a Bayes' Solution.

Thus, small support from morphology that is anything less than 0.50 probability compared to alternative support for a classical solution is *not* support but is instead refutation for the molecular analysis. It may be expected that support for most molecular clades are reduced in posterior probability every time new shared morphological traits are discovered for support. Given this, *every* molecular clade that is contrary to an accepted classical evolutionary arrangement is probably poorly supported (less than 0.95) in Bayesian probability whether apparent morphological support is offered or not because the contrary morphological arrangement is doubtless supported at greater than 0.50 probability given all data on expressed traits. This is why a pluralist approach that conciliates all methods through a theory of macro-evolution will give a higher Bayesian probability even if not precise. "There is nothing more deceptive than an obvious fact," according to Sherlock Holmes.

Multiple test problems reminds one of the birthday problem in combinatorial probability, which states that there is at least half a chance that some two out of twenty-three randomly chosen people will share the same birth date, assuming that each birthdate is equally probable. For 60 people the match is almost certain. The reason this is relevant is that when searching for support for a molecular cladogram, the entire cladogram need not show morphological support, only a part of it. This is because morphology can be homoplasious and traits can reverse themselves in morphological cladograms. Thus, it is easy to declare partial morphological support as clear support for a molecular cladogram. Because it does not matter where in the cladogram (or clade) that support occurs, or if it occurs in somewhat separated areas of the cladogram, the chance of finding this distributed support is much greater than of finding exactly the right morphological support for all relevant nodes. One must keep in mind that congruence for morphological and molecular cladograms merely indicates agreement of across the board sister groups, not necessarily of macroevolution (being a combination of some nodes with budding evolution and some with pseudoextinction).

R. Feynman (D. L. Goodstein and G. Neugebauer in Feynman et al. 2011: xxi) discussed at length the fallacy of using the same data to verify an idea as that which suggested the idea in the first place. If preselection of taxa from classical taxonomy is used as a way to select exemplars for a molecular phylogenetic analysis, it is somewhat disingenuous for a scientist to be surprised or gratified that there is generally congruence between the results.

One can ignore some data as irrelevant if it is not part of a homogeneous reference class (Salmon 1971), but contrary data cannot be eliminated by off-hand assignment to other classes like homoplasy, convergence, or incomplete lineage sorting without actually demonstrating exactly how such rejection of contrary evidence is defensible for the case at hand. We all prefer to increase statistical power of discrimination when doing preliminary analyses, and leave pursuit of reliability to later workers. This calls to mind a parallel with abduction, the generation of hypotheses that may be only suggested by some few data. After 30 years of cladistics, however, it is important to now find a better, more instructive balance between statistical power and reliability when doing decisive analyses.

Although the multiple test problem is a fairly advanced concept in statistics, the problem is tractable and should have been obvious at some point to phylogenetic researchers, many well trained in statistics. Umberto Eco (1990) had an explanation for true-believer behavior in otherwise rational persons, adapted here as follows: First, a study with a history (phylogenetics has been around 30 years, generally well funded) must therefore have cachet. People want to believe, and the more something seems to be hidden, the more they insist that something must be there. Second, there are two rules. (a) Reasoning by naïve analogy is okay, when concepts and descriptions cross. An apple is round and edible, therefore anything round is edible. A cladogram looks like a family tree, and therefore must be one. (b) Anything that "fits" is okay, even if reasoning is circular. A cladogram describes a family tree, which may be described by a cladogram. Phylogenetic analysis must yield phylogenies because it is phylogenetic analysis. (c) If one uses observations made by others, they have the force of tradition, even if suspect. An example is GenBank (see Ruedas et al. 2000: "…most specimen data in GenBank are not congruent with potential repeatability of experiments"); another is the vast literature of apparently publishable phylogenetic "results." Eco's description seems a good dissection of cognitive dissonance, and applies to modern systematics. This is why the multiple test problem is described in several places in this book, because such provides a fundamental bias in favor of phylogenetic accuracy. Apropos of misplaced certainty, Asa Gray in a letter to Charles Darwin (Burkhardt et al. 2008: 5), wrote: "It is refreshing to find a person with a new theory who frankly confesses that he finds difficulties—insurmountable, at least for the present. I know some people who never have any difficulties, to speak of."

CHAPTER 16
Summary of the Framework

There has been a misunderstanding in the present conflict between proponents of phylogenetic and classical evolutionary systematics regarding paraphyly. It has been stated (Santos & Faria 2011) that there is a "war between advocates of strictly molecular or strictly morphological systematics." Evolutionary systematics, however, promotes pluralism in analytic methods, and appreciates molecular cladograms as informative of aspects of evolution. It is how phylogeneticists interpret exemplars in molecular cladograms as representative of taxonomic units that is at least a major basis for the dispute.

The molecular cladogram can fairly accurately give a retrodiction in terms of sister-groups of the gene history of each specimen used as an exemplar for a taxon. Gene history is used here as restricted to genetic continuity and isolation events in the molecular strain ending with each *exemplar*. Branch order of the *taxa* supposedly represented by the specimen exemplars are, however, not directly modeled. Differential self-nesting ladders may scramble molecular nesting such that primitive taxa (those low in a sequence of macroevolution) may be pushed high in a molecular cladistic tree by a series of speciation events generating daughter taxa. Sequential order of taxic evolution (one taxon generating another of the same or greater rank) cannot even be modeled on a cladogram. Hörandl (2010) has pointed out that molecular data are efficient for reconstruction of descent, but commonly used DNA markers have limited value for recognizing evolutionary groups, while morphological traits that contribute to structure and function are actually involved in selection, adaptation and co-evolution, and thus may be the proper bases for evolutionary grouping in classification.

Of particular importance is the idea that a split in a molecular lineage is not necessarily a speciation event. It could signal any isolation event, followed by phenotypic stasis of isolated populations, resulting in a surviving ancestral taxon in multiple isolated populations. Identification of a surviving ancestral taxon as a kind of living fossil may be done by (1) identification of a geologic fossil with an extant taxon; (2) biosystematic and cytogenetic studies, as in the case of "quantum" or local evolution (Grant 1971; Levin 2001; Lewis 1962), the budding of a descendant species from a peripheral ancestral population, identifiable, for instance, as in the event of several apparent daughter species being all more similar to an apparent parent than to each other (originator of this idea

unknown); (3) the recent method of Theriot (1992) inferring a surviving ancestor in a group of diatoms by evaluating a morphologically based cladogram and biogeographical information (see Chapter 8 on superoptimization); (4) the somewhat more simplistic and problematic selection of a surviving ancestor as one lacking autapomorphies on a polytomous morphological clade (Wiley & Mayden 2000: 157; discussion by Zander 1998); or (5) the methods of superoptimization and determining heterophyly on molecular trees as detailed in the present book.

When exemplars of different taxa are clustered together on a molecular tree, it is impossible to satisfactorily infer the phenotype of the shared ancestor or ancestors. It could be the phenotype of any one of the exemplars or of a taxon of entirely different phenotype. When exemplars of the same taxon are clustered together on a molecular tree, it is straightforward to infer that the phenotype of the immediate shared ancestor is that of the exemplars ("homophyly"), rather than all exemplars resulting from multiple convergences from an ancestor of a different phenotype. If the exemplars are all one species, the ancestor is that species. If they are of different species of one genus, the ancestor may be inferred to be that genus; or if genera, then their family, and so on. If two such clusters are sister groups, one may infer a particular ancestor for each of both clusters, but the phenotype of the immediate shared ancestor of the two clusters is impossible to infer from phylogenetic data alone. It could be one or the other or a different extinct or unstudied taxon of perhaps intermediate phenotype.

Zander (2008, 2010a) introduced the concept of taxon mapping where heterophyly (paraphyly or phylogenetic polyphyly of undeniably the same taxon) implies a deep ancestral taxon generating two extant lineages or molecular strains of the same taxon, and this progenitor also generated one or more lineages of different, apophyletic (descendant) taxa (at the same or higher rank) at or in between the paraphyletic branches. Theoretical macroevolutionary transformations are then synthetic, emergent properties. Looking for deep ancestors linking molecular and morphological inferences is equivalent to the search for "hidden variables" in physics, such as the so far unsuccessful search for a non-obvious classical explanation for the nonsensical rules of quantum mechanics. Postulation of theoretical caulistic macroevolutionary transformations, as taxonomically named internal nodes, forms the basis for that overarching theory, as macroevolution in phylogenetics is presently a hidden variable, whether pseudoextinction or budding evolution.

The method has six elements: (1) Alpha taxonomy is a set of genetic-algorithm-based heuristics developed over 250 years, and is in part based on physical and geometric principles. (2) Cladistic analysis of morphology aids in developing a natural key to taxa based on transformations of weighted conservative characters. (3) Molecular systematics establishes genetic continuity and order of isolation events of molecular strains, but not necessarily speciation events, through deep ancestors implied by heterophyly of exemplars. The name and rank of the ancestral taxon is that inclusive of all the heterophyletic exemplars. Probabilistic support values for ancestral taxa may be calculated. (4) Taxa low in the morphological tree but high in the molecular tree are theoretically ancestral taxa of all lineages in between, while morphological analyses may be reassessed with molecular taxon mapping. (5) "Superoptimization" by maximizing theoretical ancestor-descendant hypotheses minimizes superfluous unnamed postulated shared ancestors, while biosystematic and biogeographic study through Dollo evaluation at the taxon level provides biological evidence for macroevolutionary transformations. Supergenerative taxa and there stirps are often identifiable. (6) Classification by diagnosable macroevolutionary constraints requires the generalist Linnaean classification system capable of representing to some degree all aspects of evolutionary analysis through taxon-inclusive lists (distinctions) and ranks (similarities). In this paper, the level of support for macroevolutionary transformations at the taxon level is measured, based in part from support for phylogenetic clades.

Total evidence — There is much discussion of the requirement for total evidence (Allard & Carpenter 1996; Chen et al. 2003; Eernisse & Kluge 1993; Fitzhugh 2006, 2012; Hempel 1966; Nixon & Carpenter 1996), including much confusion over what is meant by total evidence, and how different data may be combined before evaluation or separately evaluated and then combined. The present reframing of pre-classification evolutionary analysis in systematics from modeling nested OTUs to modeling serial macroevolutionary transformations of taxa allows reinterpretation of the evidence such that both morphological and molecular cladograms can be explained even when they differ.

Consider a terminal clade of a morphological cladogram ((AB)C) versus a terminal clade of the same taxa on a molecular cladogram ((AC)B). Bayesian analysis, given each is supported at 0.99 probability, says neither can be more than 0.50 probability because they contradict each other. Suppose, however, superoptimization indicates that A is the progenitor of both B and C. There two processes that may generate the two different cladograms. (1) Morphologically, reversal in C of one of the morphological traits shared by A, B and C, places C lower in the morphological cladogram. (2) A self-nesting ladder involving A generating B then C molecularly would explain the different molecular clade. Both explanations are acceptable under a macroevolutionary process theory, and the joint explanation involving total evidence is $A > (^1C, ^2B)$, that is, taxon A gives rise to first C then B. Since the processes differ, there can be no increase in Bayesian credibility, but there is no decrease, and no data need be ignored.

Points of contention — I have been prompted to write this book by several kinds of outrages over the years, some admittedly theoretic and moot, but some terribly obvious and problematic, particularly:

(1) The elimination in phylogenetic classifications of any aspect of macroevolution, resulting in synonymy and splitting of what are apparently well-founded species and higher taxa of organisms through the phylogenetic classification principle of holophyly (which reflects nothing in nature).

(2) The assumption of no surviving ancestors, or that one taxon cannot be in two different molecular lineages, though there is plenty of evidence that this may result from punctuated equilibrium followed by long stasis of taxa governed by stabilizing selection—there may be in fact up to n − 1 terminal taxa that are ancestral to at least one other terminal taxon. Using the word "taxon" rather than "species" or "population" is due to my expectation that genera and higher taxa do evolve as units through joint species' reference to the paragenetic regulatory functions of a shared particular selective regime (an "envirosome" see point 7 below).

(3) Parsimony using morphological traits and other optimization analyses should use an evolutionary model including both pseudoextinction and budding evolution. No software does this.

(4) Avoidance of weighting of traits reflects the "automatic classification" philosophy originating with pheneticists and continued by phylogeneticists as "theory-free" taxonomy, such that a parsimony analysis of morphology is in reality based on raw similarity, lacking phyletic weighting—similar prob-

lems occur with molecular traits. If cladistics is truly a theory-free discovery process then there is no hypothesis to test, since the pattern discovered is a fact. Phylogenetics is theoretically bankrupt and has always been so.

(5) Classifications based on evolutionary systematics and apprehensions of serial macroevolution cannot be directly tested by nested relationships generated via cladograms, in which evolution is modeled as nested hierarchies with unnamed, undemonstrable shared ancestors as hidden causes.

(6) The common elimination of the environment from consideration in phylogenetic analysis. E.g., following the mechanisms for macroevolution of genera and higher taxa summarized by Gould (2002) in his opus, the environment acts as a regulator and guide (external chromosome equivalent, which may be called an "envirosome") of shared change among species of the higher taxa through stabilizing and disruptive selection acting on individual species or on traits shared by all species. This shared feature is not represented in the analytic processes of phylogenetic analysis, but is part of the expected judgment involved in classification by evolutionary systematists familiar with the habitats of organisms of their specialization.

Problems in terminology — Double meanings abound in the phylogenetic literature. Monophyly now means phylogenetic holophyly not evolutionary monophyly. Evolution now means changes in traits, not changes in taxa, i.e., microevolution is substituted for macroevolution because it yields more precise results. A tree is now actually a set of parentheses. Exemplar now means some sequences from one specimen taken to represent a whole species or an even higher rank. Shared ancestor is now a space between parentheses of nested exemplars. Lineage now means a nested set of exemplars clustered in a cladogram, not a serial line of ancestry (or if so then a serial line of unnamed nodes). Speciation is now modeled by two lineages generated from an extinct ancestor, and only that, which would be true if there were only two daughter lineages and no ancestral taxon ever survived speciation. A high Bayesian posterior probability is not a Bayesian Solution, which requires all relevant information and weighs risk of loss if wrong. Similarly, "discovery" is often a result of the amplifying power of statistical discernment at the expense of reliability.

Evolutionary distance is now measured by separation of numbers of often imaginary nodes (as patristic distance, see Zander 2007e). Genetic distance can

mean any number of things, from Nei's measure which has utility, to patristic or phyletic distance which are less meaningful, to numbers of DNA base differences, which may or may not mean anything. It is worth pointing out for this last that changes in mostly non-coding sequences that are used for molecular systematics are inexorable in both ancestral and descendant taxa, which means that an extant species with fossil representation tens of millions of years old could, theoretically, include between populations of the same species a "genetic distance" similar to that between extant families. At what point are differences in molecular sequences alone sufficient to name a new taxon? Clearly there are several meaning of "genetic distance," and changes in non-coding traits in a morphologically static taxon should never be used as taxonomic characters. Double meanings and other tergiversation can mask crippling problems.

Consistent systematics — A few paragraphs here summarize salient features of the struggle for a consistent, evolutionarily based systematics:

(1) Alpha taxonomy plus biosystematic and ecological study yield well-conceived taxa as apparent results of past evolution, these correspond to what the "exemplars" of phylogenetics are supposed to represent.

(2) Phylogenetic sister-group analysis is powerful and effective in determining cladistic sister-groups, but is crippled by insistence on classification by holophyly (a simplifying classification principle corresponding to no thing in nature), and with some revision (e.g., multifurcations) can be changed to a more practical natural key.

(3) Molecular trees demonstrate molecular strain continuity and isolation but not necessarily speciation, and are often confounded by self-nesting ladders.

(4) Macroevolution is being unfortunately eliminated from classification by phylogenetic insistence on (a) classification by holophyly and (b) not naming ancestral nodes (because this would create paraphyly).

(5) Elimination of macroevolution results in phylogenetic trees without names for ancestral nodes leading to clades but no caulis, i.e., a phylogenetic Tree of Life (Pennisi 2003) may be totally replaced by a Nested Parentheses of Life. The phylogenetic Tree of Life is actually a hierarchical Tree of Classification, and is not an evolutionary tree reflecting serial transformation of taxa.

(6) One taxon may be found in two (or more) different molecular lineages, commonly as surviving

isolated molecular strains, because morphological stasis associated with punctuated evolution may also follow isolation of two populations that remain identical at some taxonomic level, followed by budding evolution, and splitting of lineages is not necessarily accompanied by speciation.

(7) Stabilizing evolution on morphology and interaction of expressed traits with the environment may be decoupled from gradual accumulation of changes in the genome, such as apparently noncoding traits that are used to track continuity and splitting of lineages. There is considerable evidence from molecular analyses that this is true, e.g. "cryptic" species, genera and families.

The redemption of systematics faced with difficulties in dealing with vast numbers of species and few broad methodological guidelines is not to follow an arbitrary classificatory system that substitutes nested monophyly for serial evolution. Evolutionary systematics recognizes phylogenetic paraphyly as a necessary phenomenon of evolution. Cladograms postulate shared ancestors as explanations of relationship, but morphological and molecular analyses are often different. The aim of the Framework presented here is to postulate one ancestral taxic structure, based on macroevolution, that is shared by classical taxonomy, morphological cladistics, and molecular analysis.

Paradigms revisited — A final word on revolution in systematics. Brinton (1952), in his classic work on the subject summarizes (1952: 277) three major conclusions, which may apply to some extent in systematics. (1) Political and cultural revolutions are alike in instructive ways. Kuhn (1970) treats such similarities as paradigm changes in science. (2) Deeds and words often differ, with revolutionaries saying one thing and doing another. In systematics one might point to the vast changes in definitions or expectations of very important operational words like monophyly (evolutionary versus strict phylogenetic), evolution (of traits versus of taxa), support (mere corroboration of morphological and molecular cladograms versus similar data-rich separate analyses), tree (caulogram versus cladogram), data (some data versus all data), polyphyletic (distant on a cladogram versus demonstrably from different ancestors), parsimony (tolerating a host of postulated unobservable unnamed shared ancestors as hidden causes versus inferentially naming as many nodes as possible), lineage (a nested set versus a branching linear transformation series), and so on. (3) An eventual post-

revolution return to previous times given humankind's culturally conservative nature that limits political change, but with some, often positive changes.

The Framework presented here details a major change in how systematics might be done, but it is to a great extent a return or instauration (renewal after decay, lapse or dilapidation) of past but still valuable manners of analyzing data and synthesizing new knowledge. Nevertheless, in its pluralistic nature, the Framework retains those parts of morphological and molecular systematics that are genuine advances in systematic method. Brinton (1952: 277) ends with the observation that no revolution has ever completely fulfilled its promises.

O'Hara (1993) cited a paragraph by Dewey (1910) that apparently predicts the future of sister-group thinking:

"Old ideas give way slowly, for they are more than abstract logical forms and categories. They are habits, predispositions, deeply engrained attitudes of aversion and preference. Moreover, the conviction persists—though history shows it to be a hallucination—that all the questions that the human mind has asked are questions that can be answered in terms of the alternatives that the questions themselves present. But in fact intellectual progress usually occurs through sheer abandonment of questions together with both of the alternatives they assume—an abandonment that results from their decreasing vitality and a change of urgent interest. We do not solve them: we get over them. Old questions are solved by disappearing, evaporating, while new questions corresponding to the changed attitude of endeavor and preference take their place." (Dewey, 1910: 19).

Although this is plausible, R. Feynman, in a filmed lecture at Cornell University, pointed out that revolutionary new ideas need to fully replace the old, and fit well into their pragmatic function. Newton's cosmology was not "gotten over," it is replaced by an Einsteinian superset when relevant. Plane trigonometry was not abandoned, it is replaced by spherical when the latter is relevant.

What now of the cladistic revolution? Brinton (1952: 238) indicated that revolutions characteristically last 35 years. It is now 35 years from the late 1970s when cladistics first burgeoned as a force in systematics. Brinton also stated (1952: 240) that there is no real return to pre-revolutionary times, but a "new equilibrium" is established something like the pre-revolutionary times but clearly different. The present Framework offers a kind of solution in which no field is slighted as wrong or irrelevant, but all contribute particular strengths to a noble undertaking. In the Nash Equilibrium sense, all players maximize their profit given the known strategies and activities of the others. Such shared knowledge of data and method is necessitated by the Framework. We can no longer both ignore each other and prosper.

Ascription of fault — Humans relieve themselves of anxiety by assigning responsibility for major problems to others. Examples of such others are scapegoats, sin eaters, those with religions more strange than one's own, and hapless leaders of once popular causes. No set of persons is, however, guilty of sustaining the intellectual bubble of phylogenetics while cognizant of its fatal flaws. Phylogenetics has been a science-wide self-delusion, 30 years of falling down an up escalator. It has been powered by the same thing that drives all victims of confidence games—blinding greed, in this case for the perquisites of Big Science.

What happens now? — If the Framework has demonstrated that there are significant problems with "tree-thinking" in both morphological and molecular cladistics, and has suggested a new way, stem-thinking or macroevolutionary transformations at the taxon level, to approach data on evolution and biodiversity, what then are we to do with the myriad phylogenetic studies published in the past 30 years in reputable journals, often financed by public funds? They, in fact, remain valuable for the illuminations they provide—no matter how obliquely—on macroevolutionary transformations. Morphological cladistics helps develop natural keys and identify primitive taxa by phylogenetic propinquity with similar taxa in basal clades. Molecular techniques may imply deep ancestral taxa through phylogenetic paraphyly, and split taxa when patristic distance is larger than reasonably expected for extinct or unsampled phylogenetic paraphyly of molecular strains. Note that many of the examples of this book are based on re-interpretation of published phylogenetic papers. Any classification changes that have been made on the basis of phylogenetic nesting must be re-considered, at least for bias on account of strict phylogenetic monophyly and self-nesting ladders, at best for a thorough search for inferable macroevolutionary transformations distinguishing pseudoextinction and budding evolution. It is quite possible, however, in even the most felicitous scenario, that systematists will be preoccupied during the next many years with salvage systematics.

GLOSSARY

There are several glossaries of phylogenetic terminology on the Web. Here are presented a small selection of these words plus new or possibly differently defined terms as used in the present volume concerning evolutionary systematics.

Abduction – Creating hypotheses.

Aleatory – Probabilistically determined, as with a throw of the dice.

Alpha taxonomy – Part of systematics concerned with collection, identification, recognition of new taxa (particularly species), preparation of regional checklists and identification manuals, and monographs or revisions; largely confined to morphological, biogeographical, and ecological analytic evaluations.

Anagenesis – Gradual evolutionary change in a lineage without splitting, or as associated with pseudoextinction.

Anastasis – Molecular parallelism (two morphologically identical descendant populations from one ancestor) or polyphyly (two morphologically identical descendant populations from two different ancestral taxa), the generation of two taxa of the same name from an ancestral taxon of a different name at the same taxonomic level.

Ancestral taxon – One or more sections of a cladogram or one section of a caulogram consisting of inferred deep ancestors of extant exemplars that are diagnosable as one particular taxon.

Apophyletic – An apophyletic branch is that branch that comes out of a paraphyletic relationship on a cladogram, being bracketed by two branches of a single taxon of the same or lower rank; evolutionarily a descendant taxon but in a cladistic context.

Autapomorphy – A unique trait uninformative of sister-group relationships but which may be informative of unique evolutionary status or direction; a distinctive trait of no use in cladistic analysis but a major element of macroevolutionary transformation.

Bayesian analysis – A statistical method of estimating phylogenetic relationships in terms of nested diagrams, using only data that is precise and amenable to such nesting; proper analysis is the Bayesian Solution, which includes the effect of all data, precise or imprecise, to give a probabilistic answer in view of risk if wrong.

BPP – Bayesian Posterior Probability (a.k.a. BP).

Budding evolution – Speciation from a static ancestor; usually associated with peripatric speciation (margins of a range) although rapid sympatric isolation should result in much the same thing.

Caulistic – Pertaining to the axis and branches of an evolutionary tree, showing serial macroevolutionary transformations, e.g. a "Besseyan cactus" or caulogram.

Caulogram – a tree with emphasis on identified stem taxa; see commagram.

CI – Confidence interval (or level), a frequentist equivalent of the BPP.

Cladogram – A cluster analysis based on synapomorphies; an expanded diagram representing a set of nested parentheses; a calculated representation of hierarchical evolutionary relationships. Commonly assumed to be equivalent to a monophylogram.

Coarse priors – Also known as "stepped priors." For estimation of reliability of the evolutionary relationships of classical taxonomy and molecular cladistics, intuitive priors are set at 0.99 (almost certain); 0.95 (just acceptable); 0.75 (some support); 0.60 (hint of support); 0.50 (equivocal). See Table 8.1.

Commagram – A Besseyan "cactus"; a tree consisting of fat tadpoles showing directions of macroevolution of taxa; a caulistic monophylogram.

Congruent – Two cladograms that agree; see *incongruent*.

Conciliate – To reconcile, to make compatible, to come half way.

Conservative traits – Traits that are refractory to adaptation and which commonly occur in different adaptive regimes associated with different taxa that are related by such traits, acting like tracking traits in molecular systematics; commonly in stasis as opposed to molecular traits which are not, although these may each have different approximate rates of mutation.

Consiliate – An induction or generalization that is obtained from two or more different sets of facts, e.g., melding logically classical taxonomy, morphological cladistics, and molecular systematics such that all three infer a single joint macroevolutionary explanatory structure.

Dissilient – Springing open, exploding apart; here referring to a genus inferred as generating many usually highly specialized descendents or stirps.

Extended paraphyly – Phylogenetic polyphyly with no evidence of clades generated by differently named ancestors; several contiguous nodes of one deep ancestral taxon may be the correct central feature of a cladogram.

Evolution – Modification through descent of taxa.

Evolutionary systematists – Systematists who accept or even celebrate phylogenetic paraphyly as a basis for classification, and by extension accept macroevolutionary transformations at the taxon level as a basis for classification.

Exemplar – A sample of one; a specimen used to represent a population, a species, a genus, a family, etc., in a molecular cladogram.

Heterophyly – Including both paraphyly and phylogenetic polyphyly, simply two exemplars of the same taxon distant on a cladogram by at least one intervening exemplar of another taxon of the same or higher rank; e.g. ((A1, B) A2) as a terminal group, where A1 and B are sister but A2 is an exemplar of the same taxon as A1; the most important evolutionary information from molecular cladograms; also known as non-monophyly. Simple heterophyly is a single clade with two separated exemplars of the same taxon, complex heterophyly is two clades generated by, for instance, two self-nesting ladders from the same ancestral taxon (also called phylogenetic polyphyly).

Holophyly – Strict phylogenetic monophyly; all members of a clade must derive from one shared ancestor and the clade can have only one name at any rank, a cladistic classification "principle."

Homoplasy – Trait similarity in cladogram lineages that lead back to different shared ancestors, in the context of holophyly.

Incongruent – Two cladograms that disagree; see *congruent*.

Macroevolution – Descent with modification of taxa, requiring explicit distinction of pseudoextinction or budding evolution; series or successively branching sequences of taxa, impossible to diagram with cladograms having unnamed nodes, therefore not critical to phylogenetics.

Mapped taxon – Nodes on a molecular cladogram between certain separated exemplars representing inferred ancient ancestors of present-day taxa and diagnosable at a particular taxonomic level (the lowest shared by the exemplars) through a kind of taxonomic uniformitarianism; the best information from molecular systematic analysis.

Microevolution – Successive changes of traits mapped on a cladogram but seldom expressly associated with changes from one taxon to another.

Monophylogram – A diagram of serial and branching evolutionary relationships; a Besseyan cactus or commagram.

Monophyly – A clade with all taxa traceable to one ancestor; monophyly as used by phylogenetists is axiomatically strict, as used by evolutionary systematists, monophyly allows nesting of taxa of the same rank.

Multiple test (or multiple comparisons – A problem in statistics in which if you look around enough you will find, by chance alone, some surprising or supportive data, e.g., finding alternative sets of traits that do support a molecular relationship that is contrary to a relationship from morphological or classical analyses.

Nesting – Hierarchical diagrams in phenetic and cladistics that show distance between taxa by multiple layers of inclusive traits or series of inferred trait changes.

Node – Where two branches diverge in a cladogram; in phylogenetics, a locus tenens for an unnamable, unobservable, shared ancestor; in evolutionary taxonomy, an often nameable, often extant progenitor of one or more exemplars.

OTU – Operational taxonomic unit, a specimen or taxon ending a branch in a phenetic or cladistic tree.

Paraphyly – Disparaging phylogenetic jargon for a cladogram's representation of a progenitor in a macroevolutionary series.

Parsimony – A method of grouping taxa, which we all tend to use as a first pass from which to *start* analysis under the rubric that the simplest causal patterns should be examined first, given theory, in absence of other information; contrarily, the phylogenetic *end* of analysis.

Phylogenetics – An advanced form of mechanical knowledge in which unexplained, unnamed, unobservable processes as hidden causes explain the relationships of progenitors and descendants; cladistics with annotations of evolutionary inference.

Phylogeneticist – A systematist who finds greater precision in modeling evolution by nested diagrams than by serial descent with modification.

Polyphyly – Evolutionary polyphyly is two lineages not reasonably derived from the same ancestral taxon but named the same. Phylogenetic polyphyly is two exemplars or lineages separated by two or more nodes; phylogenetic polyphyly can be either evolutionary polyphyly or simply extended paraphyly, the latter with an implied ancestral taxon generative of two or more descendant lineages. Complex heterophyly involves two inferred self-nesting ladders.

Primitive – First or nearly first in a series. A serial concept in macroevolution as opposed to plesiomorphic, which is a nesting concept in phylogenetics.

Pseudoconvergence – Wrong ordering or pairing of branches in a molecular cladogram due to a combination of self-nesting ladders and extinct or unsampled lineages that contribute to extended paraphyly.

Pseudoextinction – Strictly this is the changing of one species into another through anagenesis, thus the ancestral species, as such, dies out. Phylogenetic pseudoextinction is when an ancestral species goes pseudoextinct (changes into another species) after or while generating a daughter lineage or molecular strain. This is a little easier to accept than the usual assertion that an ancestral species dies out after generating two daughter species, but also rejects the common occurrence of ancestral stasis. A shared ancestor of a different species is expected at a dichotomously branching node in cladistic models of evolution.

Punctuated equilibrium – Speciation associated with, at first, bursts of rapid change, then long phenotypic stasis.

Self-nesting ladder – A portion of a cladogram in which a progenitor taxon has appeared in a tree higher than the nodes leading to its descendants, either by reversal of morphological traits or continued mutation of tracking DNA, or by both.

Stasis – Taxa remaining much the same for thousands or millions of years without apparent change in morphology and other expressed traits (at a particular rank), possibly maintained through stabilizing selection; why we can do taxonomy at all.

Stirp (plural stirps) – A line descending from a single ancestor, an English word based on Latin "stirps" (plural stirpes) as in the legal sense of distribution of a legacy equally to all branches of a family (*per stirpes*). Used for one of a cloud of descendants of a core generative species.

Superoptimize – To make a morphological or molecular cladogram even more parsimonious by attempting to name nodes as ancestral taxa of various ranks, based on considering specialized morphology, biogeography, and ecology of the exemplars involved; this minimizes unnamable, unobservable, unexplainable superfluous entities.

Taxic – Of taxa.

Tree, evolutionary – A branching representation of macroevolution, emphasizing serial, not nested, caulistic relationships.

Tree, phylogenetic – A cladogram of inferred nested evolutionary relationships; a set of nested parentheses.

BIBLIOGRAPHY

Abbott, L. A., F. A. Bisby & D. J. Rogers. 1985. Taxonomic Analysis in Biology: Computers, Models, and Databases. Columbia University Press, New York.

Adams, D. 1980. The Restaurant at the End of the Universe. Pocket Books, New York.

Adoutte, A., G. Balavoine, N. Lartillot, O. Lespinet, B. Prud'homme & R. de Rosa. 2000. The new animal phylogeny: reliability and implications. Proc. Nat'l. Acad. Sci. USA 97: 4453–4456.

Aerts, D. 2009. Quantum structure in cognition. J. Math. Psych. 53: 314–348.

Aerts, D., D'Hooghe, B. & Haven, E. 2010. Quantum experimental data in psychology and economics. *Internat. J. Theoret. Physics* 49: 2971–2990. doi: 10.1007/s10773-010-0477-0.

Ahrends, A., C. Rahbek, M. T. Bulling, N. D. Burgess, P. J. Platts, J. C. Lovett, V. W. Kindemba, N. Owen, A. N. Sallu, A. R. Marshall, B. E. Mhoro, E. Fanning & R. Marchant. 2011. Conservation and the botanist effect. Biol. Conserv. 144: 131–140.

Albers, W. 2002. Prominence theory as a tool to model boundedly rational decisions. In: G. Gigerenzer & R. Selten (eds.), Bounded Rationality: The Adaptive Toolbox. Pp. 297–317. MIT Press, Cambridge, Massachusetts.

Aldous, D. F., M. A. Krikun & L. Popovic. 2011. Five statistical questions about the Tree of Life. Syst. Biol. 60: 318–328.

Alexander, P. J. 2006. Descent and modification in evolutionary systematics. Taxon 55: 4.

Alfaro, M.E. & J. P. Huelsenbeck. 2006. Comparative performance of Bayesian and AIC-based measured of phylogenetic model uncertainty. Syst. Biol. 55: 89–96.

Allard, M. W. & J. M. Carpenter. 1996. On weighting and congruence. Cladistics 12: 183–198.

Amorós-Moya, D., S. Bedhomme, M. Hermann & I. G. Bravo. 2010. Evolution in regulatory regions rapidly compensates the cost of nonoptimal codon usage. Mol. Biol. Evol. 27: 2141–2151.

Anderson, E. 1940. The concept of the genus: II. A survey of modern opinion. Bull. Torrey Bot. Club 67: 363–369.

Angiosperm Phylogeny Group III. 2009. An update of the Angiosperm Phylogeny Group classification for the orders and families of flowering plants: APG III. Bot. J. Linnean Soc. 161: 105–121.

Applequist, W.L. and Wallace, R.S. 2001. Phylogeny of the portulacaceous cohort based on ndhF sequence data. Syst. Bot. 26: 406–419.

Arendt, J. & D. Reznick. 2007. Convergence and parallelism reconsidered: what have we learned about the genetics of adaptation? Trends Ecol. Evol. 23: 26–32.

Aris-Brosou, S. & Z. Yang. 2002. Effects of models of rate evolution on estimation of divergence dates with special reference to the metazoan 18S ribosomal RNA phylogeny. Syst. Biol. 51: 703–714.

Aron, A., E. N. Aron & E. J. Coups. 2008. Statistics for the Behavioral and Social Sciences. A Brief Course. Fourth Edition. Prentice Hall, New Jersey.

Arts, T. 2001. A revision of the Splachnobryaceae (Musci). Lindbergia 26: 77–96.

Assis, L.C.S. and Rieppel, O. 2010. Are monophyly and synapomorphy the same or different? revisiting the role of morphology in phylogenetics. Cladistics 26: 1–9.

Avise, J. C. 1994. Molecular Markers, Natural History and Evolution. Chapman and Hall, New York.

Avise, J. C. 2000. Cladists in wonderland. Evolution 54: 1828–1832.

Bachmann, K. 2001. Evolution and the genetic analysis of populations: 1950–2000. Taxon 50: 7–45.

Bakalin, V. 2001. Notes on *Lophozia* VI. Taxonomy and distribution of *Lophozia* and *Schistochilopsis* (Lophoziaceae) in North America north of Mexico. Bryologist 114: 298–315.

Balhoff, J. P., I. Mikó, M. J. Yoder, P. L. Mullins & A. R. Deans. 2013. A semantic model for species description applied to the ensign wasps (Hymenoptera: Evaniidae) of New Caledonia. Syst. Biol. 62: 639–659.

Balzer, W., C. U. Moulines, & J. D. Sneed. 1987. An Architectonic for Science: the Structuralist Approach. Dordrecht, The Netherlands: Reidel.

Barnosky, A. 2001. Distinguishing the effects of the Red Queen and Court Jester on Miocene mammal evolution in the northern Rocky Mountains. J. Vert. Paleon. 21: 172–185.

Barnosky, A. D. 2005. Effects of Quaternary climatic change on speciation in mammals. J. Mammalian Evol. 12: 247–264.

Barraclough, T. G. 2010. Evolving entities: towards a unified framework for understanding diversity at the species and higher levels. Phil. Trans. Roy. Soc. B 365: 1801–1813.

Barrett, M., M. J. Donoghue & E. Sober. 1991. Against consensus. Syst. Zool. 40: 486–493.

Barry, P. 2002. Structuralism. Beginning Theory: An Introduction to Literary and Cultural Theory. Manchester University Press, Manchester, U.K.

Bartlett, H.H. 1940. The concept of the genus. I. History of the generic concept in botany. Bull. Torrey Bot. Club 67: 349–362.

Batemen, R. M. 1996. Nonfloral homoplasy and evolutionary scenarios in living and fossil land plants. In: M. J. Sanderson & L. Hufford. Homoplasy: The Recurrence of Similarity in Evolution. Academic Press, San Diego. Pp. 91–130.

Batten, D., S. Salthe & F. Boschetti. 2008. Visions of evolution: Self-organization proposes what natural selection disposes. Biol. Theory 3: 17–29.

Baum, D. A. & A. Larson. 1991. Adaptation reviewed: a phylogenetic methodology for studying character macroevolution. Syst. Zool. 40: 1–18.

Baum, D. & S. Smith. 2012. Tree Thinking: An Introduction to Phylogenetic Biology. Roberts and Company Publishers, Greenwood Village, Colorado.

Beatty, J. 1994. Theoretical pluralism in biology, including systematics. In: Grande, L. & O. Rieppel (eds.), Interpreting the Hierarchy of Nature. Pp. 33–60. Academic Press, San Diego.

Begley, S. 2010. The Limits of Reason. Newsweek. www.newsweek.com/2010/08/05/the-limits-of-reason.print.html

Bejder, L. & B. K. Hall. 2002. Limbs in whales and limblessness in other vertebrates: mechanisms of evolutionary and developmental transformation and loss. Evol. Devel. 4: 445–458.

Benavides, E., R. Baum, D. McClellan & J. W. Sites, Jr. 2007. Molecular phylogenetics of the lizard genus *Microlophus* (Squamata: Tropiduridae), aligning and retrieving indel signal from nuclear introns. Syst. Biol. 56: 776–797.

Benford, F. 1938. The law of anomalous numbers. Proc. Amer. Phil. Soc. 78: 551–572.

Benjamini, Y. & Y. Hochberg. 1995. Controlling the false discovery rate: a practical and powerful approach to multiple testing. J. Roy. Statist. Soc. B 57: 289–300.

Bergmann, P. G. 1949. Basic Theories of Physics: Mechanics and Electrodynamics. Dover Publications, New York.

Bessey, C.E. 1915. The phylogenetic taxonomy of flowering plants. Ann. Missouri Bot. Gard. 2: 109–233.

Blanc, D. C. 2004. Statistics: Concepts and Applications for Science. Jones and Bartlett Publishers, Sudbury, Massachusetts.

Blomberg, C. L. 1987. The Historical Reliability of the Gospels. Inter-Varsity Press, Leister, U.K.

Bock, W. J. 2003. Ecological aspects of the evolutionary processes. Zool. Sci. 20: 279–289.

Bock, W.J. 2004. Explanations in systematics. In: D. M. Williams & P. L. Forey (eds.). Milestones in Systematics. Systematics Association Special Volume 67: 49–56. CRC Press, London.

Bock, W. J. & G. von Wahlert. 1965. Adaptation and the form-function complex. Evolution 19: 269–299.

Bonner, J. T. 1974. On Development: The Biology of Form. Harvard University Press, Cambridge, Massachusetts.

Boucot, A. J. & J. Gray. 1991. Extinctions. [Review.] Evolution 45: 1294–1296.

Bowler, P.J. 1989. Evolution: The History of an Idea. University of California Press, Berkeley.

Brandon, R. 1990. Adaptation and Environment. Princeton University Press, Princeton.

Breen, M. S., C. Kemena, P. K. Vlasov, C. Notredame & P. A. Kondrashov. 2012. Epistasis as the primary factor in molecular evolution. Nature 490: 535–538.

Bridgham, J.T., E. A. Ortlund & J. W. Thornton. 2009. An epistatic ratchet constrains the direction of glucocorticoid receptor evolution. Nature 461: 515–519.

Brinton, C. 1952. The Anatomy of Revolution. Revised edition. Random House, New York.

Brooks, D.R. 1985. Historical ecology: a new approach to studying the evolution of ecological associations. Ann. Missouri Bot. Gard. 72: 660–680.

Brockman, J. 1995. The Third Culture: Beyond the Scientific Revolution. Simon & Schuster, New York.

Brower, A. V. Z. 2000. Evolution is not a necessary assumption of cladistics. Cladistics 16: 143–154.

Brower, A. V. Z. 2009. [Book review:] Science as pattern. David M. Williams and Malte C. Ebach. Foundations of Systematics and Biogeography. Systematics and Biodiversity 7: 345–346.

Brown, A. 2011. Old beds, standard deviation, can openers, Trampolines, and risk. Wilmot 54: 32–35.

Brown, J. 2000. Minds, machines and the multiverse. Simon and Schuster, New York.

Brummitt, R. K. 1992. Vascular Plant Families and Genera. Royal Botanic Gardens, Kew.

Brummitt, R. K. 1996. In defense of paraphyletic taxa. In: L. J. G. van der Maesen, X. M. van der Burgt & J. M. van Medenbach van Rooy (eds.). The Biodiversity of African Plants. Pp. 371–384. Proc. XIVth AETFAT Congress, 22–27 August 1994, Wageningen. Kluwer Academic Publishers, Dordrecht.

Brummitt, R. K. 1997. Taxonomy versus cladonomy, a fundamental controversy in biological systematics. Taxon 46: 723–734.

Brummitt, R. K. 2002. How to chop up a tree. Taxon 51: 31–41.

Brummitt, R. K. 2003. Further dogged defense of paraphyletic taxa. Taxon 52: 803–804.

Brummitt, R. K. 2006. Am I a bony fish? Taxon 55: 268–269.

Burke, J. 1985. The Day the Universe Changed. Little, Brown and Co., Boston.

Burkhardt, F., S. Evans & A. M. Pearn. 2008. Evolution: Selected Letter of Charles Darwin, 1860–1870. Cambridge University Press, Cambridge.

Burleigh, G., and 27 others. 2013. Next-Generation Phenomics for the Tree of Life. PLoS Currents Tree of Life 10.1371/currents.tol.085c713acafc8711b2ff7010a4b03733.

Butler, A. 2011. [Review of:] The Paleobiological Revolution. Essays on the Growth of Modern Paleontology. D. Sepkosky & M. Ruse, eds. Systematist 33: 15–17.

Cain, S. A. 1944. Foundations of Plant Geography. Harper, New York.

Cain, S. A. 1956. The genus in evolutionary taxonomy. Syst. Zool. 5: 97–109.

Camp, W. H. 1940. The concept of the genus: V. Our changing generic concepts. Bull. Torrey Bot. Club 67: 381–389.

Camp, W. H. 1951. Biosysystematy. Brittonia 7: 113–127.

Camp, W. H. & C. L. Gilly. 1943. The structure and origin of species. Brittonia 4: 323–385.

Cano, M. J. 2011. *Pseudocrossidium adustum* (Pottiaceae) an overlooked taxon in the Southern Hemisphere. Bryologist 114: 356–361.

Cano, M. J., J. A. Jiménez & J. F. Jiménez. 2010. A systematic revision of the genus *Erythrophyllopsis* (Pottiaceae, Bryophyta). Syst. Bot. 35: 683–694.

Cantalapiedra, J. L., M. Hernández Fernández, G. M. Alcalde, B. Azanza, D. DeMiguel & J. Morales. 2012. Ecological correlates of ghost lineages in ruminants. Paleobiology 38: 101–111.

Cantino, P.D. & K. de Querioz. 2000. Phylocode: A phylogenetic code of botanical nomenclature. http://ww.ohiou.edu/phylocode (viewed 22 February 2005).

Caporale, L. H. 1999. Chance favors the prepared genome. Ann. New York Acad. Sci. 870: 1–21.

Caporale, L. H. 2003. Natural selection and the emergence of a mutation phenotype: an update of the evolutionary synthesis considering mechanisms that affect genome variation. Ann. Rev. Microbiol. 57: 467–485.

Carle, F. L. 1995. Evolution, taxonomy, and biogeography of ancient Gondwanian libelluloides, with comments on anisopteroid evolution and phylogenetic systematics (Anisoptera: Libelluloidea). Odonatologica 24: 383–506.

Carlquist. S. 1996. The biota of long-distance dispersal. II. Loss of dispensability in Pacific Compositae. Evolution 20: 30–48.

Carnap, R. 1967. The Logical Structure of the World and Pseudoproblems in Philosophy. Transl. R. A. George. University of California Press, Berkeley.

Cartwright, R.A., N. Lartillot & J. L. Thorne. 2011. History can matter: non-Markovian behavior of ancestral lineages. Syst. Biol. 60: 276–290.

Cavalier-Smith, T. 2010. Deep phylogeny, ancestral groups and the four ages of life. Phil. Trans. Roy. Soc. B. 365: 111–132.

Cavin, L. & P. L. Forey. 2007. Using ghost lineages to identify diversification events in the fossil record. Biol. Lett. 3: 201–204.

Chakravartty, A. 2004. Structuralism as a form of scientific realism. Intern. Stud. Phil. Sci. 18: 151–171.

Chapman, G. B. & E. J. Johnson. 2002. In: T. Gilovich, D. Griffin & D. Kahneman. Heuristics and Biases: The Psychology of Intuitive Judgment. Cambridge University Press, Cambridge. Pp. 120–138.

Chase, M. W., M. F. Fay & V. Savolainen. 2000. Higher-level classification in the angiosperms: new insights from the perspective of DNA sequence data. Taxon 49: 685–704.

Chen, Wei-Jen, C. Bonillo & G. Locointre. 2003. Repeatability of clades as a criterion of reliability: a case study for molecular phylogeny of *Acanthomorpha* (Teleostei) with larger number of taxa. Mol. Phylog. Evol. 26: 262–288.

Chesterton, G. K. 1956. St. Thomas Aquinas. Garden City, New York.

Christenhusz, M. J. M. 2007. Evolutionary history and taxonomy of Neotropical marattioid ferns: studies of an ancient lineage of plants. Ann. Univ. Turkuensis 216: 1–78.

Christenhusz, M. J. M., H. Tuomisto, J. S. Metzgar & K. M. Pryer. 2007. Evolutionary relationships within the Neotropical, eusporangiate fern genus *Danaea* (Marattiaceae). Mol. Phylog. Evol. 46: 34–48.

Clausen, J., D. D. Keck & W. M. Hiesey. 1940. Experimental Studies on the Nature of Species. Carnegie Inst. Wash. Publ. 520. 452 pp.

Clausen, J. & W. H. Hiesey. 1958. Experimental Studies on the Nature of Species. IV. Genetic Structure of Ecological Races. Carnegie Inst. Wash. Publ. 615. 312 pp.

Clayton, W.D. 1972. Some aspects of the genus concept. Kew Bull. 27: 281–187.

Cleland, C. E., 2001. Historical science, experimental science, and the scientific method. Geology 29: 987–990.

Cobley, M. 2012. The Ascendant Stars. Orbit, Hachette Book Group, New York.

Cohen, J. 1994. The world is round (p < .05). Amer. Psychol. 49: 997–1003.

Collin, R. & M. P. Miglietta. 2008. Reversing opinions on Dollo's Law. Trends Ecol. Evol. 23: 602–609.

Cook, T. D. & D. T. Campbell. 1979. Quasi-experimentation: Design and analysis issues for field settings. Rand McNally College Publishing Co., Chicago.

Costello, M. J., R. M. May & N. E. Stork. 2013. Can we name Earth's species before they go extinct? Science 339: 413–416.

Costello, M. J., S. Wilson & B. Houlding. 2013. More taxonomists describing significantly fewer species per unit effort may indicate that most species have been discovered. Syst. Biol. 62: 616–624.

Cox, C. J., B. Goffinet, N. J. Wickett, S. B. Boles & A. J. Shaw. 2010. Moss diversity: A molecular phylogenetic analysis of genera. Phytotaxa 9: 175–195.

Cox, D. R. 1961. Tests of separate families of hypotheses. Proceedings of the 4th Berkeley Symposium on Mathematical Statistics and Probability 1: 105–123.

Cox, D. R. 1962. Further results on tests of separate families of hypotheses. J. Roy. Statist. Soc., Ser. B 24: 406–424.

Crandall, K. A., O. R. P. Bininda-Emonds, G. M. Mace & R. K. Wayne. 2000. Considering evolutionary processes in conservation biology: an alternative to "evolutionarily significant units". Trends Ecol. Evol. 15: 290–295.

Crawford, D. J. 2000. Plant macromolecular systematics in the past 50 years: one view. Taxon 49: 479–501.

Crawford, D. J. 2010. Progenitor-derivative species pairs and plant speciation. Taxon 59: 1413–1423.

Crisp, M. D. & L. G. Cook. 2005. Do early branching lineages signify ancestral traits? Trends Ecol. Evol. 20: 122–128.

Cronin, M. A., S. C. Amstrup, G. W. Garner & E. R. Vyse. 1991. Interspecific and intraspecific mitochondrial DNA variation in North American bears (*Ursus*). Canad. J. Zool. 69: 2985–2992.

Cronquist, A. 1975. Some thoughts on angiosperm phylogeny and taxonomy. Ann. Missouri Bot. Gard. 62: 517–520.

Cronquist, A. 1978. Once again, what is a species? In: L. V. Knutson, ed. BioSystematics in Agriculture Pp. 3–20. Alleheld Osmun, Montclair, N.J.

Crosby, M. R., R. E. Magill, B. Allen & Si He. 1999. A Checklist of the Mosses. Missouri Botanical Garden, St. Louis.

Crowe, T. M. 1994. Morphometrics, phylogenetic models and cladistics: means to an end or much to do about nothing? Cladistics 10: 77–84.

Crum, H. 1985. Traditional make-do taxonomy. Bryologist 88: 121–222.

Danchin, E., A. Charmantier, F. A. Champagne, A. Mesoudi, B. Pujol & S. Blanchet. 2011. Beyond DNA: integrating inclusive inheritance into an extended theory of evolution. Nature Reviews: Genetics 12: 475–486.

Darwin, C. 1859. The Origin of Species by Means of Natural Selection, Or the Preservation of Favoured Races in the Struggle for Life. 1963 Ed. Washington Square Press, New York.

Dayrat, B. 2005. Ancestor-descendant relationships and the reconstruction of the Tree of Life. Paleobiology 31: 347–353.

Queiroz, K, de 2007. Species concepts and species delimitation. Syst. Biol. 56: 879–886.

Danchin, É, A. Charmanier, F. A. Champagne, A. Mesoudi, B. Pujol & S. Blanchet. 2011. Beyond DNA: Integrating inclusive inheritance into an extended theory of evolution. Nature Reviews: Genetics 12: 475–486.

Demaster, D. P. & I. Stirling. 1981. *Ursus maritimus.* Mammalian Species 145: 1–7.

Deméré, T.A., M. R. McGowen, A. Berta & J. Gatesy. 2008. Morphological and molecular evidence for a stepwise evolutionary transition from teeth to baleen in mysticete whales. Syst. Biol. 57: 15–37.

Denk, T. & G. W. Grimm. 2010. The oaks of western Eurasia: Traditional classifications and evidence from two nuclear markers. Taxon 59: 351–366.

Dennett, D. C. 1984. Elbow Room: The Varieties of Free Will Worth Wanting. MIT Press, Cambridge, Massachusetts.

DeSalle, R., M. G. Egan & M. Siddall. 2005. The unholy trinity: taxonomy, species delimitation and DNA barcoding. Phil. Trans. Roy. Soc. B 360: 1905–1916.

Dewey, J. 1910. The influence of Darwinism on philosophy, and other essays in contemporary thought. Henry Holt and Company, New York.

Dewey, J. 1950. Reconstruction in Philosophy, with a New Introduction. Mentor Book, New American Library, New York

Ding, G., J. Kang, Q. Liu, T. Shi, G. Pei et al. 2006. Insights into the coupling of duplication events and macroevolution from an age profile of animal transmembrane gene families. PLoS Comput. Biol. 2(8): e102, 1–7. DOI: 10.1371/journal.pcbi.0020102

Dollo, L. 1893. Les lois de l' évolution. Bulletin de la Société Belge de Géologie; Paléontologie et Hydrologie 7: 164–166.

Doyle, J. J. 1992. Gene trees and species trees: molecular systematics as one-character taxonomy. Syst. Bot. 17: 144–163.

Doyle, J. J. 1996. Homoplasy connections and disconnections: Genes and species, molecules and morphology. In: M. J. Sanderson & L. Hufford (eds.). The Recurrence of Similarity in Evolution. Pp. 37–66. Academic Press, New York.

Driscoll, C.A., M. Menotti-Raymond, A. L. Roca, K. Hupe, W. E. Johnson, E. Geffen, E. Harley, M. Delibes, D. Pontier, A. C. Kitchener, N. Yamaguchi, S. J. O'Brien & D. Macdonald. 2007. The Near Eastern origin of cat domestication. Science Express www.sciencexpress.org 28 June 2007: 1–4, 2 figs. 10.1126 science.1139518.

Eco, U. 1990. Foucault's Pendulum. Ballantine Books, New York.

Edwards, C. J. and 17 others. 2011. Ancient hybridization and an Irish origin for the modern polar bear matriline. Current Biol. 21: 1251–1258.

Eernisse, D. J. & A.G. Kluge. 1993. Taxonomic congruence versus total evidence, and amniote phylogeny inferred from fossils, molecules, and morphology. Mol. Biol. Evol. 10: 1170–1079.

Efron, B., E. Halloran & S. Holmes. 1996. Bootstrap confidence intervals for phylogenetic trees. Proc. National Acad. Sci., USA 93: 7085–7090.

Egan, M. G. 2006. Support versus corroboration. J. Biomed. Informatics 39: 72–85.

Ekeland, I. 2006. The Best of All Possible Worlds: Mathematics and Destiny. University of Chicago Press, Chicago.

Eldredge, N. 1985. Time Frames: The Rethinking of Darwinian Evolution and the Theory of Punctuated Equilibria. Simon and Schuster, New York.

Eldredge, N. 1989. Macroevolutionary dynamics. McGraw-Hill, New York.

Elmer, K. R., J. A. Davila & S. C. Lougheed. 2007. Cryptic diversity and deep divergence in an upper Amazonian frog, *Eleutherodactylus ockendeni*. BMC Evol. Biol. 7: 247. doi: 10.1186/1471-2148-7-247.

Ence, D. D. & B. C. Carstens. 2011. SpedeSTEM: a rapid and accurate method for species delimitation. Mol. Ecol. Resources 11: 473–480.

Engels, F. 1989. Socialism: utopian and scientific. In: Pp. 68–111. In: L. S. Feuer (ed.), Marx and Engels: Basic Writings on Politics and Philosophy. New Introduction by L. S. Feuer. Anchor Books, Doubleday, New York.

Farjon, A. 2007. In defense of a conifer taxonomy which recognizes evolution. Taxon 56: 639–641.

Farris, J. S. 2012. Early Wagner trees and the "cladistic redux." Cladistics 27: 1–3.

Farris, J. S, A. G. Kluge & M. Eckardt. 1970. A numerical approach to phylogenetic systematics. J. Zool. 19: 172–189.

Felsenstein, J. 1985. Phylogenies and the comparative method. Amer. Naturalist 125: 1–15.

Felsenstein, J. 2001. The troubled growth of statistical phylogenetics. Syst. Biol. 50: 465–467.

Felsenstein, J. 2004. Inferring Phylogenies. Sinauer Associates, Inc., Sunderland, Massachusetts.

Festinger, L. 1957. A Theory of Cognitive Dissonance. Stanford University Press, Stanford, California.

Feynman, R. P. 1985. Surely You Are Joking, Mr. Feynman: Adventures of a Curious Character. W. W. Norton, New York.

Feynman, R. P., R. B. Leighton & M. Sands. 2011. Six Easy Pieces: Essentials of Physics Explained by its Most Brilliant Teacher. Basic Books, New York. Fourth Edition.

Fischer, D. H. 1989. Albion's Seed: Four British Folkways in America. Oxford University Press, New York.

Fisher, K. M., D. P. Wall, K. L. Yip & B. D. Mishler. 2007. Phylogeny of the Calymperaceae in a rank-free systematic treatment. Bryologist 101: 46–73.

Fitzhugh, K. 2006. The "requirement of total evidence" and its role in phylogenetic systematics. Biology & Philosophy 21: 309--351.

Fitzhugh, K. 2012. The limits of understanding in biological systematics. Zootaxa 3435: 40–67.

Flora of North America Editorial Committee, eds. 2007a. Flora of North America North of Mexico. Magnoliophyta: Commelinidae (in part): Poaceae, part 1. Vol. 24. Oxford University Press, New York.

Flora of North America Editorial Committee, eds. 2007b. Flora of North America North of Mexico. Bryophyta, part 1. Vol. 27. Oxford University Press, New York.

Foote, M. 1996. On the probability of ancestors in the fossil record. Paleobiol. 22: 141–151.

Frey, J. K. 1993. Modes of peripheral isolate formation and speciation. Syst. Biol. 42: 373–381.

Frey, W. & M. Stech. 2009. Bryophytes and seedless plants: Marchantiophyta, Bryophyta, Anthocerotophyta. In: Frey, W., ed. Syllabus of Plant Families: 9–264.. Ed. 13. Part 3. Gebr. Borntraeger Verlagsbuchhandlung, Berlin.

Frondon, III, J. W. & H. R. Garner. 2004. Molecular origins of rapid and continuous morphological evolution. Proc. Nat. Acad. Sci., USA 101: 18058–18063.

Funk, D. J. & K. E. Omland. 2003. Species-level paraphyly and polyphyly: frequency, causes, and consequences, with insights from animal mitochondrial DNA. Ann. Rev. Ecol. Evol. Syst. 34: 397–423.

Futuyma, D. J. 1998. Evolutionary Biology. Third Edition. Sinauer Associates, Sunderland, Massachusetts.

Games, P. A. & G. R. Klare. 1967. Elementary statistics: Data analysis for the behavioral sciences. McGraw-Hill, New York.

García, L. V., 2004. Escaping the Bonferroni iron claw in ecological studies. OIKOS 105: 657–663.

Gardner, M. 1982. Aha! Gotcha: Paradoxes to puzzle and delight. W. H. Freeman, San Francisco.

Gatesy, J., C. Matthee, R. DeSalle & C. Hayashi. 2002. Resolution of a supertree/supermatrix paradox. Syst. Biol. 51: 652–664.

Giere, R. N. 2009. Essay review: Scientific representation and empiricist structuralism. Phil. Sci. 76: 101–111.

Gigerenzer, G. 2001. The adaptive toolbox. In: G. Gigerenzer & R. Selten (eds.). Bounded Rationality: The Adaptive Toolbox. Pp. 37–50. MIT Press, Cambridge, Massachusetts.

Gigerenzer, G. 2007. Gut Feelings: the Intelligence of the Unconscious. Viking Penguin: New York.

Gigerenzer, G., J. Czerlinski & L. Martignon. 2002. How good are fast and frugal heuristics? In: T. Gilovich, D. Griffin & D. Kahneman (eds.). 2002. Heuristics and biases: The psychology of intuitive judgment. Cambridge University Press, Cambridge. Pp. 559–581.

Gigerenzer, G. & R. Selten. 2001. Rethinking rationality. In: G. Gigerenzer & R. Selten (eds.). Bounded Rationality: The Adaptive Toolbox. Pp. 1–12. MIT Press, Cambridge, Massachusetts.

Gigerenzer, G., Z. Swijtink, T. Porter, L. Daston, J. Beatty & L. Küger. 1989. The Empire of Chance: How Probability Changed Science and Everyday Life. Cambridge University Press, Cambridge.

Gilder, L. 2008. The Age of Entanglement: When Quantum Physics was Born. Alfred A. Knopf, New York.

Gill, F. B., B. Slikas & F. H. Sheldon. 2005. Phylogeny of titmice (Paridae): II Species relationships based on sequences of the mitochondrial cytochrome-b gene. Auk 122: 121–143.

Gillespie, J. H. 1991. The Causes of Molecular Evolution. Oxford University Press, Oxford.

Gilmour, J. S. L. 1940. Taxonomy and philosophy. In: J. Huxley, ed. The New Systematics. Oxford Press, London. Pp. 461–474.

Gilovich, T. & D. Griffin. 2002. Introduction—heuristics and biases: then and now. In: Heuristics and biases: the psychology of intuitive judgment. Cambridge University Press, Cambridge. Pp. 1–18.

Gilovich, T., D. Griffin & D. Kahneman (eds.). 2002. Heuristics and biases: the psychology of intuitive judgment. Cambridge University Press, Cambridge.

Gishtick, A. 2006. Baraminology. Reports of the National Center for Science Education 26(4): 17–21. Available at: http://ncse.com/rncse/26/4/baraminology (September 2, 2010).

Goffinet, B., W. R. Buck & A. J. Shaw. 2008. Morphology, anatomy, and classification of the Bryophyta. In: B. Goffinet & A. J. Shaw (eds.). Bryophyte Biology. Second Edition. Pp. 55–138. Cambridge University Press: Cambridge.

Goldberg, E. E. & B. Igić. 2008. On phylogenetic tests of irreversible evolution. Evolution 62: 2727–2741.

Goldstein, G., G. Gigerenzer, R. M. Hogarth, A. Kacelnik, Y. Kareev, G. Klein, L. Martignon, J. W. Payne & K. H. Schlag. 2002. Group report: why and when do simple heuristics work? In: G. Gigerenzer & R. Selten (eds.). Bounded Rationality: The Adaptive Toolbox. Pp. 173–190. MIT Press, Cambridge, Massachusetts.

Gontier, N. 2011. Depicting the Tree of Life: The philosophical and historical roots of evolutionary tree diagrams. Evolution: Education & Outreach DOI 10.1007/s12052-011-0355-0 Online 19 August 2011.

Goremykin, V. V., S. V. Nikiforova, P. J. Biggs, B. Zhong, P. Delange, W. Martin, S. Woetzel, R. A. Atherton, P. A. McLenachan & P. J. Lockhart. 2012. The evolutionary root of flowering plants. Syst. Biol. 62: 50–61.

Gould, S. J. 1970. Dollo on Dollo's Law: irreversibility and the status of evolutionary laws. J. Hist. Biol. 3: 189–212.

Gould, S. J. 2002. The Structure of Evolutionary Theory. Belknap Press of Harvard University Press, Cambridge.

Gould, S. J. & N. Eldredge. 1993. Punctuated equilibrium comes of age. Nature 366 (6452): 223–227.

Gould, S. J. & R. C. Lewontin. 1979. The spandrels of San Marco and the Panglossian paradigm: A critique of the adaptationist programme. Proc. Roy. Soc., London B 205: 581–598.

Gould, S. J. & E. S. Vrba. 1982. Exaptation: A missing term in the science of form. Paleobiology 8:4–15.

Grant, M. 1977. Jesus, an Historian's Review of the Gospels. Charles Scribner's Sons, New York.

Grant, T. & A. G. Kluge. 2004. Transformation series as an ideographic character concept. Cladistics 20: 23–31.

Grant, V. 1971. Plant Speciation. Columbia University Press, New York.

Grant, V. 1985. The Evolutionary Process: A Critical Review of Evolutionary Theory. Columbia University Press, New York.

Grant, V. 2003. Incongruence between cladistic and taxonomic systems. Amer. J. Bot. 90: 1263–1270.

Graur, D. & W. Martin. 2004. Reading the entrails of chickens: Molecular timescales of evolution and the illusion of precision. Trends Genetics 20: 80–86.

Greenman, J. M. 1940. The concept of the genus: III. Genera from the standpoint of morphology. Bull. Torrey Bot. Club 67: 371–374.

Grehan, J. R. & J. H. Schwartz. 2009. Evolution of the second orangutan: phylogeny and biogeography of hominid origins. J. Biogeog. 36: 1823–1844.

Griffiths, P.E. 1996. Darwinism, process structuralism, and natural kinds. Phil. Sci. 63: (Proceedings): S1–S9.

Guerra, J., M. Brugués, M. J. Cano R. M. Cros. 2010. Funariales, Splachnales, Schistostegales, Bryales, Timmiales. Flora Bryifítica Ibérica. Vol. 4. Murcia, Spain: Sociedad Española de Briología, Universidad de Murcia.

Guillaumet, A., P.-A. Crochet & J.-M. Pons. 2007. Climate-driven diversification in two widespread *Galerida* larks. BMC Evol. Biol. 8: 32. doi: 10.1186/1471-2148-8-32.

Gurushidze, M., R. Fritsch & F. Blattner. 2010. Species-level phylogeny of *Allium* subgenus *Melanocrommyum:* Incomplete lineage sorting, hybridization and trnF gene duplication. Taxon 59: 829–840.

Gurushidze, M., R. Fritsch & F. Blattner. 2010. Species-level phylogeny of *Allium* subgenus

Haack, S. 1993. Evidence and Inquiry. Blackwell, Oxford.

Hailer, F., V. E. Kutschera, B. M. Halström, D. Klassert, S. R. Fain, J. A. Leanard, U. Amason & A. Janke. 2012. Nuclear genomic sequences reveal that polar bears are an old and distinct bear lineage. Science 336: 344–347.

Hall, B.K. 2003. Descent with modification: The unity underlying homology and homoplasy as seen through an analysis of development and evolution. Biol. Rev. 78: 409–433.

Hanc, J., S. Tuleja & M. Hancova. 2003. Simple derivation of Newtonian mechanics from the principle of least action. Amer. J. Physics 71: 386–391.

Hebert, P. D. N. & T. Gregory. 2005. The promise of DNA barcoding for taxonomy. Syst. Biol. 54: 842.

Hebert, P. D. N., E. H. Penton., J. M. Burns, D. H. Janzen & W. Hallwachs. 2004. Ten species in one: DNA barcoding reveals cryptic species in the neotropical skipper butterfly *Astraptes fulgerator*. Proc. Nat. Acad. Sciences 101: 14812–14817.

Hedenäs, L. & P. Eldenäs. 2007. Cryptic speciation, habitat differentiation, and geography in *Hamatocaulis vernicosus* (Calliergonaceae, Bryophyta). Plant Syst. Evol. 268: 131–145.

Hedenäs, L. 2008. Molecular variation and speciation in *Antitrichia curtipendula* s.l. (Leucodontaceae, Bryophyta). Bot. J. Linn. Soc. 156: 341–354.

Hegde, S. G., J. D. Nason, J. M. Clegg & N. C. Ellstrand. 2006. The evolution of California's wild radish has resulted in the extinction of its progenitors. Evolution 60: 1187–1197.

Hempel, C. 1966. Recent problems of induction. In: R. G. Colodny, ed. Mind and Cosmos: Essays in Contemporary Science and Philosophy. University of Pittsburgh Press, Pittsburgh.

Hempel, C. 1988. Provisoes: A Problem Concerning the Inferential Function of Scientific Theories. Erkenntnis 28: 147–164.

Hennig, W. 1966. Phylogenetic Systematics. Transl. D. D. Davis. University of Illinois Press, Chicago. 1979 reissue.

Hernández-Maqueda, R., D. Quandt, O. Werner & J. Muñoz. 2008. Phylogeny and classification of the

Grimmiaceae/Ptychomitriaceae complex (Bryophyta) inferred from cpDNA. Mol. Phylog. Evol. (2008), doi:10.1016/j.ympev.

Hey, J. 2009. On the arbitrary identification of real species. In: R. K. Butlin, J. Bridle & D. Schluter (eds.). Speciation and Patterns of Diversity. Pp. 15–28. Cambridge University Press, Cambridge.

Hey, J. & Kliman, R. M. 1993. Population genetics and phylogenetics of DNA sequence variation at multiple loci within the *Drosophila melanogaster* species complex. Mol. Biol. Evol. 10: 804–822.

Heywood, V. H. 1963. The "species aggregate" in theory and practice. In: V. H. Heywood & Á. Löve, eds. Symposium on Biosystematics. Regnum Vegetabile 27: 26–37.

Heywood, V. H., R. K. Brummitt, A. Culham & O. Seberg. 2007. Flowering Plant Families of the World. Roy. Botanic Gardens, Kew.

Heywood, V. H. & J. McNeill. 1964. Phenetic and Phylogenetic Classification. Systematics Association Publ. 6: 1–164.

Hochberg, Y. 1988. A sharper Bonferroni procedure for multiple tests of significance. Biometrika 75: 800–802.

Hofbauer, J. & K. Sigmund. 2003. Evolutionary game dynamics. Bull. Amer. Math. Soc. 40: 479–519.

Holm, S. 1979. A simple sequentially rejective multiple test procedure. Scandinavian J. Statist. 5: 65–70.

Holyoak, D. T. 2010. Notes on taxonomy of some European species of *Ephemerum* (Bryopsida: Pottiaceae). J. Bryol. 32: 122–132.

Hopkins, W. G. 2001. Clinical versus. Statistical Significance. Physiology and Physical Education, University of Otago, Dunedin 9001, New Zealand. Sportscience 5(3), html://sportsci.org/jour/0103/inbrief.htm#clinical. 2001.

Hopkins, W. G. 2003. A New View of Statistics. Sportscience: a Peer-reviewed Site for Sports Research Web site. http://www.sportsci.org/resource/stats/index.html. May 19, 2004.

Hörandl, E. 2006. Paraphyletic versus monophyletic taxa—evolutionary versus cladistic classifications. Taxon 55: 564–570.

Hörandl, E. 2007. Neglecting evolution is bad taxonomy. Taxon 56: 1–5.

Hörandl, E. 2010. Beyond cladistics: Extending evolutionary classifications into deeper time levels. Taxon 59: 345–350.

Hörandl, E. & K. Emadzade. 2012. Evolutionary classification: A case study on the diverse plant genus *Ranunculus* L. (Ranunculaceae). Perspectives Pl. Ecol. Evol. Syst. 14: 310–324.

Hörandl, E. & T. F. Stuessy. 2010. Paraphyletic groups as natural units of biological classification. Taxon 59: 1641–1653.

Hubbell, S. P. 2005. The neutral theory of biodiversity and biogeography and Stephen Jay Gould. Paleobiology 31: 122–132.

Hudson, R. R. & J. A. Coyne. 2002. Mathematical consequences of the genealogical species concept. Evolution 56: 1557–1565.

Hudson, R. R. 1992. Gene trees, species trees and the segregation of ancestral alleles. Genetica 131: 509–512.

Huelsenbeck, J. P., B. Larget, R. E. Miller & F. Ronquist. 2002. Potential applications and pitfalls of Bayesian inference of phylogeny. Syst. Biol. 51: 673–688.

Hughes-Wilson, J. 1999. Military Intelligence Blunders. Carroll & Graf Publishers, New York.

Hull, D. L. 1979. The limits of cladism. Syst. Zool. 28: 416–440.

Huse, S. M. 1983. The Collapse of Evolution. Baker Book House, Grand Rapids, Michigan.

Hutchinson, G. E. 1957. Concluding remarks. Cold Spring Harbor Symposium. Quant. Biol. 22: 415–427.

Hutchinson, J. M. C. & G. Gigerenzer. 2005. Simple heuristics and rules of thumb: Where psychologists and behavioural biologists might meet. Behavioural Processes 69: 97–124.

Jablonski, D. 1986. 1986. Background and mass extinctions: The alternation of macroevolutionary regimes. Science 231:129–133.

Jablonski, D. 2007. Scale and hierarchy in macroevolution. Palaeontology 50: 87–109.

Jablonski, D. & D. J. Bottjer. 1990. The origin and diversification of major groups. Environmental patterns and macroevolutionary lags. In: P. D. Taylor & G. P. Larwood (eds.). Major Evolutionary Radiations. Pp. 17–57. Clarendon Press, Oxford.

Jackman, W. R. & D. W. Stock. 2006. Transgenic analysis of Dlx regulation in fish tooth development reveals evolutionary retention of enhancer function despite organ loss. Proc. National Acad. Sci., USA 103: 51: 19390–19395.

Jameel, S. A. 2009. Fuzzy Set and Logic Lectures. University of Technology, Baghdad, Computer and Information Technology, Engineering Department. http://www.itswtech.org/Lec/Adaptive_Systems/01 Fuzzy%20Set%20Theory%20and%20Fuzzy%20Logic%20Lecture%20One.pdf Viewed May 9, 2012.

Jansson, R., G. Rodriquez-Casteñada & L. E. Harding. 2013. What can multiple phylogenies say about the latitudinal diversity gradient? A new look at the tropical conservatism, out of the tropics, and diversification rate hypotheses. Evolution 67: 1741–1755.

Jardin, N. & R. Sibson. 1971. Mathematical Taxonomy. John Wiley & Sons, New York.

Jenner, R. A. 2004. Accepting partnership by submission? Morphological phylogenetics in a molecular millennium. Syst. Biol. 53: 333–342.

Jiménez, J. A. 2006. Taxonomic revision of the genus *Didymodon* Hedw. (Pottiaceae, Bryophyta) in Europe, North Africa and Southwest and Central Asia. J. Hattori Bot. Lab. 100: 211–292.

Jiménez, J. A., M. J. Cano & J. F. Jiménez. 2012. Taxonomy and phylogeny of *Andina* (Pottiaceae, Bryophyta): a new moss genus from the tropical Andes. Syst. Bot. 37: 293–306.

Kalinowski, S.T. 2005. Do polymorphic loci require large sample sizes to estimate genetic distances? Heredity 94: 33–36.

Kårched, J. & Bremer, B. 2007. The systematics of Knoxieae (Rubiaceae)--molecular data and their taxonomic consequences. Taxon 56: 1051–1076.

Kearney, M. & O. Rieppel. 2006. Rejecting "the given" in systematics. Cladistics 22: 369–377.

Kelley, H. 1973. The processes of causal attribution. Amer. Psychologist 28: 107–128.

Kendall, M. G. & W. R. Buckland. 1971. A Dictionary of Statistical Terms, 3rd Edition. Oliver & Boyd, Edinburgh.

Kimura, M. 1968. Evolutionary rate at the molecular level. Nature 217: 624–626.

Kimura, M. 1983. The Neutral Theory of Molecular Evolution. Cambridge University Press, Cambridge.

Kipling, R. 1966. Just So Stories. Illustrated by the Author. Airmont Edition, New York.

Kitcher, P. 1988. Explanatory unification. In: J. C. Pitt (ed.), Theories of Explanation. Oxford University Press, New York. Pp. 167–187.

Kliman, R. M. & Hey, J. 1993. DNA sequence variation at the *period* locus within and among species of the *Drosophila melanogaster* complex. Genetics 133: 375–387.

Kline, M. 1980. Mathematics: The Loss of Certainty. Oxford, New York: Oxford University Press.

Kline, M. 1985. Mathematics and The Search for Knowledge. Oxford University Press, Oxford.

Kluge, A. G. & J. S. Farris. 1969. Quantitative phyletics and the evolution of anurans. Syst. Zool. 18: 1–32.

Knox, E. B. 1998. The use of hierarchies as organizational models in systematics. Biol. J. Linnean Soc. Lond. 63: 1–49.

Kolecki, J. C. 2002. An Introduction to Tensors for Students of Physics and Engineering. National Aeronautics and Space Administration, Glenn Research Center, Cleveland, Ohio. Available electronically at: http://gltrs.grc.nasa.gov Accessed July 2011.

Koonin, E. V. 2009. Darwinian evolution in the light of genomics. Nucleic Acids Res. 37: 1011–1034.

Korn, D. & W.-E. Reif. 2003. Problems of cladistics: The phylogenetic relevance of cladograms and the hierarchical structure of organic diversity. N. Jb. Geol. Paläont. Mh. 2003: 683–704.

Kosko, B. 1993. Fuzzy Thinking: The New Science of Fuzzy Logic. Hyperion, New York.

Kosnik, M. A., D. Jablonski, R. Lockwood & P. M. Novack-Gottshall. 2006. Quantifying molluscan body size in evolutionary and ecological analyses: Maximizing the return of data-collection efforts. Palaios 21: 588–597.

Kolaczkowski, B. & J. W. Thornton. 2009. Long-branch attraction bias and inconsistency in Bayesian phylogenetics. PLOS One 4(12): e7891.

Kress, W. J., K. J. Wurdack, E. A. Zimmer, L. A. Weigt & D. H. Janzen. 2005. Use of DNA barcodes to identify flowering plants. Proc. National Acad. Sciences, USA 102: 8369–8374.

Kubicek, F. C. 1993. Evolution—Guilty as Charged. Destiny Image, Shippensburg, Pennsylvania.

Kučera, J., J. Kušnar & O. Werner. 2013. Partial generic revision of *Barbula* (Musci: Pottiaceae): re-establishment of *Hydrogonium* and *Streblotrichum*, and the new genus *Gymnobarbula*. Taxon 62: 21–39.

Kucera, M. & B. A. Malmgren. 1998. Differences between evolution of mean form and evolution of new morphotypes: An example from Late Cretaceous planktonic formainifera. Paleobiology 24: 49–63.

Kuhn, T. 1970. The Structure of Scientific Revolutions. Edition 2. University of Chicago Press, Chicago.

La Farge, C., A. J. Shaw & D. H. Vitt. 2002. The circumscription of the Dicranaceae (Bryopsida) based on the chloroplast regions trnL-trnF and rps4. Syst. Bot. 27: 435–452.

Lam, J. J. 1959. Taxonomy: general principles and angiosperms. In: W. B. Turrill (ed.). Vistas in Botany. Pp. 3–75. Pergamon Press, London & New York.

Lauder, G. V. 1981. Form and function: Structural analysis in evolutionary morphology. Paleobiology 7: 430–442.

Laurin, M. 2010. Assessment of the relative merits of a few methods to detect evolutionary trends. Syst. Biol. 59: 689–704.

Lee, M .S. Y. 2003. Species concepts and species realities: Salvaging a Linnaean rank. J. Evol. Biol. 15: 179–188.

Lee, M. S. Y. 2006. Morphological phylogenetics and the universe of useful characters. Taxon 55: 5–7.

Lee, M. S. Y. & A. B. Camens. 2009. Strong morphological support for the molecular evolutionary tree of placental mammals. J. Evol. Biol. 22: 2243–2257.

Legendre, P. 1972. The definition of systematic categories in biology. Taxon 21: 381–406.

Leschen, R. A. B., T. R. Buckley, H. M. Harman & J. Shulmeister. 2008. Determining the origin and age of the Westland beech (Nothofagus) gap, New Zealand, using fungus beetle genetics. Mol. Ecol. 17: 1256–1276.

Levin, D.A. 1993. Local speciation in plants: The rule, not the exception. Syst. Bot. 18: 197–208.

Levin, D.A. 2001. 50 years of plant speciation. Taxon 50: 69–91.

Levinton, J. 1988. Genetics, Paleontology, and Macroevolution. Cambridge University Press: Cambridge.

Lewis, H. 1962. Catastrophic selection as a factor in speciation. Evolution 16: 257–271.

Lewis, H. 1965. The taxonomic problem on inbreeders or how to solve any taxonomic problem. In: V. H. Heywood & Á. Löve, eds. Symposium on Biosystematics. Regnum Vegetabile 27: 37–44.

Lewis, H. 1966. Speciation in flowering plants. Science 3152: 167–172.

Lewis, H. & M. R. Roberts. 1956. The origin of *Clarkia lingulata*. Evolution 10: 126–138.

Li, Cong-jun. 2013. DNA demethylation pathways: recent insights. Genetics & Epigenetics 5: 43–49.

Lim, G. S., M. Balke & R. Meier. 2012. Determining species boundaries in the world full of rarity: singletons, species delimitation methods. Syst. Biol. 61: 165–169.

Lindley, J. 1853. Vegetable Kingdom, or The Structure, Classification, and Uses of Plants. Bradbury and Evans, London.

Lipscomb, D., N. Platnick & Q. Wheeler. 2003. The intellectual content of taxonomy. Trends Ecol. Evol. 18: 65–66.

Littlejohn, S. W. 1978. Theories of Human Communication. Charles E. Marrill Publishing Company: Columbus, Ohio.

Liu, J. 1993. Discounting initial population sizes for prediction of extinction probabilities in patchy environments. Ecol. Modelling 70: 51–61.

Livio, M. 2002. The Golden Ratio: The Story of Phi, The World's Most Astonishing Number. Random House, New York.

Lönnig, W.-E., K. Stüber, H. Saedler & J. H. Kim. 2007. Biodiversity and Dollo's Law: To what extent can the phenotypic differences between *Misopates orontium* and *Antirrhinum majus* be bridged by mutagenesis? Bioremediation, Biodiversity and Bioavailability 1: 1–30.

Löve, A. 1963. Cytotaxonomy and Genetic Delimitation. Regnum Vegetabile 27: 45-51.

Lowery, R. 2012. VassarStats: Web Site for Statistical Computation. Department of Psychology, Vassar

College, Poughkeepsie, New York. http://vassarstats.net/ Viewed: June 15, 2012.

Lyons-Weiler, J. & M. C. Milinkovitch. 1997. A phylogenetic approach to the problem of differential lineage sorting. Mol. Biol. Evol. 14: 968–975.

Mace, G. M., J. L. Gittleman & A. Purvis. 2003. Preserving the Tree of Life. Science 300: 1707–1709.

Maddison, W. P. 1996. Molecular approaches and the growth of phylogenetic theory. In: J. D. Ferraris & S. R. Palumbi (eds.). Molecular Zoology: Advances, Strategies, and Protocols. Pp. 47–63. Wiley-Liss, New York.

Mailath, G. J. 1998. Do people play Nash equilibrium? Lessons from evolutionary game theory. J. Econ. Lit. 36: 1347–1374.

Mallet, J. 1995. A species definition for the modern synthesis. Trends Ecol. Evol. 10: 294–299.

Mallet, J. & K. Willmott. 2003. Taxonomy: renaissance or Tower of Babel? Trends Ecol. Evol. 18: 57–59.

Manzotti, R. 2011. Machine free will: Is free will a necessary ingredient of machine consciousness? Adv. Experim. Medical Biol. 718: 181–191.

Marshall, C. R. 1997. Statistical and computational problems in reconstructing evolutionary histories from DNA data. Computing Sci. Statist. 29: 218–226.

Martignon, L. 2001. Comparing fast and frugal heuristics and optimal models. In: G. Gigerenzer & R. Selten (eds.). Bounded Rationality: The Adaptive Toolbox. Pp. 147–171. MIT Press, Cambridge, Massachusetts.

Martin, J., D. Blackburn & E. O. Wiley. 2010. Are node-based and stem-based clades equivalent? Insights from graph theory. Publ. Libr. Sci. Currents Tree of Life November 18, 2010: 1–12. [modified April 4, 2012] doi: 10.1371/currents.RRN1196.

Matthews, P. 2001. A Short History of Structural Linguistics. Cambridge, Cambridge University Press.

Mayer, M. S. & L. Beseda. 2010. Reconciling taxonomy and phylogeny in the *Streptanthus glandulosus* complex (Brassicaceae). Ann. Missouri Bot. Gard. 97: 106–116.

Maynard Smith, J. & G. R. Price. 1973. The logic of animal conflict. Nature 246: 15–18.

Mayr, E. 1954. Geographic speciation in tropical echinoids. Evolution. 8: 1–18.

Mayr, E. 1969. Principles of Systematic Zoology. McGraw-Hill, New York.

Mayr, E. 1982. Speciation and macroevolution. Evolution 36: 1119–1132.

Mayr, E. 1983. How to carry out the adaptationist program? Amer. Naturalist 121: 324–334.

Mayr, E. & W. J. Bock. 2002. Classifications and other ordering systems. J. Zool. Evol. Res. 40: 169–194.

Mayr, E. & W. B. Provine (eds). 1980. The Evolutionary Synthesis: Perspectives on the Unification of Biology. Harvard University Press, Cambridge, Mass.

Mazur, J. 2006. Euclid in the Rainforest: Discovering Universal Truth in Logic and Math. Plume Books, New York.

McShea, D. W. 2005. The evolution of complexity without natural selection, a possible large-scale trend of the fourth kind. Paleobiology 31: 146–156.

Mercier, H. & D. Sperber. 2011. Why do humans reason? Arguments for an argumentative theory. Behavioral and Brain Sciences 34: 57–111.

Millstein, R. L. 2002. Chance and macroevolution. Phil. Sci. 67: 603–624.

Millstein, R. L., R. A. Skipper, Jr. & M. R. Dietrich. 2009. (Mis)interpreting mathematical models: Drift as a physical process. Philos. Theor. Biol. 1: 1–13.

Mishler, B. D. 1999. Getting rid of species? In: R. Wilson, ed. Species: New Interdisciplinary Essays. Pp. 307–315. MIT Press, Cambridge, Massachusetts.

Mooi, R. D. & A. C. Gill. 2010. Phylogenies without synapomorphies—A crisis in fish systematics: Time to show some character. Zootaxa 2450: 26–40.

Moran, M. D. 2003. Arguments for rejecting the sequential Bonferroni in ecological studies. OIKOS 100: 403–405.

Moritz, C., C. J. Schneider & D. B. Wake. 1992. Evolutionary relationships within the *Ensatina eschscholtzii* complex confirm the ring species interpretation. Syst. Biol. 41: 273–291.

Morrison, D. A. 2013. {Review:] Evolutionary Genomics: Statistical and Computational Methods. Syst. Biol. 62: 348–350.

Morrison, P. 1963. Fermi questions. Amer. J. Physics 31: 626–627.

Morrone, J. J., L. Katinas & J. V. Crisci. 1996. On temperate areas, basal clades and biodiversity conservation. Oryx 30: 187–194.

Mort, M. E., C. P. Randle, R. T. Kimball, M. Tadess & D. J. Crawford. 2008. Phylogeny of Coreopsideae (Asteraceae) inferred from nuclear and plastid DNA sequences. Taxon 57: 109–120.

Nagel, E. & J. R. Newman. 2008. Gödel's Proof, Revised Edition. D. R. Hofstadter, ed. New York University, New York.

Nakagawa, S. 2004. A farewell to Bonferroni: The problems of low statistical power and publication bias. Behavioral Ecol. 15:1044–1045.

Nei, M. 2005. Selectionism and neutralism in molecular evolution. Mol. Biol. Evol. 22: 2318–2342.

Nei, M., S. Kumar & K. Takahashi. 1998. The optimization principle in phylogenetic analysis tends to give incorrect topologies when the number of nucleotides or amino acids is small. Proc. Nat'l Acad. Sci. USA 95: 12890–12397.

Neisser, U. & R. Becklen. 1975. Selective looking: Attending to visionally specified events. Cogn. Psych. 7: 480–494.

Nelson, G. 2004. Cladistics: Its arrested development. In: D. M. Williams & P. L. Forey (eds.). Milestones in Systematics. Systematics Association Special Volume Series 67: 127–147.. CRC Press: London.

Newman, D. & D. Pilson. 1997. Increased probability of extinction due to decreased genetic effective population size: Experimental populations of *Clarkia pulchella*. Evolution 51: 354–362.

Newmaster, S. G, A. J. Fazekas & S. Ragupathy. 2006. DNA barcoding in land plants: evaluation of *rbcL* in a multigene tiered approach. Canad. J. Bot. 84: 335–341.

Newman, D. & D. Pilson. 1997. Increased probability of extinction due to decreased genetic effective population size: experimental populations of *Clarkia pulchella*. Evolution 51: 354–362.

Nicklas, K. J. 1992. Plant Biomechanics. University of Chicago Press, Chicago.

Nicklas, K. J. & H.-C. Spatz. 2012. Plant Physics. University of Chicago Press, Chicago.

Niiniluoto, I. 1998. Defending abduction. Phil. Sci. 66: S436–S451.

Nixon, K. C. & J. M. Carpenter. 1996. On simultaneous analysis. Cladistics 12: 221–241.

Noorbergen, R. 1977. Secrets of the Lost Races: New Discoveries of Advanced Technology in Ancient Civilizations. Barnes & Noble Books, New York.

Nordal, I. & B. Stedje. 2005. Paraphyletic taxa should be accepted. Taxon 54: 5–6.

Nowack, M. A. & K. Sigmund. 2004. Evolutionary dynamics of biological games. Science 303: 793–799.

O'Hara, R. J. 1993. Systematic generalization, historical fate, and the species problem. Syst. Biol. 42: 231–246.

Ohta, T. 1992. The nearly neutral theory of molecular evolution. Ann. Rev. Ecol. Syst. 23: 263–286.

O'Keefe, F.R. and Sander, P.M. 1999. Paleontological paradigms and inferences of phylogenetic pattern: a case study. Paleobiology 25: 518–533.

O'Leary, M.A. & S. Kaufman. 2007. MorphoBank 2.5: web application for morphological systematics and taxonomy. www.morphobank.

O'Leary, M. & S. Kaufman. 2011. MorphoBank: Phylophenomics in the "cloud." Cladistics 27: 529–537.

Olmstead, R.G. 1995. Species concepts and plesiomorphic species. Syst. Bot. 20: 623–630.

Omland, K. E., S. M. Lanyon & S. J. Fritz. 1999. A molecular phylogeny of the New World Orioles (*Icterus*): The importance of dense taxon sampling. Mol. Phylog. Evol. 12: 224–239.

Otnyukova, T. N. 2002. A study of the *Didymodon* species (Pottiaceae, Musci) in Russia. I. Species with caducous leaf apices. Arctoa 11: 337–349.

Overton, W. F. 1975. General systems, structure and development. In: K. Riegel & G. C. Rosenwald (eds.). Structure and Transformation: Developmental and Historical Aspects. Vol. 3: 61–81. John Wiley & Sons, New York.

Padial, J. M., A. Miralles, I. De la Riva. & M. Vences. 2010. The integrative future of taxonomy. Frontiers in Zoology 7: 1–16. http://www.frontiersinzoology.com/content/7/1/16 Accessed Dec. 19, 2011

Page, R. D. M. 1996. TREEVIEW: An application to display phylogenetic trees on personal computers. Computer Applications in the Biosciences 12: 357–358.

Pamilo, P. & M. Nei. 1988. Relationships between gene trees and species trees. Mol. Biol. Evol. 5: 568–583.

Pap, A. 1962 An Introduction to the Philosophy of Science. Free Press of Glencoe, Macmillan Company, New York.

Paradis, E. 2005. Statistical analysis of diversification with species traits. Evolution 59: 1–12.

Parker, B. 1988. Creation: The Story of the Origin and Evolution of the Universe. Basic Books, Cambridge, Massachusetts.

Parker, T. 1983. Rules of Thumb. Houghton Mifflin Co., Boston.

Paul, G. S. 2002. Dinosaurs of the Air: The Evolution and Loss of Flight in Dinosaurs and Birds. Edition 2. Johns Hopkins University Press, Baltimore.

Peirce, C. S. 1903. Pragmatism as a Principle and Method of Right Thinking. The 1903 Harvard Lectures on Pragmatism. P.A. Turisi (ed.). State University of New York Press: Albany, New York.

Pelser, P. B., B. Nordenstam, J. W. Kadereit & L. E. Watson. 2007. An ITS phylogeny of tribe Senecioneae (Asteraceae) and a new delimitation of *Senecio* L. Taxon 56: 1077–1104.

Pennisi, E. 2003. Modernizing the Tree of Life. Science 300: 1692–1697.

Peterson, I. 1988. The Mathematical Tourist: Snapshots of Modern Mathematics. W. H. Freeman and Company, New York.

Philippe, H., G. Lecointre, L. Hoc Lanh Vân Lê & H. Le Guyader. 1996. A critical study of homoplasy in molecular data with the use of a morphologically based cladogram, and its consequences for character weighting. Mol. Biol. Evol. 13: 1174–1186.

Pianka, E. R. 2000. Evolutionary Ecology. Edition 6. Addison Wesley Longman, Inc., San Francisco.

Pielou, E. C. 1966. Shannon's formula as a measure of specific diversity: its use and misuse. American Naturalist 100: 463–465.

Pitman, N. C. & P. M. Jørgensen. 2002. Estimating the size of the world's threatened flora. Science 298: 989.

Podani, J. 2013. Tree thinking, time and topology: Comments on the interpretation of tree diagrams in evolutionary/phylogenetic systematics. Cladistics 29: 315–327.

Poole, M. 1990. A Guide to Science and Belief. Lion Publishing, Oxford.

Popper, K. R. 1959. The Logic of Scientific Discovery. Basic Books: New York.

Porter, D. M. & P. W. Graham, eds. 1993. The Portable Darwin. Penguin Books, New York. [Pp. 321–360, Excepts from the Descent of Man, and Selection in Relation to Sex, 1871.]

Posada, D. & T. R. Buckley. 2004. Model selection and model averaging in phylogenetics: Advantages of Akaike Information Criterion and Bayesian approaches over likelihood ratio tests. Syst. Biol. 53: 793–808.

Price, J. J. and S. M. Lanyon. 2002. A robust phylogeny of the oropendolas: polyphyly revealed by mitochondrial sequence data. Auk 119: 335–348.

Pullen, W. D. 2011. Maze Classification. Think Labyrinth. http://www.astrolog.org/labyrnth/algrithm.htm Viewed September 8, 2011.

Purvis, A., J. L. Gittleman & T. Brooks. 2001. Phylogeny and Conservation. Cambridge University Press, Cambridge.

Robosky, D. L. 2010. Extinction rates should not be estimated from molecular phylogenies. Evolution 64: 1816–1824.

Radinsky, L.B. 1985. Approaches in evolutionary morphology: A search for patterns. Ann. Rev. Ecol. Syst. 16: 1–14.

Rajakumar, R., D. San Sauro, M. B. Dijkstra, M. H. Huang., D. E. Wheeler, F. Hiou-tim, A. Khila, M. Cournoyea & E. Abouheif. 2012. Ancestral development potential facilitates parallel evolution in ants. Science 335: 79–82.

Ramdhani, S., N. P. Barker & H. Baijnath. 2009. Rampant non-monophyly of species in *Kniphofia*

Moench (Asphodelaceae) suggests a recent Afromontane radiation. Taxon 58: 1141–1152.

Ramdhani, S., N. P. Barker & R. M. Cowling. 2011. Revisiting monophyly in *Haworthia* Duval (Asphodelaceae): incongruence, hybridization and contemporary speciation. Taxon 60: 1001–1014.

Rannala, B. & Z. Yang. 2003. Bayes estimation of species divergence times and ancestral population sizes using DNA sequences from multiple loci. Genetics 164: 1645–1656.

Raup, D. M. 1981. Extinction: bad genes or bad luck? Acta Geol. Hisp. 16: 25–33.

Redelings, B. D. & M. A. Suchard. 205. Joint Bayesian estimation of alignment and phylogeny. Syst. Biol. 54: 401–418.

Reiners, W. A. & J. A. Lockwood. 2010. Philosophical Foundations for the Practices of Ecology. Cambridge University Press, Cambridge.

Ren, Fengrong, H. Tanaka & Ziheng Yang. 2005. An empirical examination of the utility of codon-substitution models in phylogeny reconstruction. Syst. Biol. 54: 808–818.

Reynolds, A. 2007. Pushing Ice. Ace, New York.

Ricklefs, R. E. and S. S. Renner. 2012. Global correlations in tropical tree species richness and abundance reject neutrality. Science 335: 464–467.

Riddihough, G. & L. M. Zahn. 2010. What is epigenetics? Science 330: 611. [And following articles.]

Ridley, M. 1996. Evolution. Edition 2. Blackwell Science: Cambridge, Massachusetts.

Rieppel, O. 2011. Willi Hennig's dichotomization of nature. Cladistics 26: 103–112.

Rieppel, O. 2012. Othenio Abel (1875–1946) and "the phylogeny of the parts." Cladistics 29: 328–335.

Rieppel, O. & L. Grande. 1994. Summary and comments on systematic pattern and evolutionary process. In: L. Grande & O. Rieppel (eds.). Interpreting the Hierarchy of Nature. Pp. 227–255. Academic Press, San Diego.

Rieseberg, L. H. & L. Brouillet. 1994. Are many plant species paraphyletic? Taxon 43: 21—32.

Rieseberg, L. H. & J. M. Burke. 2001. A genic view of species integration. J. Evol. Biol. 14: 883–886.

Rindal, E. & A. V. Z. Brower. 2011. Do model-based phylogenetic analyses perform better than parsimony? A test with empirical data. Cladistics 27: 331–334.

Robinson, H. 1986. A key to the common errors of cladistics. Taxon 35: 309–311.

Robinson, R. L., J. C. Kilbride & S. Nagy. 1992. Elders Living Alone: Frailty and the Perception of Choice. Aldine de Gruyter, New York.

Robson, A. J. 1990. Efficiency in evolutionary games: Darwin, Nash and the secret handshake. J. Theor. Biol. 144: 379–396.

Rockman, M. V., M. W. Hahn, N. Soranzo, F. Zimprich, D. B. Goldstein & G. A. Wray. 2005. Ancient and recent positive selection transformed opioid *cis*-regulation in humans. PLoS Biology 3(12): 1–12.

Rolf, F. J. 1990. Morphometrics. Ann. Rev. Ecol. Syst. 21: 299–316.

Roseman, C. C. 2004. Detecting interregionally diversifying natural selection on modern human cranial form by using matched molecular and morphometric data. Proc. National Acad. Sci., USA. 101: 12824–12829.

Rosenberg, N. A. 2003. The shapes of neutral gene genealogies in two species: Probabilities of monophyly, paraphyly, and polyphyly in a coalescent model. Evolution 57: 1465–1477.

Ross, H. H. 1972. The origin of species diversity in ecological communities. Taxon 21: 253–259.

Ross, S. 2009. A First Course in Probability. Eighth Edition. Posts and Telecom Press, Beijing.

Rothwell, G. W. 1999. Fossils and ferns in the resolution of land plant phylogeny. Bot. Rev. 65: 188–218.

Rozin, P. & C. Nemeroff. 2002. Sympathetic magical thinking: The contagion and similarity "heuristics." In: T. Gilovich, D. Griffin & D. Kahneman. Heuristics and Biases: The Psychology of Intuitive Judgment. Campride University Press, Cambridge. Pp. 201–216..

Rucker, R. 1983. Infinity and the Mind: The Science and Philosophy of the Infinite. Bantam Books, New York.

Ruedas, L. A., J. Salazar-Bravo, J. W. Dragoo & T. L. Yates. 2000. The importance of being earnest: What, if anything, constitutes a "specimen examined?" Mol. Phylog. Evol. 17: 129–132.

Russo, C. A. M., N. Takezaki & M. Nei. 1996. Efficiencies of different genes and different tree-building methods in recovering a known vertebrate phylogeny. Mol. Biol. Evol. 13: 525–636.

Salmon, W. C. 1971. Statistical Explanation and Statistical Relevance. Univ. Pittsburgh Press, Pittsburgh, Pennsylvania.

Sanderson, M. J. & M. G. Wojciechowski. 2000. Improved bootstrap confidence limits in large-scale phylogenies, with an example from *Neoastragalus* (Leguminosae). Syst. Biol. 49: 671–685.

Santos, L. & L. R. R. Faria. 2011. The taxonomy's new clothes: A little more about the DNA-based taxonomy. Zootaxa 3025: 66–68.

Sasaki, K. & S. F. Fox & D. Duvall. 2008. Rapid evolution in the wild: Changes in body size, life-history traits, and behavior in hunted population of the Japanese mamushi snake. Conservation Biol. 23: 93–102.

Schmidt-Lebuhn, A. N. 2011. Fallacies and false premises—a critical assessment of the arguments for the recognition of paraphyletic taxa in botany. Cladistics 28: 174–187.

Schneider, H., A. R. Smith & K. M. Pryer. 2009. Is morphology really at odds with molecules in estimating fern phylogeny? Syst. Bot.. 34: 455–475.

Schuettpetz, E. & K. M. Pryer. 2007. Fern phylogeny inferred from 400 leptosporangiate species and three plastid genes. Taxon 56: 1037–1050.

Schwartz, J. H. & B. Maresca. 2006. Do molecular clocks run at all? A critique of molecular systematics. Biol. Theory 1: 357–371.

Scotland R. W., R. G. Olmstead & J. R. Bennett. 2003. Phylogeny reconstruction: the role of morphology. Syst. Biol. 52: 539–548.

Scott-Elliot, W. 1968. The Story of Atlantis and the Lost Lemuria. Theosophical Publishing House, London. Revised edition.

Scott-Ram, N. R. 1990. Transformed Cladistics, Taxonomy and Evolution. Cambridge University Press, Cambridge.

Seberg, O., C. J. Humphries, S. Knapp, D. W. Stevenson, G. Petersen, N. Scharff & N. M. Andersen. 2003. Shortcuts in systematics? A commentary on DNA-based taxonomy. Trends Ecol. Evol. 18: 63–65.

Seberg, O., G. Petersen & C. Baden. 1997. Taxonomic incongruence—a case in point from plants. Cladistics 13: 180.

Semple, C. 2007. Hybridization networks. In: O. Gascuel & M. Steel. Reconstruction Evolution: New Mathematical and Computational Advances. Pp. 277–314. Oxford University Press, New York & Oxford.

Shaban-Nejad, A. & V. Haarslev. 2008. Ontology-inferred phylogeny reconstruction for analyzing the evolutionary relationships between species: ontological inference versus cladistics. Proceedings of the 8th IEEE International Conference on Bioinformatics and Bioengineering, BIBE 2008, Oct. 8–10, 2008, Athens, Greece, IEEE, 2008, pp. 1–7.

Shanahan, T. 2011. Phylogenetic inertia and Darwin's higher law. Studies in Hist. Phil. Biol. Biomed. Sci. 42: 60–68.

Shannon, C. & W. Weaver. 1963. The Mathematical Theory of Communication. University of Illinois, Urbana, Ill.

Shaw, A. J., I. Holz, C. J. Cox & B. Goffinet. 2008. Phylogeny, character evolution, and biogeography of the Gondwanic moss family Hypopterygiaceae (Bryophyta). Syst. Biol. 33: 21–30.

Shaw, J. 2001. Biogeographic patterns and cryptic speciation in bryophytes. J. Biogeog. 28: 253–261.

Shaw, J. & R. L. Small. 2005. Chloroplast DNA phylogeny and phylogeography of the North American plums (*Prunus* subgenus *Prunus* section *Prunocerasus,* Rosaceae). Amer. J. Bot. 92: 2011–2030.

Shen, Bing, Lin Dong, Shuhai Xiao & M. Kowalewski. 2008. The Avalon explosion: Evolution of Ediacara morphospace. Science 319 (5859): 81–84.

Sherff, E. E. 1940. The concept of the genus: IV. The delimitations of genera from the conservative point of view. Bull. Torrey Bot. Club 67: 375–380.

Shrader-Frechette, K. 2008. Statistical significance in biology: Neither necessary nor sufficient for hypothesis acceptance. Biol. Theory 3: 12–16.

Sidor, C. A. & J. A. Hopson. 1998. Ghost lineages and "mammalness": Assessing the temporal pattern of character acquisition in the Synapsida. Paleobiol. 24: 254–273.

Simons, D. J. & C. F. Chabris. 1999. Gorillas in our midst: Sustained inattentional blindness for dynamic events. Perception 28: 1059–1074.

Simpson, C. 2013. Species selection and the macroevolution of the coral coloniality and photosymbiosis. Evolution 67: 1607–1621.

Simpson, G. G. 1953. The Major Features of Evolution. Columbia University Press, New York.

Simpson, G. G. 1961. Principles of Animal Taxonomy. Columbia University Press, New York.

Sites, J. W., Jr., S. K. Davis, T. Guerra, J. B. Iverson & H. L. Snell. 1996. Character congruence and phylogenetic signal in molecular and morphological data sets: a case study in the living iguanas (Squamata, Iguanidae). Mol. Biol. Evol. 13: 1087–1105.

Sivarajan, V. V. 1991. Introduction to the Principles of Plant Taxonomy. Ed. 2. N. K. P. Robson, ed. Cambridge University Press, Cambridge.

Smith, A. J. E. 1990. The Liverworts of Britain and Ireland. Cambridge University Press, Cambridge.

Smith, J. & J. L. Clark. 2013. Molecular phylogenetic analyses reveal undiscovered monospecific genera in the tribe Episcieae (Gesneriaceae). Syst. Biol. 38: 451–463.

Smith, Z. R. & Wells, C. S. 2006. Central Limit Theorem and Sample Size. Paper presented at the annual meeting of the Northeastern Educational Research Association, Kerhonkson, New York, October 18–20, 2006. http://www.umass.edu/remp/Papers/Smith&Wells_NERA06.pdf Accessed Dec. 19, 2011.

Smyth, M. B. & G. D. Platkin. 1982. The category-theoretic solution of recursive domain equations. SIAM J. on Computing 11: 761–783.

Sneath, P. H. A. 1976. Phenetic taxonomy at the species level and above. Taxon 25: 437–450.

Sneath, P. H. A. 1995. Thirty years of numerical taxonomy. Syst. Biol. 44: 281–298.

Sober E. 1991. Core Questions in Philosophy: A Text with Readings. Macmillan Library Reference, New York.

Sokal, R. R. & P. H. A. Sneath. 1963. Principles of Numerical Taxonomy. W. H. Freeman & Co., San Francisco.

Solé, R. V. & S. C. Manrubia. 1996. Extinction and self-organized criticality in a model of large-scale evolution. Physical Rev. E 54: R42–R45.

Sonneborn, T. M. 1957. Breeding systems, reproductive methods, and species problems in Protozoa. In: E. Mayr, ed. The Species Problem. Pp. 39–80.

Sosef, M. S. M. 1997. Hierarchical models, reticulate evolution and the inevitability of paraphyletic supraspecific taxa. Taxon 46: 75–85.

Springer, S. P. & G. Deutsch. 1993. Left Brain, Right Brain. Fourth Edition. W. H. Freeman and Company, New York.

Stadler, T. 2013. How can we improve accuracy of macroevolutionary rate estimates? Syst. Biol. 62: 321–329.

Stanley, S. M. 1981. The New Evolutionary Timetable: Fossils, Genes, and the Origin of Species. Basic Books, New York.

Stanovich, K. E., M. E. Toplak & R. F. West. 2008. The development of rational thought: A taxonomy of heuristics and biases. In: R. V. Kail (ed.), Advances in child development and behavior. Pp. 251–285. Elsevier B. V., Amsterdam, The Netherlands.

Stat Trek. 2012. Statistics, probability, and survey sampling. http://stattrek.com/online-calculator/-binomial.aspx Viewed April 23, 2012.

Stebbins, G. L. 1959. Genes, chromosomes, and evolution. In: W. B. Turrill (ed.). Vistas in Botany. Pp. 258–290. Pergamon Press, London & New York.

Stech, M. & D. Quandt. 2013. 20,000 species and five key markers: The status of molecular bryophyte phylogenetics. Phytotaxa 9: 196–228.

Stegemann, S., M. Keuthe, S. Greiner & R. Bock. 2012. Horizontal transfer of chloroplast genomes between plant species. Proc. National Acad. Sciences, USA 109: 2434–2438.

Stenseth, N. C. & J. M. Smith. 1984. Coevolution in ecosystems: Red Queen evolution or stasis? Evolution 38: 870–880.

Stevens, P. F. 1985. The genus concept in practice: But for what practice? *Kew Bull.* 40: 457–465.

Stevens, P. F. 1994. The Development of Biological Systematics: Antoine-Laurent de Jussieu, Nature, and the Natural System. Columbia University Press, New York.

Stevens, P. F. 2000. Botanical systematics 1950–2000: change, progress, or both? Taxon 49: 635–659.

Stevens, P. F. 2008. Angiosperm Phylogeny Website. Version 9. http://www.mobot.org/MOBOT-/research/APweb/. Viewed June 23, 2012.

Stewart, I. 2011. Mathematics of Life. Basic Books, New York.

Stewart, I., T. Elmhurst & J. Cohen. 2000. Symmetry breaking as an origin of species. In: J. Buescu et al., eds. Bifurcations, Symmetry and Patterns. Birkhäuser, Basel. Pp. 3–54.

Stuessy, T. F. 2008. Plant Taxonomy: the Systematic Evaluation of Comparative Data. Columbia University Press. New York.

Stuessy, T. F. 2009. Paradigms in biological classification (1707–2007): Has anything really changed? Taxon 58: 68–76.

Stuessy, T. F. 2010. Paraphyly and the origin and classification of angiosperms. Taxon 59: 689–693.

Stuessy, T. F. & C. König. 2008. Patrocladistic classification. Taxon 57: 594–601.

Stuessy, T. F. & H. W. Lack. 2011. Monographic Plant Systematics: Fundamental Assessment of Plant Biodiversity. Regnum Vegetabile 153. A. R. G. Gantner Verlag: Ruggell, Liechstenstein.

Suchard M. A., R. E. Weiss, K. S. Dorman & J. S. Sinsheimer. 2002. Oh brother, where art thou? A Bayes factor test for recombination with uncertain heritage. Syst. Biol. 51:715–728.

Swanson, T. M. 1994. The economics of extinction revisited and revised: A generalised framework for the analysis of the problems of endangered species and biodiversity loss. Oxford Econ. Pap., n.s. 46: 800–821.

Syring, J., K. Farrell, R. Businsky, R. Cronn & A. Liston. 2007. Widespread genealogical nonmonophyly in species of *Pinus* subgenus *Strobus*. Syst. Biol. 55: 163–181.

Szalay, F. S., A. L. Rosenberger & M. Dagosto. 1987. Diagnosis and differentiation of the order Primates. Amer. J. Physical Anthrop. 30: 75–105.

Sznajd-Weron, K. & R. Weron. 2001. A new model of mass extinctions. Physica A 293: 559–565.

Talbot, S. & Shields, G. 1996. Phylogeography of Brown Bears (*Ursus arctos*) of Alaska and paraphyly within the Ursidae. Mol. Phylog. Evol. 5: 477–494.

Taleb, N. N. 2008. Fooled by Randomness: The Hidden Role of Chance in Life and in the Markets. Second edition. Random House, New York.

Taleb, N. N. 2010. The Black Swan: Second Edition: The Impact of the Highly Improbable: With a New Section: On Robustness and Fragility. Random House, New York.

Templeton, A. 1986. Relation of humans to African apes: a statistical appraisal of diverse types of data. In: S. Karlin & E. Nevo (eds.). Evolutionary Processes and Theory. Pp. 365–388. Academic Press. New York.

Theriot, E. 1992. Clusters, species concept, and morphological evolution of diatoms. Syst. Biol. 41: 141–157.

Thomason, J. J., ed. 1995. Functional Morphology in Vertebrate Paleontology. Cambridge University Press, Cambridge.

Thomas, C. D. and 18 others. 1994. Extinction risk from climate change. Nature 427: 145–148.

Thorne, R. F. 2007. An updated classification of the class Magnoliopsida (Angiospermae). Bot. Rev. 73: 67–182. [nomenclatural additions by James L. Reveal.]; also http://www.rsabg.org/content/-view/50/144/

Tilman, D., R. M. May, C. L. Lehman & M. A. Nowak. 1994. Habitat destruction and the extinction debt. Nature 371: 65–66.

Tobias, J A., N. Seddon, C. N. Spottiswoode, J. D. Pilgrim, L. D. C. Fishpool & N. J. Collar. 2010. Quantitative criteria for species delimitation. IBIS, Intern. J. Avian Sci. 152: 724–746.

Turner, B. M. 2002. Cellular memory and the histone code. Cell 111: 285–291.

Tverksy, A. & D. Kahneman. 1974. Judgment under uncertainty: heuristics and biases. Scienc*e* 185: 1124–1131.

Van Deemter, K. 2004. Towards a probabilistic version of bidirectional OT syntax and semantics. J. Semantics 21: 251–280.

Van der Pijl, L. 1982. Principles of Dispersal in Higher Plants. Edition 2. Springer-Verlag, New York.

Van Fraasen, B. D. 2007. Structuralism(s) about science: Some common problems. Proc. Aristotelian Soc., Suppl. 81: 45–61.

Vanderpoorten, A. & A. J. Shaw. 2010. The application of molecular data to the phylogenetic delimitation of species in bryophytes: a note of caution. Phytotaxa 9: 229–237.

Van Valen, L. 1973. A new evolutionary law. Evol. Theory 1: 1–30.

Van Valen, L. 1976. Ecological species, multispecies, and oaks. Taxon 25: 233–239.

Vasek, F. C. 1968. The relationships of two ecologically marginal sympatric *Clarkia* populations. Amer. Naturalist 102: 25–40.

Vernon, K. 1993. Desperately seeking status: Evolutionary systematics and the taxonomists' search for respectability 1940–60. Brit. J. Hist. Sci. 26: 207–227.

Verhoeven, K. J. F., K. L. Simonsen & L. M. McIntyre. 2005. Implementing false discovery rate control: Increasing your power. OIKOS 108: 643–647.

Vernon, K. 1993. Desperately seeking status: Evolutionary Systematics and the taxonomists' search for respectability 1940–60. Brit. J. Hist. Sci. 26: 207–227.

Vavillov, N. I. 1951. The origin, variation, immunity and breeding of cultivated plants. K. S. Chester, transl. Chronica Botanica 13: i–xvii, 1–364. Waltham, Massachusetts.

Vrba, E. S. 1980. Evolution, species and fossils: How does life evolve? So. Afr. J. Sci. 76: 61–84.

Vrba, E. S. 1984. Evolutionary pattern and process in the sister-group Alcelaphini-Aepycerotini (Mammalia: Bovidae). In: N. Eldredge & S. M. Stanley (eds.). Living Fossils. Pp. 62–79. Springer-Verlag, New York.

Vrba, E. S., ed. 1985. Species and speciation, Transvaal Museum Monograph; no. 4. Transvaal Museum, Pretoria, South Africa.

Wagner, W. H., Jr. 1952. The fern genus *Diellia:* its structure, affinities and taxonomy. Univ. Calif. Publ. Bot. 26: 1–212, pl. 1–21.

Walsh, B. 2009. *Lithophane leeae* (Lepidoptera, Noctuidae, Xyleninae), a striking new species from southeastern Arizona. ZooKeys 10: 11–16.

Walsh, P. D. 2000. Sample size for the diagnosis of conservation units. Conservation Biology 14: 1533–1537.

Walton, D. N. 1989. Informal Logic: A Handbook for Critical Argumentation. Cambridge University Press, Cambridge.

Watzlawick, P. 1976. How Real is Real? Confusion, Disinformation, Communication. Vintage Books, New York.

Weaver, W. 1949. The mathematics of communication. Sci. Amer. 181: 11–15.

Weinstein, L. & J. A. Adam. 2008. Guesstimation: Solving the World's Problems on the Back of a Cocktail Napkin. Princeton University Press, Princeton.

Weisstein, E. W. 1999a. Golden ratio. MathWorld—A Wolfram Web Resource. http://mathworld.wolfram.com/GoldenRatio.html Viewed Nov. 15, 2010.

Weisstein, E.W. 1999b. Geometric distribution. MathWorld—A Wolfram Web Resource. http://mathworld.wolfram.com/GeometricDistribution.html Accessed Nov. 15, 2010.

Werner, O., R. M. Ros, M. J. Cano & J. Guerra. 2004. Molecular phylogeny of Pottiaceae (Musci) based on chloroplast rps4 sequence data. Pl. Syst. Evol. 243: 147–164.

Werner, O., J. A. Jiménez, R. M. Ros, M. J. Cano, & J. Guerra. 2005a. Preliminary investigation of the systematics of *Didymodon* (Pottiaceae, Musci) Based on nrITS Sequence Data. Syst. Bot. 30: 461–470.

Werner, O., R. M. Ros & M. Grundmann. 2005b. Molecular phylogeny of Trichostomoideae (Pottiaceae, Bryophyta) based on *nrITS* sequence data. Taxon 54: 361-368.

Werner, O, R. M. Ros & B. Goffinet. 2007. A reconsideration of the systematic position of *Goniomitrium* (Funariaceae) based on chloroplast sequence markers. Bryologist 110: 108–114.

Whalen, T. 2012. Problem Solving and Decision Making Processes. http://www2.gsu.edu-/~dscthw/x130/04susyll.htm. Also: Heuristics & Biases. http://www2.gsu.edu/~dscthw/x130-/Heuristics-biases.html. Viewed Sept. 3, 2012.

Whittaker, D. J. 2009. Phylogeography of Kloss's Gibbon (*Hylobates klossii*): Populations and implications for conservation planning in the Metawai Islands. In: S. Lappan & D. J. Whittaker (eds.). Pp. 73–89. The Gibbons: Developments in Primatology: Progress and Prospects. Springer, New York.

Whittaker, R. H. 1972. Evolution and measurement of species diversity. Taxon 21: 213-251.

Wiens, J. J. 2007. Species delimitation: new approaches for discovering diversity. Syst. Biol. 56: 875–878.

Wiley, E. O. 1981. Phylogenetics: The Theory and Practice of Phylogenetic Systematics. John Wiley and Sons, New York.

Wiley, E. O. & Mayden, R. L. 2000. The evolutionary species concept. In: Q. D. Wheeler & R. Meier (eds.). Species Concepts and Phylogenetic Theory: A Debate. Pp. 70–89; 146–158; 198–208. Columbia University Press, New York.

Wiley, E. O., D. Siegel-Causey, D. R. Brooks & V. A. Funk. 1991. The Compleat Cladist: A Primer of Phylogenetic Procedures. Museum of Natural History, University of Kansas, Lawrence, Kansas.

Wilkinson, L., R. Rosenthal, R. Abelson, J. Cohen, L. Aiken, M. Appelbaum, G. Boodoo, D. A. Kenny, H. Kraemer, D. Rubin, B. Thompson & H. Wainer. 1999. Statistical methods in psychology journals: guidelines and explanations. Amer. Psychologist 54: 594–604.

Will, K. W., B. D. Mishler & Q. D. Wheeler. 2005. The perils of DNA barcoding and the need for integrative taxonomy. Syst. Biol. 54: 844–851.

Williams, D. M. 2002. Precision and parsimony. Taxon 51: 143–149.

Wilson, J. J., R. Rougerie, J. Schonfeld, D. H. Janzen, W. Hallwachs, M. Hajibabael, I. Kitching, J. Haxaire & P. D. N. Hebert. 2011. When species matches are unavailable are DNA barcodes correctly assigned to higher taxa? An assessment using sphingid moths. BMC Ecol. 11: 18.

Winkler, R. L. 1972. An Introduction to Bayesian Inference and Decision. Holt, Rinehart and Winston, New York.

Winkler, R. L. & W. L. Hays. 1975. Statistics: Probability, Inference, and Decision. Second Edition. Holt, Rinehart and Winston: New York.

Witmer, L. M. 1995. The extant phylogenetic bracket and the importance of reconstructing soft tissues in fossils. In: J. J. Thomason (ed.). Functional Morphology in Vertebrate Paleontology. Pp. 19–33. Cambridge University Press, Cambridge, U.K.

Witmer, L. M. 1998. Application of the extant phylogenetic bracket (EPB) approach to the problem of anatomical novelty in the fossil record. J. Vert. Paleont. 18(3: Suppl.): 87A.

Wright, R. 2001. Nonzero: the Logic of Human Destiny. Vintage, New York.

Yablokov, A. V. 1986. Phenetics: Evolution, Population, Trait. Transl. M. J. Hall. Columbia University Press, New York.

Yamane, T. 1967. Statistics: An Introductory Analysis. Harper & Row, New York.

Yoshimura, J., T. Hayashi, Y. Tanaka, K. Tainaka & C. Simon. 2008. Evolution 63: 288–294.

Yoon, C. K. 2009. Naming Nature: The Clash between Instinct and Science. W. W. Norton, New York.

Zander, R. H. 1972. Revision of the genus *Leptodontium* (Musci) in the New World. Bryologist 75: 213–280.

Zander, R. H. 1982. Various thoughts on characters in moss taxonomy. Beihefte zur Nova Hedwigia 71: 81–85.

Zander, R. H. 1993. Genera of the Pottiaceae: mosses of harsh environments. Bull. Buffalo Soc. Nat. Sci. 32 1–378.

Zander, R. H. 1995. Phylogenetic relationships of *Hyophiladelphus* gen. nov. (Pottiaceae, Musci) and a perspective on the cladistic method. Bryologist 98: 363–374.

Zander, R. H. 1998a. A phylogrammatic evolutionary analysis of the moss genus *Didymodon* in North America North of Mexico. Bull. Buffalo Soc. Nat. Sci. 36: 81–115.

Zander, R. H. 1998b. Phylogenetic reconstruction, a critique. Taxon 47: 681–693.

Zander, R. H. 1999. Randset generates random data files for phylogenetic analysis. Res Botanica Web site, Missouri Botanical Garden. http://www.mobot.org/plantscience/ResBot/Phyl/Rand/Randset.htm.

Zander, R. H. 2001. A conditional probability of reconstruction measure for internal cladogram branches. Syst. Biol. 50: 425–437.

Zander, R. H. 2003a. Reliable phylogenetic resolution of morphological data can be better than that of molecular data. Taxon 52: 109–112.

Zander, R. H. 2003b. Silk Purse Spreadsheet. Res Botanical Web site, Missouri Botanical Garden. http://www.mobot.org/plantscience/ResBot/phyl/silkpursesspreadsheet.htm.

Zander, R. H. 2004. Minimal values for reliability of bootstrap and jackknife proportions, decay index, and Bayesian posterior probability. Phyloinformatics 2: 1–13.

Zander, R. H. 2005. Unaccounted Assumptions. Res Botanica Web Site, Missouri Botanical Garden. <http://www.mobot.org/plantscience/ResBot/phyl/unaccounted.htm>

Zander, R. H. 2006. The Pottiaceae s.str. as an evolutionary Lazarus taxon. J. Hattori Bot. Lab. 100: 581–602.

Zander, R. H. 2007a. Nine easy steps for constructing reliable trees from published phylogenetic analyses. Ann. Missouri Bot. Gard. 94: 691–709.

Zander, R. H. 2007b. When biodiversity study and systematics diverge. Biodiversity (Tropical Conservancy) 8: 43–48.

Zander, R. H. 2007c. Paraphyly and the species concept, a reply to Ebach et al. Taxon 56: 642–644.

Zander, R. H. 2007d. When biodiversity study and systematics diverge. Biodiversity 8: 43–48.

Zander, R. H. 2007e. Neutralist evolution and strict monophyly adversely affect biodiversity study. Anales Jardín Bot. Madrid 64: 107–108.

Zander, R. H. 2008a. Evolutionary inferences from non-monophyly of traditional taxa on molecular trees. Taxon 57: 1182–1188.

Zander, R. H. 2008b. Statistical evaluation of the clade "Rhabdoweisiaceae." Bryologist 111: 292–301.

Zander, R. H. 2009. Evolutionary analysis of five bryophyte families using virtual fossils. Anales Jardín Bot. Madrid 66: 263–277.

Zander, R. H. 2010a. Taxon mapping exemplifies punctuated equilibrium and atavistic saltation. Pl. Syst. Evol. 286: 69–90.

Zander, R. H. 2010b (2011). Structuralism in phylogenetic systematics. Biol. Theory 5: 383–394.

Zander, R. H., J. A. Jiménez & T. Sagar. 2005. *Didymodon bistratosus* (Pottiaceae) in the New World. Bryologist 108: 540–543.

Zhang, C., D.-X., Zhang, T. Zhu. & Z. Yang. 2011. Evaluation of a Bayesian coalescent method of species delimitation. Syst. Biol. 60: 747–761.

INDEX